Tropical Deforestation

Methods and Cases in Conservation Science
Mary C. Pearl, Editor

Biology and Resouce Management in the Tropics
Michael J. Balick, Anthony B. Anderson,
and Robert Bailey, Editors

Methods and Cases in Conservation Science
Mary C. Pearl, Editor
Christine Padoch and Douglas Daly, Advisers

Tropical Deforestation: Small Farmers and Land Clearing in the Ecuadorian Amazon
Thomas K. Rudel with Bruce Horowitz

Bison: Mating and Conservation in Small Populations
Joel Berger and Carol Cunningham

Wildlife Conservation: International Case Studies of Education and Communication Programs
Susan Jacobson

Population Management for Survival and Recovery
Jonathan D. Ballou, Michael Gilpin, and Thomas J. Foose

Remote Sensing: A Tool for Biodiversity Conservation
David S. Wilkie and John Finn

At the End of the Rainbow? Gold, Land, and People in the Brazilian Amazon
Gordon MacMillan

Biology and Resource Management
Michael J. Balick, Anthony B. Anderson, and Kent Redford, Editors
Robert Mendelsohn and Charles Peters, Advisers

Alternatives to Deforestation: Steps Toward Sustainable Use of the Amazon Rain Forest
Anthony B. Anderson, editor

Useful Palms of the World: A Synoptic Bibiliography
Michael Balick and Hans T. Beck, editors

The Subsidy from Nature: Palm Forests, Peasantry, and Development on an Amazon Frontier
Anthony B. Anderson, Peter H. May, and Michael J. Balick

Contested Frontiers in Amazonia
Marianne Schmink and Charles Wood

Conservation of Neotropical Forests
Kent H. Redford and Christine Padoch, editors

Footprints of the Forest: Ka'apor Ethnobotany
William Balée

Tropical Biodiversity and Human Health
Michael J. Balick, Elaine Elisabetsky, and Sarah Laird

Pricing the Planet
Peter H. May and Ronaldo Seroa da Motta

Tropical Deforestation

The Human Dimension

Edited by:
Leslie E. Sponsel
Thomas N. Headland
and Robert C. Bailey

with a foreword by Jeffrey A. McNeely

 Columbia University Press   *New York*

Columbia University Press
New York   Chichester, West Sussex
Copyright (c) 1996 Columbia University Press
All rights reserved

Library of Congress Cataloging-in-Publication Data
Tropical Deforestation : the human dimension / edited by Leslie E.
  Sponsel, Thomas N. Headland, and Robert C. Bailey ; with a foreword
  by Jeffrey A. McNeely.
      p.   cm. — (Methods and cases in conservation science)
  (Biology and resource management in the tropics)
    Includes bibliographical references and index.
    ISBN 0–231–10318–2 (cloth). — ISBN 0–231–10319–0 (pbk.)
    1. Deforestation—Social aspects—Tropics.   2. Forest ecology—
Tropics.   3. Forest management—Tropics.   4. Forests and forestry—
Tropics.   I. Sponsel, Leslie E. (Leslie Elmer), 1943–   .
II. Headland, Thomas N.   III. Bailey, Robert Converse.   IV. Series.
V. Series: Biology and resource management in the tropics series.
SD418.2.T76T75   1996
304.2'8—dc20   95–47256
                                                      CIP

Casebound editions of Columbia University Press books are printed
on permanent and durable acid-free paper.

Printed in the United States of America

c 10 9 8 7 6 5 4 3 2 1

Dedicated to a better future for all forest-dwelling people struggling with deforestation in the tropics

Contents

Illustrations

Tables

Foreword

I was walking through a primeval forest in Kalimantan, Indonesia, several years ago, looking for orangutans. With my eyes cast toward the tree tops I paid insufficient attention to where I was putting my feet, so I stumbled. It turned out that what I had stumbled over was an old brick wall. I subsequently learned that it was part of an important archaeological site bearing witness to the ancient civilization of Srwijaya, which was centered in Palambang, Sumatra, in the sixth to the fourteenth centuries A.D. Orangutans were living in a forest reclaimed from a human civilization that was long gone.

This observation spurred me to look more carefully at the concept of virgin forest, convincing me to join many ecologists in concluding that virtually all forests have been significantly influenced by humans. The old idea of primary forests, much less virgin forests, needs to be replaced by a much more sophisticated view of the dynamic relationship between forests and people.

Tropical Asia, for example, was one of the heartlands of shifting cultivation, a practice that has had a profound influence on habitats throughout the region over the past ten thousand years. Under traditional systems of shifting cultivation wildlife flourishes, with elephants, wild cattle, deer, and wild pigs all feeding in the abandoned fields. Tigers, leopards, and other predators are in turn attracted by the herbivores. The older fields contained a high proportion of fruit trees, which are attractive to primates, squirrels, hornbills, and a variety of other animals. Those wishing to see wildlife in Asian forests are therefore well advised to seek abandoned fields, where the forest's productivity is down on the ground, where it can be reached by hungry browsers.

While civilizations ebbed and flowed across the Asian countryside over the centuries, the forests too ebbed and flowed, providing a sort of balance between people and the forest. This book shows that a similar pattern can be found in tropical America and Africa as well.

In precolonial times forest management was primarily in the hands of the people who lived in the forests. But the colonial era brought forests into the global market system, leading to the nationalization of many forests, the importation of forest-management technology from Europe, and the loss of many traditional resource-management practices that had the effect of maintaining biodiversity in forests (even if they were not explicitly designed to do so). Throughout the tropics conflicts between the policies of the colonial governments and the traditional forest use systems has been a major cause of the mismanagement of forest land. These radical changes in tenure rights, and resulting lack of clarity over ownership of products of the forest, are keys to understanding the speed with which tropical forests have been depleted. Perhaps now is the time to consider how to recapture some techniques and benefits of the age-old forest-management systems developed by the people who actually live in the tropical forests.

Further, modern technological innovations—such as plantation forests or industrial logging—tend to favor overexploitation of forests and the weakening of traditional approaches to forest management that had developed in response to historical experience. Today technology—through processing, transport, and marketing—enables the global consumer society to harvest forest resources from alternative locations when local resources are exhausted. The sustainable, conservative, and cyclical use that tended to characterize the groups that lived in balance with their resources are no longer viable, because the market-driven economy feeds most benefits of the forest into the global system but pays few of the local environmental costs. These costs remain with the local people, who must live with the consequences of resource-management decisions imposed upon them from outside, losing both their resources and their knowledge of traditional resource management.

Throughout human history local societies have ebbed and flowed as their wisdom was tested against the criterion of sustainability. Those societies that were able to develop the wisdom, technology, and knowledge to live within the limits of their environment were able to survive. Others overexploited their resources, so they flourished only briefly, giving up sustainability and adaptability for a flash-in-the-pan enjoyment of immediate wealth. This book presents a number of case studies that demonstrate that the people who live closest to the forest are likely to have the greatest vested interest in ensuring that the forest is used in a sustainable manner to provide long-term benefits to local people. Their lives depend on how well they manage their forests.

Jeffrey A. McNeely
Chief Biodiversity Officer
International Union for the Conservation of Nature—The World
    Conservation Union
Gland, Switzerland

Preface

Deforestation in the tropics is an environmental crisis that has increasingly drawn the attention of professionals, the media, the public, and governments since the 1980s (Goldschmidt, Hildyard, and Bunyard 1987). However, most of the discussion has concentrated on topics such as the catastrophic extinction of species and reduction of biodiversity, the wasteful loss of valuable natural resources like plant species that may have important medical applications, and the potential for serious climatic disruptions at the regional and global levels (Myers 1992; Park 1992). Considerations of deforestation usually neglect indigenous and other local people who inhabit the forests, except, for example, when so-called slash-and-burn farmers are blamed for deforestation or when the media briefly pick up a story about a group such as the Penan in Sarawak who resist external agents of deforestation. This relative neglect is the major reason for publishing this book. Through a representative set of anthropological case studies this book aims to provide an alternative perspective on the causes, consequences, and solutions of deforestation.

Whereas other disciplines usually approach deforestation at the regional, national, or global levels, anthropology, through its unique methods of fieldwork, emphasizes the more immediate human contexts and aspects of the phenomena and processes of deforestation at the local community level. This is what we mean by the subtitle of this book—the human dimension.

Although many tend to treat forests and humans as separate and independent phenomena, to do so is not only artificial and misleading but can be counterproductive and sometimes even dangerous for environmental conservation and economic development (Gomez-Pompa and Kaus 1992). This view is reflected in, among other places, Principle 22 of the Rio Declaration on Environment and Development from the famous United Nations world conference of June 3–14, 1992:

> Indigenous people and their communities, and other local communities, have a vital role in environmental management and development because of their knowledge and traditional practices. States should recognize and duly support their identity, culture, and interests, and enable their effective participation in the achievement of sustainable development.

This book focuses on the changing niche of human populations in tropical forest ecosystems across the major geographical regions. Most of the essays herein discuss some aspects of political economy in the context of environmental or forest history. Several authors marshal arguments and data to suggest that government management of forests often results in deforestation, whereas local community management of forests is usually more likely to contribute to forest conservation, although the latter do influence their ecosystems, sometimes profoundly. The crisis of deforestation is also placed in the context of recent trends and future needs in basic and applied research, and specific policy recommendations are implicit if not always explicit in most articles.

In examining the phenomena and processes of deforestation the authors in this volume touch on a wide range of related themes in the tropical world, including biological ecology, forest ecology, forest history, conservation biology, human ecology, anthropology, political economy, economic development, and environmental conservation. Although no single book could ever cover all the numerous topics and countries relevant to the complex process of tropical deforestation, together the fourteen essays in this book provide a representative sample of current anthropological research on the human context and aspects of this subject.

Although this book is written primarily by anthropologists, it is designed to be useful to scholars and practitioners working in a variety of fields, including those in the natural sciences as well as the social sciences. We hope that the many environmentalists, conservation biologists, and economic development experts working on issues related to deforestation will profit from the lessons in these essays. Moreover a major aim of this book is to provide background as well as examples of anthropological case studies to inform and stimulate further basic and applied research by anthropologists and others, especially by new generations of students who may well be at the battle front of deforestation in the future.

References

Goldschmidt, E., N. Hildyard, and P. Bunyard, eds. 1987. Save the forests: Save the planet—A plan for action. *Ecologist* 17 (4/5): 129–204.

Gomez-Pompa, A. and A. Kaus. 1992. Taming the wilderness myth. *BioScience* 42 (4): 271–279.

Myers, Norman. 1992. *The Primary Source: Tropical Forests and Our Future.* New York: W. W. Norton.

Park, Chris. 1992. *Tropical Rainforest.* London: Routledge.

Contributors

Elliot Abrams has a Ph.D. in anthropology from Pennsylvania State University. He is an associate professor at Ohio University. Abrams has conducted archaeological research in Honduras, Belize, and Mexico, as well as in the United States, especially Ohio. His research interests include human ecology, cultural evolutionary theory, and architectural energetics.

Janis Alcorn has a Ph.D. in botany with a minor in anthropology from the University of Texas at Austin. She is director for Asia and the Pacific in the Biodiversity Support Program, a consortium of the World Wildlife Fund, the Nature Conservancy, and the World Resources Institute funded by the U.S. Agency for International Development. Her research interests include ethnobotany, indigenous forest management, and international development. She is author of *Huastec Mayan Ethnobotany, Ethnobiology: Culture and Biodiversity* (forthcoming) and coeditor with Margery L. Oldfield of *Biodiversity: Culture, Conservation, and Ecodevelopment.*

Robert Bailey has a Ph.D. in anthropology from Harvard University and is an associate professor of anthropology at the University of California, Los Angeles. He is a National Institutes of Mental Health National Research Service Award recipient in the HIV/AIDS Training Program at Emory University's School of Public Health and the Centers for Disease Control and Prevention in Atlanta, Georgia. His research interests are human behavioral ecology, growth, nutrition and physical activity, and

the interface between conservation ecology and indigenous peoples'
needs. In 1980 he founded the Ituri Project, a long-term interdisciplinary
research and community development project in northeast Zaire. His
publications include *The Behavioral Ecology of Efe Men in the Ituri Forest,
Zaire,* and, with Nadine Peacock, *Efe: Investigating Food and Fertility in the
Ituri Forest.*

Eduardo Bedoya has a Ph.D. in anthropology from the State University of
New York at Binghamton. He is an associate professor at
FLACSO–Ecuador (Facultad Latinoamericana de Ciencias Sociales). He
has conducted long-term research on the social and economic causes of
deforestation in Peru. Bedoya is a specialist in peasant economy, politi-
cal ecology, and development studies. He has worked as a consultant for
various organizations such as the U.S. Agency for International
Development, Institute for Development Anthropology, and the World
Health Organization. His recent book is titled *La deforestacion en la
Amazonia Peruana: Un problema estructural* (Deforestation in the Peruvian
Amazon: A structural problem).

Alfonso Peter Castro has a Ph.D. in anthropology from the University of
California at Santa Barbara. He is an associate professor at Syracuse
University and an adjunct faculty member in forestry at the State
University of New York in Syracuse. He has served as a consultant on
community forestry issues for several agencies, including the United
Nations Development Program, CARE, and the UN Food and
Agriculture Organization. His publications include *Facing Kirinyaga: A
Social History of Forest Commons in Southern Mount Kenya.*

Carolyn Cook has a Ph.D. in anthropology from the University of Hawaii
and an M.A. in anthropology from Washington State University. She
lived in Irian Jaya from 1977 to 1983 and conducted formal fieldwork
with the Amung-me in the Timika area during the summers of 1985 and
1987 for her master's thesis on traditional land tenure and the effects of
development. Her dissertation fieldwork in 1992 was supported by a
grant from the Program on the Environment of the East-West Center.
She recently received a National Science Foundation postdoctoral fellowship to work with forest ecologist Mathius Kilmaskossu at
Cenderawasih University in Irian Jaya.

Billie DeWalt has a Ph.D. in anthropology from the University of
Connecticut. He is director of the Center for Latin American Studies,
professor in the Graduate School of Public and International Affairs,
and professor of anthropology at the University of Pittsburgh. His
research focuses on natural resource, agricultural, land tenure, and food
policies as well as political ecology, economic anthropology, and anthropological theory and methods. His books include *Modernization in a
Mexican Ejido: A Study in Economic Adaptation, Food Crops Versus Feed
Crops: Global Substitution of Grains in Production* (written with David

Barkin and Rosemary Batt), and *The End of Agrarian Reform in Mexico: Past Lessons and Future Prospects* (written with Martha Rees, with the assistance of Arthur Murphy).

James Eder has a Ph.D. in anthropology from the University of California, Santa Barbara. He is a professor of anthropology at Arizona State University. His research interests include ecological anthropology, economic development, Southeast Asia, and the Philippines. He is author of *Who Shall Succeed? Agricultural Development and Social Inequality on a Philippine Frontier,* and *On the Road to Tribal Extinction: Depopulation, Deculturation, and Adaptive Well-Being Among the Batak of the Philippines.*

AnnCorinne Freter has a Ph.D. in anthropology from Pennsylvania State University. She is an associate professor of anthropology and directs the Mesoamerican Obsidian Hydration Laboratory at Ohio University. She is an archaeologist whose primary research interests include human ecology, Mayan regional settlement patterns, and chronological demographic reconstructions. She has worked extensively at the sites of Copán in Honduras and Quiché in Guatemala.

Glen Green has a Ph.D. in earth and planetary sciences from Washington University in St. Louis, Missouri. He is affiliated with the Anthropological Center for Training and Research on Global Environmental Change at Indiana University in Bloomington. His research interests include mapping vegetation cover and deforestation using remote sensing, modeling anthropogenic land-cover change, landscape ecology, and evaluation of conservation program effectiveness.

Thomas Headland has a Ph.D. in anthropology from the University of Hawaii. He is an international anthropology consultant with the Summer Institute of Linguistics and adjunct professor of linguistics at the University of Texas in Arlington. His primary research interests are hunter-gatherer societies and human ecology in tropical forest ecosystems. He spent more than eighteen years doing fieldwork among Agta Negritos in the Philippines. He is editor of *The Tasaday Controversy: Assessing the Evidence* and with Kenneth Pike and Marvin Harris co-edited *Emics and Etics: The Insider/Outsider Debate.*

Lorien Klein has worked as an assistant investigator in research on the processes of environmental destruction caused by coca expansion in the upper Huallaga Valley of the Peruvian Amazon. She is enrolled at New York Chiropractic College in Seneca Falls.

Jeffrey McNeely is chief scientist at the International Union for the Conservation of Nature–The World Conservation Union. Before going to IUCN in 1980, he worked in Asia for twelve years on a wide range of conservation activities. He is author or editor of more than twenty books, including *Soul of the Tiger, Economics and Biological Diversity, Mammals of Thailand, Culture and Conservation,* and *Conserving the World's*

Biological Diversity. He was secretary-general of the fourth World Congress on National Parks and Protected Areas, held in Caracas, Venezuela, in 1992.

Brien Meilleur has a Ph.D. from the University of Washington at Seattle. He is president and executive director of the Center for Plant Conservation headquartered at the Missouri Botanical Garden in St. Louis. Formerly he was the manager of the Amy B. H. Greenwell Ethnobotanical Garden, a distinct unit of the Bishop Museum, located on Hawaii Island. His recent research centers on the ethnobiology and ethnoecology of indigenous Polynesian farmers, especially the traditional management of crop and wild economic plants in Hawaii and the Marquesas.

Augusta Molnar is a natural resource management specialist in the Agriculture Division of the Central American Department of the World Bank. Previously, she worked in the Asia Region Environmental Division at the World Bank on the design and supervision of social forestry projects in India, Nepal, and Bangladesh and participated in studies of India's forestry experience.

Emilio Moran has a Ph.D. in anthropology from the University of Florida. He is a professor of anthropology at Indiana University, where he also directs the Anthropological Center for Training and Research on Global Environmental Change. His research interests include cultural ecology, migration and rural development, social and environmental impact assessment, agrarian systems, applied anthropology, remote sensing, and geographical information systems. He is the author of *Human Adaptability: An Introduction to Ecological Anthropology, Developing the Amazon,* and *Through Amazonian Eyes: The Human Ecology of Amazonian Populations* and editor of *The Dilemmas of Amazonian Development* and *The Ecosystem Approach in Anthropology: From Concept to Practice.*

David Rue has a Ph.D. in anthropology from Pennsylvania State University. He is a senior archaeologist with Archaeological and Historical Consultants, Inc., an archaeological firm in central Pennsylvania. His research interests include the prehistory, cultural ecology, and palynology of Mesoamerica and eastern North America.

Leslie Sponsel has a Ph.D. in anthropology from Cornell University. He is an associate professor at the University of Hawaii where he directs the Ecological Anthropology Program. His research interests include human ecology in tropical forests, Buddhist environmental ethics, anthropological aspects of peace studies, and advocacy anthropology. He is chair of the Commission for Human Rights of the American Anthropological Association. Sponsel has conducted extensive fieldwork in the Venezuelan Amazon and more recently in southern Thailand. He is editor of *Indigenous Peoples and the Future of Amazonia: An Ecological Anthropology of an Endangered World,* and co-editor with Thomas Gregor of *The Anthropology of Peace and Nonviolence.*

Susan Stonich has a Ph.D. in anthropology from the University of Kentucky. She is an associate professor of anthropology and environmental studies at the University of California, Santa Barbara. Her research interests include the relationships among social processes, demographic change, economic development, food security, human rights, and environmental destruction. She is author of *I Am Destroying the Land: The Political Ecology of Poverty* and *Environmental Destruction in Central America.*

Linda Sussman has a Ph.D. in sociology from Washington University. She is a research instructor at the Center for Health Behavior Research at the School of Medicine and a research associate in the Department of Anthropology at Washington University in St. Louis, Missouri. Her research interests include medical belief systems, ethnomedicine, ethnobotany, and local resource use. She is also interested in the effects of conservation and development projects on the health status, resource use, and economics of local inhabitants. She has conducted fieldwork in Madagascar, Mauritius, and the United States.

Robert Sussman has a Ph.D. in anthropology from Duke University. He is professor of anthropology at Washington University in St. Louis, Missouri. His research interests focus on the nature and evolution of the relationships between primate habitat preferences, morphology, and social organization. He is also interested in conservation and development policy. In 1978 he co-founded the Beza Mahafaly Reserve in Madagascar, a long-term project focusing on ecology, conservation, and development.

John Vandermeer is a professor of biology at the University of Michigan. He has been involved in research on tropical biology for the past two decades, concentrating his work in Costa Rica, Nicaragua, and southern Mexico. His most recent books are *The Ecology of Intercropping* and the co-edited volume *Agroecology.* His current research is in the lowland rain forests of Nicaragua, where he is studying forest successional patterns after catastrophic damage from Hurricane Joan.

David Wilkie earned a B.S. in biology from Stirling University and a Ph.D. in conservation ecology from the University of Massachusetts. Since 1990 he has taught the United Nations Environment Programme/Tufts environmental management postgraduate program for professionals from developing countries. When not at Tufts, Wilkie provides technical assistance to multilateral, government, and nongovernmental organization conservation and development projects within Africa and continues his research on the impact of human land-use practices on tropical forest ecosystems in central Africa and Central America. Wilkie is author with J. T. Finn of *Remote Sensing Imagery for Natural Resources Monitoring: A First Time User's Guide.*

John Wingard has an M.S. in agricultural economics and a Ph.D. in anthropology from Pennsylvania State University as well as an M.S. in anthropology from the University of Oregon. He is an anthropologist with the National Marine Fisheries Service. His research interests include assessing social effects of human or naturally induced changes in the resource base and the long-term impacts on societies of major changes in resource availability and distribution. He has conducted field research in Micronesia, Honduras, and the United States.

Tropical Deforestation

I

The Big Picture

In part 1, The Big Picture, chapter 1 by the coeditors, Leslie E. Sponsel, Robert C. Bailey, and Thomas N. Headland, provides an overview of tropical forest ecology and the causes and consequences of deforestation and its solutions, emphasizing the human dimension and an anthropological perspective. This first essay provides the background for the subsequent essays, which discuss selected aspects in greater detail, usually in the format of a specific case study.

1

Anthropological Perspectives on the Causes, Consequences, and Solutions of Deforestation

Leslie E. Sponsel, Robert C. Bailey,
and Thomas N. Headland

After four billion years of organic evolution on this planet, life reached its greatest diversity and complexity in the tropical rain forests (Myers 1992; Signor 1990; Solbrig 1991). Yet in recent decades this biome as a whole has become endangered. Although many processes contributing to deforestation have been at work for centuries (Perlin 1991), the recent rate of deforestation in the tropics is unprecedented (Park 1992:31–42).

Most deforestation has transpired in this century, especially since the 1970s. By 1989 the annual rate of deforestation had reached 142,200 square kilometers, which represents 1.8% of the 8 million square kilometers of remaining forest, and the rate of deforestation is even accelerating (Myers 1992:xvii–xviii). Current rates of deforestation exceed 0.4 hectares per second (Repetto 1990), and each hectare may contain millions of individuals representing thousands of species (Uhl and Parker 1986). As a result of habitat destruction as many as 10,000 species may become extinct each year, a level unprecedented in all of geological history (Raup 1988; Simberloff 1986; Whitemore and Sayer 1992; Wilson 1992).

What are the causes and consequences of deforestation, and what are the solutions? The purpose of this book is to demonstrate through a series of case studies that anthropology has a special contribution to make in helping to understand and resolve this crisis, one that ultimately involves every human being on this planet.

This essay provides a background and framework for the detailed studies that follow; all were conducted by anthropologists or by researchers and practitioners using a distinctly anthropological approach and emphasizing the human contexts of deforestation. First, we outline some characteristics of tropical forests that present challenges to human inhabitants. We discuss some ways in which humans have responded to those challenges, how they have used resources to meet their needs, and how they have influenced their habitat. Second, we review some causes of deforestation, focusing principally on relatively recent efforts to exploit the resources of the fragile ecosystems of tropical forests. Third, we discuss the consequences of deforestation for the forests themselves, for the people who now live within them, and for the planet as a whole. Fourth, we touch on some potential solutions to deforestation. And finally, we review some attributes of the anthropological approach to deforestation as illustrated in subsequent essays. (Especially useful surveys of deforestation include Allen and Barnes 1985; Clay 1988; Colchester and Lohmann 1993; Denslow and Padoch 1988; Dinerstein and Wikramanayake 1993; Durning 1989; Hurst 1990; Klein and Perkins 1987; LeTacon and Harley 1990; Lundberg 1984; Mather 1987; Myers 1992, 1993b; Park 1992; Prance 1993; Ranjitsinh 1979; Repetto 1990; Singh 1994; Solo 1993; Thapa and Weber 1990; *Ecologist* 1987; Vanclay 1993; Williams 1989, 1990; Wilson 1989.)

Forest Ecosystems as Human Habitats

> What has taken tens of millions of years to evolve can be wiped out by human activities within a matter of a few human generations. Once it is gone, it is lost forever.   —*Park 1992:11*

Richness in Diversity

Tropical forests are located on either side of the equator at latitudes as great as 23.5°. About 61% of the forest is in Latin America, 23% in Asia, and 16% in Africa (Allen and Barnes 1985:168). Some thirty-three countries have tropical forest, but those with the largest blocks and greatest biological diversity (megadiversity countries) include Brazil, Zaire, and Indonesia. Brazil embraces about 60% of the Amazon. It also gets the most publicity about deforestation in the scientific literature and mass media; unfortunately, not as much is known about deforestation in the other seven Amazonian nations. However, size alone is not a sufficient measure of the significance of an area of tropical forest for biodiversity conservation, because many small areas have a high percentage of endemic species (species found nowhere else), such as the Atlantic Forest of Brazil and forests in Australia, Hawaii, and many other Pacific islands (Whitehouse 1991; Padua 1994; chapters 3 and 13 of this volume).

Equatorial regions typically provide excellent climatic conditions for plant growth—high temperature, rainfall, and humidity throughout or during most of the year. Beyond the equator plant communities are generally susceptible to greater stresses such as seasonal drought or cold (Osmond 1987; Whitmore, Flenley, and Harris 1982). Thus nearly half the biomass (organic weight) on earth is contained in tropical rain forests, yet they cover a mere 6 to 8% of the terrestrial surface of the planet. The luxuriance of the forest vegetation is deceptive, however; it does not necessarily reflect high soil fertility, because the climatic conditions are also ideal for chemical weathering and leaching. Accordingly, the Amazon in particular has been characterized as a counterfeit paradise (Meggers 1971; Moran 1993). At the same time tropical soils have great diversity, and some areas are quite fertile, as, for example, volcanic slopes in Indonesia.

Another characteristic of tropical rain forests that can only be described with superlatives is biological diversity (Huston 1994). Generally, at least ten times as many species occupy a hectare of tropical rain forest as a hectare of temperate forest. Biological diversity also occurs beyond the species level. For instance, in the vertical gradient from the forest floor up through the canopy, sunlight, temperature, and wind increase and humidity decreases. Such differences in microclimate create variations in microenvironment and associated species. At another level of diversity an area of forest is actually a mosaic of plant and associated animal communities at different stages of development (succession) in response to storms that fell trees, cause rivers to shift course, mud to slide from steep slopes, and so on. Moreover tropical forests are not alike; various classifications identify a dozen or more types related to differences in soil, drainage, seasonality, altitude, and so on (e.g., Whitmore 1990:13; Moran 1993:24–31; Park 1992:2–3).

In general, tropical rain forests are the oldest terrestrial biome on the planet, having evolved over a period of 60 to 100 million years, whereas many temperate forests, such as the primary successions that followed the retreat of glaciers, are only a few thousand years old. However, climatic change during the Pleistocene epoch, with the advance and retreat of continental glaciers from polar regions and mountains, affected the climate of tropical regions as well. One controversial position is that the forests contracted and fragmented, forming refugia, and savanna and intermediate plant communities expanded simultaneously. Subsequently, as rainfall increased, forests expanded and coalesced and savannas contracted. Several such cycles occurred during the Pleistocene (Flenley 1979; Prance 1982). Thus deforestation is nothing new; it occurred under natural conditions and independent of human influences but usually over geological time periods of thousands of years or more.

Human Colonization

Some changes in climate and biomes may have facilitated the human exploration of and adaptation to forests from the savannas via gallery (riverine) forest and other wooded patches. However, the human colonization of this biome appears to be a relatively recent phase of human prehistory, even though our closest living relatives in the animal kingdom, the primates, are mostly limited to tropical forests, and even though the tropical rain forest is the traditional home and source of livelihood for many human societies (Milton 1993; Coppens 1994). However, because archaeological research in tropical forests is limited, little is known about when humans moved into the rain forests and all the ways by which prehistoric humans exploited tropical rain forests (Bailey et al. 1989; Eggert 1993; Glover 1992; Headland 1987; Hutterer 1988; Meggers 1988; Roosevelt 1989, 1992).

The apparent recency of the human occupation of tropical rain forests, the low density of human populations in their interior, and other phenomena may reflect the relative scarcity of food for humans, with their comparatively large body and group sizes. High biological mass, productivity, and diversity do not necessarily mean an abundance of food for humans. About 95% of the plant biomass is wood, which most animals, except for specialists like termites, find inedible. The leaves are mostly toxic as chemical defenses against herbivore pressure. Fruits, nuts, tubers, and other edible portions of plants are sporadic in time and space and usually appear in only small amounts. Thus locating, acquiring, and processing such foods may require considerable time and energy. For instance, in the neotropics a spider monkey, which is primarily frugivorous (fruit eating) has a home range about twenty-five times larger than that of a howler monkey, which is mainly herbivorous (leaf eating) (Milton 1981, 1993). Likewise, the distribution of individuals within an animal species tends to be of low density and patchy. Moreover, at least in Amazonia, most mammalian species are nocturnal, solitary, small, arboreal, and well camouflaged (Sponsel 1986). As a consequence of these and other ecological factors subsistence from wild plants and animals is at best difficult for human foragers (hunter-gatherers) in tropical rain forests. Plant carbohydrates may be more efficiently and reliably obtained by farming or at least by trade with farmers, as in the exchange between Pygmy foragers and Bantu farmers in the forests of central Africa (Bailey 1991; Wilkie 1988), or Negrito foragers and lowland farmers in the Philippines (Peterson 1978; Headland and Reid 1989, 1991). Agriculture probably facilitated human colonization and adaptation in tropical rain forests, especially in the Amazon (Sponsel 1989), and some argue that humans could not survive in this biome by foraging alone, without either farming or trading with farmers (Bailey et al. 1989;

Bailey 1990; Bailey and Headland 1991; Headland 1987; Headland and Bailey 1991; for counter arguments see Brosius 1991; Endicott and Bellwood 1991; Dwyer and Minnegal 1991; Bahuchet, McKey, and de Garine 1991; Milton 1991; Stearman 1991).

Sustainability

In any case, under traditional circumstances slash-and-burn cultivation, which is better called swidden, or shifting horticulture, is the only type of farming that has proved sustainable in the long term in tropical rain forest ecosystems. (*Sustainable* refers to practices that do not irreversibly deplete resources and/or degrade the habitat.) (See Dovers and Handmer 1993; Goodland, Daly, and Serafy 1993; Robinson 1993; Wagner and Cobbinah 1993 for recent discussions of sustainability.) This type of farming effectively recognizes and takes advantage of some key characteristics of forest ecology. Cutting a section of the forest allows the penetration of sunlight for photosynthesis by the crop plants. Burning dried cuttings fertilizes the soil, which is usually of low fertility. Burning also temporarily eliminates wild plants that would compete with crops for nutrients and insects that would graze on the crops. Both lower crop productivity. Small, discontinuous, and roughly circular garden plots encourage the regeneration of the forest from seeds carried in by wind and animals from the adjacent forest. Some trees may be purposefully retained in the garden or even planted, which also encourages forest regeneration (Balée 1994; Conklin 1957; Carneiro 1988; Clay 1988; Kundstadter 1988).

Under traditional conditions—a relatively mobile population of low density, a subsistence economy, polycropping, adequate fallow periods, and holding an ample area of forest in reserve for future gardens—swidden cultivation can be efficient, productive, and sustainable (Clarke 1976; Denevan 1992b; Posey and Balée 1989). Indeed swiddens are rather similar to the natural gaps caused by tree falls from wind storms (Walschburger and von Hildebrand 1991). Traditional indigenous practices, when used at moderate or low population densities, are usually consistent with the regenerative capacity of the forest. In many cases traditional swiddening even enhances the biological diversity of the forest (Bailey 1990; Balée 1994; Blankespoor 1991; Brookfield and Padoch 1994; Gadgil, Berkes, and Folke 1993; Park 1992:48; Sponsel 1992a; Uhl et al. 1989; Wilkie and Finn 1990). Like other small-scale disturbances swiddens can contribute to a diverse and healthy forest with a mosaic of plant and associated animal communities in different stages of succession (Sponsel 1992a; see also Fimbel 1994 and Salafsky 1993).

Swiddening has endured for centuries or even millennia in many areas of the tropics, and so has some kind of forest. For instance, the Maya used an intricate system of forest management to support an unusually high degree of sedentariness, a population both large and dense, and a complex culture (Atran 1993; Gómez-Pompa and Kaus 1992; Gómez-Pompa, Flores, and Sosa 1987; Nations and Nigh 1980; Rice and Rice 1984). Nevertheless some prehistoric societies, including some Maya, were responsible for deforestation. (See chapters 2 and 3 in this volume.) Another consideration is that before Westerners systematically depopulated the rain forests through epidemics, warfare, and slavery, indigenous populations were large in many areas (Balée 1994; Beckerman 1979; Denevan 1992a; Newson 1985; Roosevelt 1989; Wood 1992:59), and accordingly must have used, managed, and affected the forests extensively.

Interdependence

Because the forest and humans have profoundly influenced each other for millennia, neither can be fully understood in isolation. (See chapter 4 in this volume.) Only recently have scientists begun to understand the degree to which tropical forests may be anthropogenic (Alcorn 1981; Balée 1989, 1992, 1993; Denevan 1992a, 1992b; Gómez-Pompa and Kaus 1992; Posey and Balée 1989; Yoon 1993). For instance, William Balée, who defines *anthropogenic forests* as those having "a biocultural origin that would not have existed without past human interference" (1993:231), identifies several types of anthropogenic forest in the Amazon and estimates that some 12% of the interior forest is anthropogenic (Balée 1989). For at least thousands of years humans have played important roles in prey selection and coevolution, seed dispersal, plant domestication, triggering biotic community succession through swiddening, and altering soil composition in forest ecosystems (Balée 1994; Blankespoor 1991; Hecht and Posey 1989; Sponsel 1992a). Many tropical forest plant communities are the outcome of hundreds of generations of shifting horticulturists' clearing and cultivating plots of forest (Pelzer 1968). Thus tropical forests today are a patchwork of various stages of successional growth interspersed with mature forest. From this perspective alone biologists and others concerned with the ecology and conservation of forests and wildlife must be concerned with the humans who reside and historically have resided in and adjacent to these ecosystems (e.g., Robinson and Redford 1991; Redford and Padoch 1992). This has become clear at the La Selva Biological Research Station in Costa Rica, which has been studied for many years as pristine tropical forest but where charcoal and other evidence of prehistoric human disturbances were only recently reported (McDade et al. 1993). The articles in this book underscore the seminal importance of such a human-centered approach to the ecology, history, and conservation of forests, as well as to natural resource use and management in these ecosystems. The importance of considering the human dimension is emphasized in chapter 9 by James Vandemeer. Although a biologist Vandemeer finds it necessary to compare the political economies of Costa Rica and Nicaragua in the context of their environmental history in order to understand deforestation.

Causes of Deforestation

> Atomic warfare could make our planet uninhabitable very quickly. But we are increasingly aware of another danger, just as apocalyptic but more insidious. We may be making the planet uninhabitable gradually, without even being sure that we are doing so. The signs are all around us but we are not sure how to read them.　　　—*Maybury-Lewis 1992:35*

Shifting and Shifted Farmers

When conditions for sustainable swiddening change, such as a shift from subsistence to cash cropping or an increase in population density, swiddening can contribute to deforestation. Also, the effects of different modes of swiddening are different. For instance, some hill peoples of Thailand like the Lua' usually practice sustainable swiddening, whereas others like the Hmong usually do not and contribute instead to deforestation (Kunstadter 1988). Government policy makers are often either unaware of or ignore such differences, and their policies tend to characterize all swidteners as primitive, inefficient, unproductive, destructive, and so on (Brookfield and Padoch 1994; Cultural Survival 1987:21; Dove 1983, 1992). Such pretenses often serve as justification for nonrecognition of land rights and either forced or de facto resettlement, as Carolyn D. Cook describes in her essay (chapter 12) in this book. The result may amount to genocide, ethnocide, and/or ecocide. For whatever reasons the traditional swiddeners are more often the victims rather than the villains in deforestation. However, in some cases the swiddeners may be both victims and villains, as when they practice forms of forest use for short-term survival because they have no choice, although they realize that these patterns lead to deforestation and undermine their society in the long term. (See the case studies in chapters 4, 9, 11, and 13 in this volume.)

There are many examples of cultures that have changed because contact with others rendered a previously sustainable swidden system unsustainable (Burkhalter and Murphy 1989; Hires and Headland 1977; Milton 1992). One factor that has not been studied but that may have contributed to an increase in deforestation is the introduction of steel axes into societies that used only stone axes. Because steel axes are several times more efficient than stone axes (Carneiro 1979a, 1979b; Townsend 1969; see review in

Denevan 1992b), it is reasonable to suppose that deforestation increases with their introduction (Denevan 1992b). However, this has yet to be documented, although other aspects of this technology have been considered (Toth, Clark, and Ligabue 1992). The introduction of cash cropping and the use of chain saws are additional factors that may contribute to deforestation (Ekanade 1987).

The distinction between shifting and "shifted" farmers is also important, the latter usually being both victims and villains in deforestation (Myers 1992:363–365). Because they cannot deal effectively with the potentially explosive problems of population growth, poverty, and landlessness, governments have allowed, and often even encouraged, massive migrations to the forests, as in Madagascar, (see chapter 13) and transamazon colonization, as in Brazil (Moran 1983; Smith 1981; see chapters 6 and 7 in this volume). The Amazon has often been viewed as a "land without men for men without land" (Fearnside 1986; Hecht and Cockburn 1989). Similarly, Indonesia launched its transmigration project to relieve population, economic, and political pressures in the central islands by settling people on the outer islands (*Ecologist* 1986; Kartawinata et al. 1981; Secrett 1986). Indonesia also has used transmigration as a military weapon, invading and conquering Irian Jaya (Cultural Survival 1987:157; Nietschmann 1985; chapter 12 in this volume). Commercial logging operations are infamous for opening new areas of forest and either allowing or introducing new immigrants to the forest. (See Wilkie in chapter 10 of this book.) These and other shifted farmers have triggered deforestation on a scale that is usually way beyond that of any acculturated long-term residents of forests.

Tropical countries are seeing an increase in the number of poor and landless who migrate to farm in forests or in deforested areas. Shifted cultivators, in contrast to shifting cultivators, now account for about half the deforestation in the tropics (Myers 1992). The total number of swiddeners (shifted and shifting farmers combined) is estimated at 200 to 500 million people. Today about 1 in 10 humans is a swidden farmer (Myers 1992:364). (See chapters 4, 9, and 11 in this book.)

At current rates world population will double by the year 2035. Most of this will occur in countries with substantial areas of tropical forest, and most of these populations are young (with a median age of less than fifteen years). Thus an explosive potential for growth is ahead (Rudel 1989). In developing countries that depend on oil imports the rural poor can barely afford even the least expensive source of energy for cooking—wood from local trees—and the situation worsens with inflation of oil prices, economic decline, and population growth. Women and children may sacrifice disproportionate amounts of time and energy just gathering fuelwood. The number of cooked meals may decline, contributing to malnutrition. Where the wood supply is inadequate, crop refuse and animal dung may be burned for fuel, but this sacrifices valuable nutrients that would otherwise be returned to the soil for agricultural production. Thus in much of the Third World the ongoing energy crisis of fuelwood is a major cause of deforestation and linked to many other problems (Dewees 1989; Eckholm, Foley, and Bernard 1994). In parts of Africa and the Indian subcontinent estimates are that 60 to 90% of the energy is supplied by wood (Baidya 1984; Kalapula 1989; Witte 1992). Swiddening and fuelwood will both contribute to even more deforestation in many regions as the local human populations grow. (See chapters 4, 8, and 13 in this volume.) In this book Robert C. Bailey notes in chapter 14 that population increase in the Ituri Forest of Central Africa comes not from residents but from immigrants and other outsiders. However, in many areas much of the population growth may be concentrated in urban centers rather than in the forest itself (Volbeda 1986). For example, in the Brazilian Amazon population growth was 3.86% in the rural regions and 10.85% in urban centers during 1960–1980 (Godfrey 1990:107). Still, cities and towns make demands on the resources of nearby forests that can lead to deforestation (Bowonder, Prasad, and Unni 1987; Lugo 1991; Perlin 1991). (See chapters 4, 5, and 8 in this book.)

Extractive Consumption

Entities outside tropical countries also contribute to deforestation. For instance, logging for export is a major factor in the deforestation of Southeast Asia and Africa (Horta 1991; Headland 1988; Hurst 1987; Rush 1991; Wilson and Johns 1982; cf. Vincent 1992). Even selective logging, where less than 10% of the trees are cut, may damage 60% or more of the trees in the forest. In the rain forest trees may be connected with other vegetation such as lianas. Also, large trees damage smaller ones when they fall. Removing the logs damages vegetation and soil (Park 1992:61–62). Selective logging harvests the best commercial trees and may cause genetic erosion of forest resources over time (see chapter 9 in this book). Accordingly, logging can cause a decline in the quality as well as the quantity of trees and forests. Logging also has contributed to extensive forest fires in Kalimantan, Borneo, and elsewhere (Mackie 1984; Malingreau, Stephens, and Fellows 1985). In addition, as David S. Wilkie's essay in this volume (chapter 10) documents so well for central Africa, the loggers do more than cut trees. They create the conditions for depletion of wildlife through hunting, and they kill fish by using streambeds as their roads. Some loggers even fish with insecticides or dynamite, as in the Philippines. Other consequences of logging are indirect, such as the roads that are conduits for migrants and colonists like shifted cultivators because they open up previously isolated sections of forest. Thus logging is just the first step in a chain reaction of deforestation (Gillis 1988). (Also see chapters 6, 8, 9, and 10 in this volume.)

Logging has intensified as such technology as chain saws and logging trucks has become available (Williams 1990:184). As developed countries have harvested their own forests and launched conservation programs, they have turned to the developing countries for timber; the developing countries have responded eagerly because they need the foreign exchange income. Whereas deforestation through nontraditional swidden and fuelwood use is a matter of need, logging is often a matter of greed linked with materialism and consumerism. Some countries, such as the Philippines, Malaysia, and Thailand, have already logged their forests so extensively that they are being forced to shift from net exporters to net importers of wood (Brookfield 1988). Japan is the largest importer of wood from the tropical forests of Southeast Asia, and as that region is logged the Japanese may target the Amazon next (Colchester 1994; Johnstone 1987; Swinbanks 1987). (See Institute of Social Analysis [INSAN] 1989, and chapter 6 by Emilio F. Moran in this book.)

Yet another factor contributing to deforestation is forest conversion for growing commercial monocrops, such as the coffee plantations in Côte D'Ivoire and the coconut and rubber tree plantations in Southeast Asia. In Central America large areas of forests have been converted to pasture for ranching to supply inexpensive beef for the fast-food industry in the United States (Nations and Komer 1982, 1983; Myers and Tucker 1967). (See chapters 8 and 9 in this volume.) In Brazil ranching is also an important cause of deforestation, but the main cause is land speculation, encouraged by government subsidies such as tax credits and special loans. About 85% of the deforestation in the Brazilian Amazon is caused by some five thousand ranch owners. The five hundred biggest ranches range in size from about 4,000 hectares to 40,000 hectares or more (Buschbacher 1986; Hecht 1993; Hecht and Cockburn 1989; Loker 1993). Actually, in most cases government policy effectively recognizes deforestation as land improvement and gives land titles accordingly (Buschbacher 1986; Hecht 1989; Poelhekke 1986). (For forest conversion to pasture in Latin America see Downing et al. 1992; in this book see the discussions of ranching in chapters 6, 8, and 9.)

Cattle ranching and hydroelectric dams are the biggest causes of deforestation in Brazil. Moreover the wood is simply wasted rather than harvested—burned on ranches and left standing behind dams to rot (Fearnside 1989a; Gribel 1990). Development banks, such as the World Bank and the Interamerican Development Bank, which loan money to the government for the construction of highways, transamazon colonization, hydroelectric dams, and so on, indirectly finance such deforestation (*Ecologist* 1985; Lutzenberger 1985; Rich 1990).

Mining is yet another cause of the degradation and destruction of forests. Although most mineral deposits are relatively concentrated, the activities associated with mining are not necessarily so. In the Brazilian

Amazon iron mining involves using charcoal produced from the surrounding forest to smelt the ore (Bunker 1989; Fearnside 1989b). Also, since the 1980s about half a million gold miners have invaded various parts of the Amazon. They use mercury to amalgamate the ore, which contaminates the ecosystem. Mercury washes into the soil and waters and is magnified as it goes up the food chain, reaching humans as well as fish. It will take centuries for the mercury to wash out of the ecosystem (Greer 1993; Malm et al. 1990; Martinelli et al. 1989). In Brazil and Venezuela the Yanomami are just the latest victims of this gold mining (Albert 1994; Berwick 1992; Sponsel 1994; Tierney 1995).

Politics and the Military

Politics and even the military are connected to deforestation in several ways (Giaimo 1988; Ledec 1985; Pasca 1988; Stowe 1987). In many countries, such as the Philippines and Malaysia, logging and politics are bedfellows, because politicians often hold large logging concessions and provide contracts in return for political support and other favors (Hurst 1987; Ledec 1985; Marshall 1990; Rush 1991).

Increasingly, wars in developing countries find the government fighting its indigenous peoples (Nietschmann 1987). Environmental interdiction, and scorched earth policy in particular, are military strategies. During the Vietnam War the American military sprayed Agent Orange and other defoliants along the Ho Chi Minh Trail, the major supply route of the Viet Cong, and on mangrove forests along the coast and estuaries to expose enemy troops for bombing runs (Stevens 1993). More recently, scorched earth techniques have been used in El Salvador and other countries where the enemy used the forest for refuge, supply routes, and guerrilla tactics.

The war against drugs and the drug trade in the Amazon are linked with deforestation, especially in Peru and Colombia. The case of the Huallaga River valley in Peru is becoming infamous; there the war on drugs has also become a war on peasants and forests (Morales 1989). (See chapter 7 in this volume.) Opium production in the Golden Triangle region of Myanmar (Burma), Thailand, and Laos by some hill peoples such as the Hmong has been a target of the U.S. drug war and often negatively affected these people and their environment. Often the forest that remains is along border zones and thus becomes a matter of strategic concern. Villages of hill peoples in the monsoon forests along the border between Myanmar and Thailand have been forcibly relocated for security reasons (Cultural Survival 1987).

The Brazilian military has long been concerned with infrastructural and other development in the Amazon to control its frontiers and integrate the country for national security (Hecht and Cockburn 1989:95–128; Pinto

1989; Santilli 1989; Treece 1989). The Indonesian military has used transmigration and the ensuing deforestation as part of its strategy to conquer and control Irian Jaya (Nietschmann 1985). (See chapter 12 in this volume.) Also, as Vandemeer describes in chapter 9 here, during the cold war the different policies of the U.S. government toward Nicaragua and Costa Rica had dramatically different effects on the political economies of deforestation and forest conservation in these two countries.

In summary, deforestation is a complex phenomenon with multiple causes; the specific combinations of proximate causes and their proportionate contributions vary from country to country and even regionally within a country (Myers 1993b; Park 1992:80; Plumwood and Routley 1982). However, the ultimate causes are usually the same—need and/or greed. Both proximate and ultimate causes need to be considered in developing solutions to the crisis.

Consequences of Deforestation

> The largest tragedy of this century for humanity as a whole, will be irreversible destruction of tropical forests.   —*Bowonder 1986:186*

Environmental Functions

Deforestation has many far-reaching consequences. The environmental functions and services of the forest ecosystem are reduced or even lost, depending on the extent of deforestation. Forests protect the soils from the tropical rainfall and its effects on erosion. Deforestation increases soil erosion and the siltation of rivers and streams and can endanger hydroelectric dams, agricultural irrigation systems, and other technological and economic facilities (Bruijnzeel 1991; Goodland, Juras, and Pachauri 1993; Randrianarijaona 1983). Some of the worst deforestation and soil erosion has occurred in Madagascar (Randrianarijaona 1983; chapter 13 in this volume). Even the operation and longevity of the Panama Canal are threatened by extensive deforestation in its drainage basin (Jukofsky 1991; Whelan 1988). In the Amazon about half the water in the hydrological cycle is internal, a result of daily rainfall, evaporation, and transpiration. Deforestation could disrupt this cycle, leading to a serious reduction of the rainfall in Amazonia and adjacent regions (Bunyard 1985). This in turn would lead to serious consequences for agricultural production.

Deforestation also changes the albedo of the land surface in the region, and that can alter local climate. The rough surface of the forest canopy reduces winds, which otherwise would be stronger and contribute to soil erosion. In short, extensive deforestation can trigger a chain reaction of negative climatic and ecological consequences (Clark 1992; Holloway 1993; Myers 1992; Salati 1990).

Global warming is related to the increased emissions of greenhouse gases into the atmosphere as a result of human activities (Gentry and Lopez-Parodi 1980; Janetos and Hinds 1990; Jones and Henderson-Sellers 1990; Jones and Wigley 1990; Lovelock 1985; Michaels 1990; Molion 1989; Palm et al. 1986). Although fossil fuel combustion and industrial pollutants comprise most emissions, about 30% of the carbon dioxide released into the atmosphere results from trees burned in connection with deforestation in the tropics (Bunyard 1985; Detwiler and Hall 1988; Goreau and de Mello 1988; Meehl 1987; Myers 1992; Post et al. 1990). The average global temperature, which is predicted to increase two to five degrees centigrade within the next century, could trigger melting of the polar ice caps and other glaciers, thereby raising the sea level. Because almost two-thirds of the human population lives on one-third of the world's land adjacent to coasts and along estuaries (this includes most of the world's great cities), the rise in sea level could be catastrophic (Myers 1993b; Westing 1992). Moreover many small coral island societies and nations in the Pacific, Caribbean, and elsewhere would be endangered (Stoddart 1990). The economic cost would be astronomical. For instance, it would cost $75 to $111 billion just to safeguard the eastern seacoast of the United States against rising sea level. On the other hand, the cost of compensating countries for conserving tropical rain forest would be about $20 billion per year, and planting deforested areas with fast-growing trees to reduce carbon dioxide would mean a one-time cost of about $40 billion (Myers 1992:370–372; Postel and Heise 1988; Sedjo 1989).

Deforestation has many other climatic effects. In parts of Africa and Madagascar, deforestation is an important contributing factor in desertification. Deforestation in the Himalayas has been related to disastrous floods in Bangladesh (Myers 1986). Many linked the severe flooding in southern Thailand in November 1988 to deforestation from logging; as a result Thailand became the first country to establish a nationwide ban on logging, although it has turned to its neighbors for timber (Lohmann 1989, 1993; Project for Ecological Recovery 1992). Already, peninsular Malaysia's rice production has declined by more than a quarter, at least partly the result of decreased rainfall, soil erosion, and other phenomena believed to be connected with deforestation (Aiken and Moss 1975; Ekachai 1990; Myers 1992; Pushparajah 1985).

Biodiversity and Germ Plasm

Deforestation reduces biological diversity. Forests contain numerous species of plants and animals, many of which are endemic to relatively small areas (e.g., Pomeroy 1993; Tangley 1988). Species are becoming extinct before biologists have even had a chance to classify and describe them, let alone understand their natural history. Because all species are interconnected and interdependent, the extinction of one species is likely to have a synergetic or multiplier effect on several others. Extinction is irreversible, and because each species is a closed genetic system, the loss of any species is the loss forever of unique genetic information that evolved over thousands or even millions of years (McNeely et al. 1990a, 1990b; Wilson 1988, 1989, 1992; World Resources Institute et al. 1992). Deforestation's repercussions on fauna include its effects on birds, which are intercontinental migrants from temperate zones during the winter season (Hutts 1988; Lovejoy 1983).

The reduction of biodiversity not only impoverishes life on this unique planet but reduces the possibilities for future evolution (Mannion 1992). Genetic engineering, a promising technology for the future welfare of humankind, cannot create new genetic material but only rework existing material. From a human utilitarian viewpoint the loss of biodiversity is also the loss of potential natural resources of economic and social value, including medicinal plants that might cure cancer, acquired immune deficiency syndrome (AIDS), or other diseases (Blum 1993; Rasoanaivo 1990; Reid et al. 1993; Schultes 1991; Shiva et al. 1991). If tropical rain forests are destroyed, cures for some diseases may never be found (Park 1992:89). The Amazon alone contains more than thirteen hundred plant species that have medicinal value (Plotkin and Famolare 1992; Posey and Overall 1988; Schultes 1991).

Erosion of biodiversity also affects the world's food supply because it involves the loss of germ plasm for genetic engineering, including wild counterparts of domestic crops that could infuse the latter with greater resistance to diseases and pests. The latter is particularly critical because most modern agriculture is based on monocropping of just a few species of cultigens, and an infusion of fresh genetic material from wild varieties is sometimes critical for their survival and productivity. Moreover, except for ornamentals, less than 1% of the total number of species of higher plants have been domesticated during and since the Neolithic (Barrau 1982; Clement 1989; Oldfield and Alcorn 1991; Pimentel et al. 1992; Smith and Schultes 1990; and Smith, Williams, and Plucknett 1991).

Natural Capital or Interest

Deforestation also means the loss of other renewable natural resources. Forests contain more than just stands of trees or board feet of timber. Many valuable nontimber forest products have been mislabeled as minor forest products (Godoy and Lubowski 1992). Most were originally discovered and used by indigenes (Plotkin and Famolare 1992).

In some regions such as Southeast Asia nontimber products have been collected by forest indigenes and traded to middlemen for an export market for centuries or longer (Dunn 1975; Headland and Reid 1989, 1991; Hoffman 1986; McNeely and Wachtel 1988; Ryan 1992; Schwartzman 1989; Stiles 1992; Turnbull 1983). Peoples of the Amazon and Andes exchanged nontimber products in prehistoric times. Today rattan for furniture manufacture is one of the most important nontimber products from the forests of the Philippines, Indonesia, and other countries of Southeast Asia. Rattan exports from Southeast Asia are worth more than $60 million, whereas the value of the finished products is about $4 billion (Myers 1992:232). Wildlife, another important renewable natural resource in tropical rain forests, could be harvested on a sustainable basis for the long term (Adeola 1992; Robinson and Redford 1991; Redford and Padoch 1992). The flesh of wild animals provides a major source of quality protein for many Africans (see chapter 10 in this volume).

Forest wildlife also supplies the fur, skin, pet, zoo, and other trades, and many species are important for biomedical research and experimentation (Adeola 1992; McNeely and Wachtel 1988; Robinson and Redford 1991). One indication of the magnitude of wildlife exploitation is that researchers recorded 68,654 native birds of 225 species (most supposed to be protected) for sale at the Sunday market in Bangkok during a twenty-five-week period (Round 1990). Indeed the economic value of nontimber forest products actually exceeds that of the timber (Peters, Gentry, and Mendelsohn 1989) and depends on the conservation of the forest ecosystem rather than its destruction or degradation, as in the cases of clear-cut and selective logging.

The tropical rain forest also has significant economic value for other nonconsumptive uses such as ecotourism, which is the fastest-growing component of the world tourism industry. Ecotourism can generate major foreign income, thereby encouraging and helping to finance conservation in developing countries such as Costa Rica, Panama, and Thailand (Boza 1993; Brockelman and Dearden 1990; Burnie 1944; Paaby, Clark, and Gonzalez 1991; Simons 1988; Tobias and Mendelsohn 1991; Wallace 1993). On the other hand, deforestation decreases a country's natural capital for future economic development like ecotourism.

Disease.

Diseases and deforestation are often linked in many different ways (Independent Commission on International Humanitarian Issues [ICIHI] 1986). New migrants like shifted farmers introduce alien diseases to the long-term residents and in turn contract local diseases from their new environment. Cutting trees brings canopy mosquitoes to the ground, where they may feed on humans and transmit malaria, whereas formerly they preyed on canopy animals (Yuill 1983; chapter 6 in this volume). Brazil now has the highest number of new cases of malaria in the world (Moran 1993:151, 160).

Violence.

Deforestation also has direct and indirect sociopolitical consequences. It is linked at least in part to the growing security problems of places like Guatemala, Haiti, El Salvador, Ethiopia, India, and Pakistan (Colchester 1991; Myers 1992:xxiv—xxv). Among the more serious human aspects of deforestation is its connection with warfare and other forms of violence (Homer-Dixon, Boutwell, and Rathjens 1993; Westing 1992). In this book Moran predicts that violent conflict will increase as land and resource competition intensify with continued deforestation in the Brazilian Amazon. (Also see Pace 1992.)

Deforestation is indirectly related to the genocide, ethnocide, and ecocide of indigenous societies. Jason W. Clay states the case most concisely:

> Virtually all the world's tropical forests are populated, usually by indigenous peoples. In order for local, state or international interests to exploit forest resources, the rights of indigenous groups must be denied and the groups themselves displaced. It is no accident, therefore, that indigenous peoples are disappearing at an even faster rate than the tropical forests upon which they depend. Their own survival is intricately linked with that of their forests. They also represent our best first line of defense against the destruction of the forests.   (1989:1)

Since the 1970s indigenes have increasingly organized politically from the village to the regional, national, and international levels to defend their rights and interests (Berger 1991; Bodley 1990:152–178; Burger 1987; Carneiro da Cunha 1989; Durning 1993; Hildyard 1989; Lutzenberger 1987; Moody 1988; Nyoni 1987; Posey 1989a; Schwartz 1989). Likewise, advocacy groups among the anthropology and human rights organizations, such as Cultural Survival, International Work Group for Indigenous Affairs, and Survival International, are collaborating with indigenes to promote the survival, welfare, identity, rights, self-determination, and freedom of indigenous groups. For instance, for several years the Penan in Borneo have been struggling to defend their forest homelands against loggers supported by the government (Langub 1988). Indigenous people have a unique status and deserve special attention because of their historical claims to the territory of their ancestors, their use and knowledge of natural resources, and their vulnerability to disease and disruptive change as they come into contact with the so-called modern world.

Although many indigenous societies like the Yanomami in Brazil (Sponsel 1992a, 1994) and the Agta of the Philippines (Headland 1988, 1993) are threatened with extinction, others have avoided ethnocide, experienced a population rebound after initial depopulation with Western contact, accepted Western culture selectively, and maintained their ethnic identity and territorial integrity. It is a myth that all indigenous peoples are inevitably destined to extinction (Bodley 1990:179–207). The Kayapó in Brazil, Shuar in Ecuador, Ye'kuana in Venezuela, and the Kuna in Panama are among those that have thus far been successful at maintaining control over their forest lands and their ethnic integrity and identity (e.g., Fisher 1994).

Solutions to Deforestation

In many ways the problems of the rainforest are microcosms of the problems of the world, because they reflect the tensions and interplay between the powerful and the powerless, the rich and the poor, the north and the south. As a result some critics of existing attitudes and policies toward the forests have joined the call for a new world economic order, which would be more sustainable, less environmentally damaging and more equitable. 　　　　　　　　　　　　　　　　*—Park 1992:161*

Gravity and Urgency

Perhaps the greatest unanswered question is whether enough can be done soon enough to substantially reduce the rate of deforestation, conserve the forests that remain, and protect the rights of indigenes. Is it already too late? The next few years may well prove decisive. The task ahead is monumental because of the large numbers of people, diverse and sometimes conflicting interests, and multiple causes, but fixing the problems is a matter of survival for life and humanity as a whole. Governments in the tropics and beyond need to become far more responsive and responsible in regard to deforestation—we believe the problem is most serious and urgent, requiring an internationally coordinated effort on a massive scale (see Ehrlich and Wilson 1991; Farnworth and Golley 1974; Lugo 1991; Myers 1992; National Research Council 1980, 1992; Parker 1992; Whitmore 1989; World Resources Institute et al. 1992).

Just as the causes and consequences of deforestation are many, it has numerous solutions (Budowski 1976; Elfring 1984; Head and Heinzman 1990; LeTacon and Harley 1990; McNeely et al. 1990a, 1990b; Murray 1990; Park 1992; Rubinoff 1983). However, any solutions ultimately must effectively deal with both elemental factors, need and greed.

Curing the Cancer of Society

The population explosion, now mainly in the developing countries, must be reduced as much as possible through, among other means, the effective implementation of adequate family-planning programs such as that in Thailand (Lappe and Schurman 1990). Much more must be done to satisfy the basic needs of the world's poor, to improve their quality of life, and to provide them with economic alternatives that are environmentally and socially sustainable in the long term (see essays 8 and 11 in this volume).

Simultaneously, the relatively affluent nations must abandon their ethos of materialism, consumerism, and growth mania, learn to consume much less and with greater efficiency, to recycle more, and in many other ways to fundamentally alter their lifestyles to also be more environmentally and socially sustainable in the long term. This will happen inevitably, if not voluntarily then more painfully, forced by such practical circumstances as the diminishing resources of the planet (Gunn 1994; Myers 1993b; Upreti 1994). Persuasive studies argue that a change in lifestyle would also improve the quality of life of the developed societies (Bodley 1994, chaps. 5, 10, 13; Head 1990; Myers 1992; Silver and DeFries 1990).

John H. Bodley defines the culture of consumption as a "culture whose major economic, social, and ideological systems are geared to nonsustainable levels of resource consumption and to continual, ever-higher elevation of those levels on a per capita basis" (1985:67). For example, the average American contributes about five times more than most other human beings to the global greenhouse effect, through fossil fuel combustion in cars and other waste products connected directly or indirectly to excessive and irresponsible consumption (Myers 1992:372). Paul Ehrlich argues that one American does twenty to one hundred times more damage to the planet than one person in an underdeveloped country (Ehrlich and Ehrlich 1991).

Basic changes in consumer behavior in developed countries would go a long way toward relieving some pressures on tropical forests (Gunn 1994; Head 1990). Japan, which imports about a quarter of the logs cut from tropical rain forests, uses much of the wood only once and for a short time, such as for molds in concrete construction for high-rise buildings and in disposable chopsticks. Environmental activist organizations like Rainforest Action Network advocate the boycott of selected tropical forest products (Hayes 1990). In recent years changes in consumer values and behavior have successfully reduced a significant portion of the pressure on African elephant populations for the ivory trade. The same could be done for the tropical timber trade. Solutions must come at many levels, from the government to the community and in terms of education, science, technology, economics, politics, religion, and values.

Impeding these concerns and measures are certain cultural values

(Ehrlich and Ehrlich 1992). In particular, the frontier mentality has dominated exploitation of tropical rain forests, especially by outsiders. That is, this biome has been viewed as empty of people, or underpopulated and unused or underused, yet rich in almost infinite resources that are free for the taking for the maximization of quick profits—in short, a place to conquer, master, and exploit to the fullest (Dickenson 1989; Shrader-Frechette 1981). Western economic development in tropical rain forests is predicated on ignorance as well as greed and operated for external interests rather than the benefit of residents, as demonstrated by the emphasis on the extraction of such raw resources as timber, wildlife, minerals, and energy (hydroelectric dams) for consumptive uses. In the process social and economic justice and equity for the people of the forest are disregarded, as are the rights of future generations of humanity (Hecht and Cockburn 1989:193–210).

Elite Solutions

However, more specific solutions to deforestation exist, and these are not simply future projections; many are actually already under way. These involve people from the leaders of government, business, and industry at the international and national levels to the grassroots level (Alpert 1993; Burwell, Helin, and Joyce 1994; Pasca 1988). Increasingly, people are recognizing that national security includes environmental security and that many environmental problems transcend national borders. People also are recognizing that resource depletion and environmental degradation have many hidden costs and that a society can be no healthier than its economy, which is rooted in its environment (see chapters 6, 8, 9, and 13). These and other considerations have been emerging as important factors in the political consciences of many nations since the first Earth Day and the Stockholm Conference in the early 1970s and received widespread publicity during the 1992 Earth Summit in Rio and in many other contexts and forms (Haas, Levy, and Parson 1992; Parson, Haas, and Levy 1992). However, the Food and Agriculture Organization of the United Nations (FAO) and other programs on forestry and deforestation have received mixed reviews (Lanly, Singh, and Janz 1991; Marshall 1991; Winterbottom 1990). Such programs as the Costa Rican government's collaboration in an international research and conservation effort to preserve that country's rich biodiversity have been better received (Blum 1993; Boza 1993; Reid et al. 1993; cf., chapter 9 in this volume). Also, the mass media have helped publicize the plight of indigenes and forests in the tropics, such as through Hollywood movies, including *The Emerald Forest, Medicine Man, Fern Gully,* and *The Mission.*

Community Solutions

The community level shows promising developments as well. Forest people are organizing to defend their human rights and to protect their habitats and conserve their natural resources, which they depend on for their survival and well-being (Bodley 1990; Clay 1993; Fisher 1994; Lutzenberger 1987). One of the most dramatic and successful cases is the Chipko movement, a nonviolent political strategy in the foothills of the Himalayas of India in which village women embrace trees to prevent logging (Haigh 1988; Morris 1986; Shiva 1990; Shiva and Bandyopadhyan 1987). In Thailand many Buddhist monks have become environmental activists, combating deforestation and championing reforestation (Sponsel and Natadecha-Sponsel 1988, 1993). In Kenya the ecofeminist Wangari Maathai has been the catalyst for the successful Greenbelt Movement in which thousands of women and children planted millions of trees (O'Keefe and Hosier 1983). Native Hawaiians and other residents of the island of Hawai'i have voiced substantial opposition to deforestation through geothermal and other development projects (Hannah 1990; Holden 1985; Sakai 1988). In the Brazilian Amazon, following the martyrdom of Chico Mendes, the Forest People's Alliance was successful in getting the government to establish several extractive reserves (Allegretti 1990; Browder 1992; Fearnside 1989c; Hecht and Cockburn 1989; Hyman 1988; Salafsky, Dugelby, and Terborgh 1993; Schwartzman 1989). In Haiti, which has suffered severe deforestation, tree-planting programs have achieved some success, thanks to applied anthropology (Kurlansky 1988; Lewis and Coffey 1985; Lugo, Schmidt, and Brown 1981; Pellek 1990; Murray 1987; Stevenson 1989). (Also see Bray and Irvine 1993; Croll and Parkin 1992; and Solo 1993.)

Another hopeful development is the hundreds of nongovernmental organizations (NGOs) concerned with deforestation and related issues (Hendee and Pitstick 1994; Kaufman and Mallory 1993; Mohd and Laarman 1994). For example, the Rainforest Action Network has nearly two hundred local offices in the United States, and a new one opens about every ten days (Hayes 1990). Local NGOs have succeeded where government policies have failed in India, according to Alcorn and Molnar (chapter 4 in this book). Conservation organizations have also contributed money to reduce the debts of countries in return for the establishment of protected areas (Gullison and Losos 1993; Visser and Mendoza 1994). However, considerable controversy surrounds the socioeconomic and environmental advantages and disadvantages of NGO projects for developing commercial extraction enterprises among local communities (Carr, Pedersen, and Ramaswamy 1993; Cory 1993; Dove 1993; Stiles 1992).

So far, the West has yet to demonstrate that it knows how to develop the tropical rain forest without seriously degrading or even destroying it. Economic dreams have become nightmares in virtually every respect (environmental, economic, medical, social, and political). Earlier in this century the Fordlandia rubber plantation in the Brazilian Amazon established a common pattern—huge initial capital investment, subsequent project failure, economic loss, and abandonment. Fordlandia cost $10 million and was sold for only $500,000. The Jari paper pulp plantation in the Brazilian Amazon is just the latest nightmare. This transpired despite the billionaire status of its owner, the late Daniel Ludwig. Had the plantation used them fully, Ludwig had at his disposal considerable technological and scientific expertise, as well as enormous capital (Hecht and Cockburn 1989; McNabb, Borges, and Welker 1994; Schmink 1988; Stone 1992).

On the other hand, a large group of scientists has marshaled arguments and evidence to suggest that many indigenes, like the Bora and Desana in Amazonia and the Mentawaians in Indonesia, are effective conservationists; their societies have used and managed the natural resources of their habitat in sustainable ways for centuries or even millennia (Anderson 1990; Clay 1991, 1993; Denevan and Padoch 1988; Dufour 1990; Gadgil, Berkes, and Folke 1993; Hecht 1982; Hyndman 1994; Lamb 1991; Posey 1989b; Posey et al. 1984; Reichel-Dolmatoff 1971, 1976; Schefold 1988; Sponsel 1992a; Treacy 1982; Winterbottom and Hazlewood 1987). This was recognized recently when the president of Colombia awarded indigenes the rights to their ancestral lands in Colombia's Amazon territory (Bunyard 1989). Southeast Asia also has ancient traditions of agroforestry and other forms of forest management (Conklin 1957; Poffenberger 1990; Rush 1991). The extent to which indigenes may know and use the forest is suggested by botanical studies of sample plots that reveal that a local community may identify and use 49 to 79% of the tree species in the surrounding forest (Prance 1990:58–59). For example, the Ka'apor recognize 768 plant species, from the stages of seed to reproductive adult (Balée 1994; see also Denevan and Padoch 1988; Plotkin and Famolare 1992). In chapter 12 Cook recommends the intensification of traditional staple crops as a source of economic development for the Amung-me and other peoples of Irian Jaya and Papua New Guinea.

Others have marshaled arguments and evidence against a conservation ethic or sustainable practices for some indigenous groups (Alcorn 1993; Alvard 1993; Dei 1990; Diamond 1992; Edgerton 1992; Hames 1991; Headland 1994; McGlone 1983; McLarney 1993; Parker 1993; Rambo 1985; Redford 1991; Redford and Robinson 1985; Redford and Stearman 1993). Nevertheless there is growing recognition that the values, knowledge, wisdom, and skills of indigenes in the use, management, and conservation of land and resources can be of considerable practical value for planning, implementing, and monitoring sustainable economic development and the conservation of forest ecosystems and biodiversity. Also, to be successful developers and conservationists must consider the interest, needs, and concerns of local people who live in and adjacent to forests. For example, when the Khao Yai Park was established in Thailand, officials initially encountered serious problems with local "poachers" who had traditionally used the forest resources, until they were given economic alternatives and assistance (Brockelman 1987). Dian Fossey's active conservation on behalf of the mountain gorillas (1983), which included aggressively combating poachers in the vicinity of her famous Karisoke Research Center, most tragically cost her her life. Had the local human population, including the poachers, been provided with environmental education, practical training, and economic alternatives, the conservation of the gorillas and their habitat might have been more successful and less costly, as subsequent initiatives appear to have proved (Mowat 1987).

Increasingly, experts in such fields such as wildlife management (Eltringham 1985; Holdgate 1993; Redford and Padoch 1992), tropical forestry and agroforestry (Anderson 1990; Bonnicksen 1991; Lamb 1987; Lugo 1991; Parker 1993), environmental and biodiversity conservation (Cohn 1988; Gradwohl and Greenberg 1988; Kemp 1993; McNeely et al. 1990a, 1990b; Oldfield and Alcorn 1991), economic development (Bodley 1988; Croll and Parkin 1992; McNeely and Pitt 1984; Oldfield and Alcorn 1991), and ethnobotany (Plotkin and Famolare 1992) are becoming aware of the crucial importance of considering the local people and collaborating with them. Describing the usual approach to economic development, George Appell writes, "A development act is any act by an individual who is not a member of a local society that devalues . . . their relationship with their natural and social world" (1975:33).

Further, Appell observes that "every act of development involves, of necessity, an act of destruction. This destruction— social, ecological, or both—is seldom accounted for in development projects, despite the fact that it may entail costs that far outweigh the benefits arising from development" (p. 31).

On the other hand, conservation education for local communities may be helpful (Godoy 1994; Padua 1994; Stapp and Polunin 1991).

Forest Management

In the past, state management of forests that alienates local communities has often contributed to deforestation, whereas community management is more likely to contribute to forest conservation, although not always (Lohmann 1993; Poffenberger 1990; see also chapters 4, 5, and 9). Today more and more development workers are recognizing that economic development and environmental conservation must work together instead of against each other to create sustainable uses of forests that benefit local as well as external societies (Clay 1991). The interests of the indigenes and other peoples of the forest and the interests of environmentalists are converging.

The long-term costs of the short-term benefits of the exploitation of tropical forests by outsiders have yet to be realistically calculated and fully appreciated. Part of the solution is to recognize the value of "externalities"—the environmental services such as climate regulation performed by the forest ecosystem and the sustainable harvesting of nontimber forest products (Godoy and Lubowski 1992; Plotkin and Famolare 1992). These are nondestructive uses that do not depend on converting forest to other land uses. They harvest the natural interest rather than the capital of the forest. The new field of ecological economics holds promise in this regard (Barbier, Burgess, and Markandya 1991; Cartwright 1989; Farnworth et al. 1983; Katzman and Cale 1990; Leonard 1985; McNeely and Dobias 1991; McNeely 1989; Patterson 1990; Raloff 1988; Randall 1991; Repetto 1987, 1992; Ryan 1992; United Nations Eduation [UNESCO] 1991).

Scientists now recognize a whole spectrum of silvicultural systems (Gómez-Pompa and Burley 1991:5). Examples are the strictly protected core areas of biosphere reserves, extraction reserves such as those in Brazil where selective logging and the exploitation of nontimber forest products are allowed, forest conversion or replacement with commercial tree plantations such as the rubber, coconut, and eucalyptus monocrops in Thailand and elsewhere in Southeast Asia, and restoration of forests such as in the Guanacaste area of Costa Rica (also see Gómez-Pompa, Whitmore, and Hadley 1991; Posey and Balée 1989).

Many scientists and environmentalists may perceive that nothing is more valuable than a natural wilderness area that has not suffered significant anthropogenic disturbances or transformations (Hannah et al. 1994; McCloskey and Spalding 1989; Oelschlaeger 1991). However, in areas such as the Amazon the appearance of wilderness may be part of the aftermath of the massive depopulation of indigenous societies through introduced epidemic diseases and other effects of Western contact (Denevan 1992a; Wood 1992). In reality it is improbable that many if any pristine forests exist (Balée 1994; Gómez-Pompa and Kaus 1992; Heizer 1955; Pelzer 1968). In any event the concept of a wilderness area in which no person (other than a few biologists) may ever tread is increasingly an impractical Western luxury, given the practical needs of many societies in the tropics, especially with their rapid population growth and expansion into forest zones. Thus managing forests wisely, rather than simply demarcating and guarding supposed wilderness areas, is the real challenge for future forest conservation (see chapter 14 in this volume).

Anthropology's Role

Radical ecology emerges from a sense of crisis in the industrialized world. It acts on a new perception that the domination of nature entails the domination of human beings along lines of race, class, and gender. Radical ecology confronts the illusion that people are free to exploit nature and to move in society at the expense of others, with a new consciousness of our responsibilities to the rest of nature and to other humans. It seeks a new ethic of the nurture of nature and the nurture of people. It empowers people to make changes in the world consistent with a new social vision and a new ethic.   —*Merchant 1992:1*

People

Anthropology can make a significant contribution to understanding the causes and determining the consequences of deforestation, and it will be integral to most efforts to develop solutions (Denslow and Padoch 1988; F. B. Lamb 1987; D. Lamb 1991). Anthropologists also need to develop a greater role in tropical forestry education (Lamb 1987; Putz 1989). What anthropologists have always done best is work closely with people, and it is people who are responsible for deforestation (although not necessarily all the people with whom anthropologists work). Thus, even though contemporary anthropologists may take advantage of new technologies such as satellite imagery, information derived from anthropological and other kinds of field observations remains indispensable (see chapter 13 in this book).

The hallmark of cultural anthropology is its special method of field research, participant observation, which ideally includes gaining competence in the local language (Spradley 1979, 1980). Anthropologists study people and their culture and ecology by participating as well as by observing in their communities on an intimate daily basis for months or even years. This involves establishing rapport in a host community, learning to act so that the people go about their business as usual, and recording and analyzing the findings daily (Bernard 1994). For the most part anthropologists in tropical forest areas have used this approach to study people in small-scale societies. (Among the more important ethnographic case studies of forest societies are Chernela 1993; Dwyer 1990; Graeve 1989; Hayano 1990; Lizot 1985; McGee 1990; Rai 1990; Reichel-Dolmatoff 1971; Stearman 1989; Turnbull 1961, 1983.)

Accordingly, anthropologists are able to provide a unique human perspective on the phenomena and processes of deforestation. They can see and portray deforestation from the people's viewpoint, a perspective that is seldom if ever described or considered by experts from other fields (e.g., Murray 1987). Also, in many cases the anthropologist adopts the point of view of the victims of deforestation. For example, because he has worked closely for years with forest people in Central Africa, Wilkie is able to convey in chapter 10 the meaning of commercial logging operations for the subsistence and social relations of both foragers and farmers. Sometimes anthropologists even go further than observing, analyzing, and reporting field data; they may become advocates for the survival, welfare, identity, rights, and self-determination of a community (Bodley 1990; Downing and Kushner 1988; Johnston 1994; Sponsel 1992b, 1994; Wright 1988). (In this volume Alcorn and Molnar, Bailey, and Wilkie all mention human rights, especially land rights, as a factor in curbing deforestation and promoting appropriate management and conservation of forests.)

Anthropologists can help to empower local forest communities, as Bailey suggests in chapter 14. In addition, in the arena of the interaction of forest peoples, environmentalists, and developers, where conflicts of interest often arise, the anthropologist may be in a strategic position to serve as translator and mediator.

Although they have tended to work with foragers and swiddeners in the forest, in principle anthropologists are just as capable and should be just as relevant in studying and interpreting the ideas, behavior, and culture of other participants, including the foresters, loggers, ranchers,politicians, and bankers (Dove 1983, 1992). Two relatively new but rapidly growing fields are development and policy anthropology, which seek to understand the forces involved in sociocultural change as well as economic development (Partridge 1984; van Willigen and DeWalt 1985). Increasingly, these two new fields, together with economic and ecological anthropology, can provide insights to the phenomena and processes of deforestation that other approaches do not provide, as illustrated by most case studies in this book. (For policy statements in this volume, see chapters 6, 8, and 14 in particular.)

Cross-Cultural Approach

Anthropology also contributes to studies of the causes, consequences, and solutions of deforestation through its cross-cultural approach to problems. Because anthropologists are familiar with a broad spectrum of human societies across time and in various geographical settings, they can adopt a broader view of problems. By virtue of having learned of a variety of responses to similar problems, the anthropologist often can see beyond the solution used by one institution or culture to offer an approach borrowed from some distant circumstance (see Castro 1992). Furthermore, controlled comparisons can reveal powerful insights, as illustrated especially well in the essays by Cook and Vandemeer.

A cross-cultural perspective also offers a tolerance for differences, indeed even an enthusiasm for considering differences as attributes rather than as obstacles or faults to be minimized or eliminated altogether. (Readers will notice this perspective in all essays in this book.) Anthropologists' special appreciation of cultural diversity is important in relation to tropical forests and deforestation, because countries such as Brazil, Indonesia, and Zaire are also especially high in cultural diversity (Clay 1993; Gadgil, Berkes, and Folke 1993; Hyndman 1994; McNeely 1993; Nelson 1992; Oldfield and Alcorn 1991; Shiva et al. 1991).

Diachronic Approach

Yet another dimension of comparison is temporal. The diachronic approaches of anthropology (that is, those that include the historical antecedents when studying a subject) include prehistoric and historic archaeology, ethnohistory, historical ecology, and historical linguistics. They can all be important in understanding deforestation as well as the ecology, land and resource use, management, and conservation of forests. People have lived in and adjacent to tropical forests for centuries or even millennia. Good examples come from the studies of the Maya in Central America (Abrams and Rue 1988; Gómez-Pompa and Kaus 1992; Murdy 1990; Nations and Nigh 1980; Rice and Rice 1984), as well as other groups in the Amazon (Balée 1989, 1992, 1994; Moran 1983, 1993; Roosevelt 1980, 1989, 1992; Sponsel 1992a), central Africa (Bailey et al. 1989; Vansina 1990), and Negritos in the Philippines (Headland 1987, 1988; Headland and Reid 1989, 1991). Through the study of the remains and contexts of human activities archaeology provides lessons for the present and even for the future. For example, deforestation may be at least one factor in the collapse of many complex prehistoric societies, including the Maya, the Harrapan of the Indus River valley in India, the Borobudor in Java, and the Khmer of Angkor, Cambodia (e.g., Diamond 1992). (Several essays in this volume incorporate diachronic approaches, especially those in parts 2 and 3 and chapters 7 and 13.)

Diachronic approaches are also important for understanding the ecology and cultures of refugee and other societies in the forest (e.g., Headland and Reid 1989, 1991). Societies do not change only in response to their ecosystems but also in reaction to other societies, including contact societies, as in the context of colonialism and neocolonialism (Bodley 1990; Crosby 1972, 1986; Wolf 1982). Indeed for many indigenous societies in the forest the greatest adaptive challenge or hazard has become the effects of outside influences, including the external agents of deforestation (Bahuchet and Maret 1993; Clay 1988, 1989; Gray 1991; Hecht and

Cockburn 1989; ICIHI 1986; Solo 1993; Sponsel 1994, 1995). Studying and understanding how cultures respond to agents of change provides insights into how some mistakes may be avoided. This is revealed, for example, by Castro's discussion (essay 5) of state-sponsored farm forestry in one region of Kenya from 1912 to 1963, a case that provides lessons useful for planning afforestation efforts today (see also Poffenberger 1990).

Anthropology is distinctive in its research repertoire, which can combine several special approaches: holistic (biocultural and systems), evolutionary, comparative (cross-cultural and diachronic), ethnographic (participant observation and interviewing), and emic (folk or native viewpoint). By marshaling this variety of approaches anthropology is unique among all the basic and applied fields working on economic development and environmental conservation in tropical forest zones. Experts in development, planning, economics, tropical ecology, tropical forestry, conservation, and the like do not enjoy the combined tool kit of the anthropologist, which provides special insights into the lives and viewpoints of the people of the forest as they are embedded in ecology, culture, and history. In recent years ecologically oriented studies in anthropology have increasingly embraced political economy in the context of environmental history, as the essays in this book demonstrate. For these and other reasons we believe that anthropology can and should take a larger and more central role in tackling the difficult issues entailed in solving the growing crisis of deforestation.

Examples abound of how anthropology can and has contributed to understanding the phenomena and processes of deforestation, including its effects on people and their habitat. We have cited many here. However, they are spread across many years and throughout numerous sources. One goal in writing this essay was to provide a fairly thorough review of the pertinent literature, an extensive bibliography as background for the remaining essays, and a resource for future research and teaching. The goal of this book is to provide a representative sample of such case studies in a single volume, although topical and regional gaps inevitably exist because of space limitations. However, the coeditors and authors of this volume hope that it will help stimulate more research and teaching on the human ecology of forests and deforestation in the tropics and that future generations of students, among others, will help fill some gaps remaining in this book and the literature beyond.

Acknowledgments   We would like to express our appreciation to Ursula Thiele for her meticulous bibliographic and editorial assistance with this essay. Any deficiencies are ours.

References Cited

Abrams, E. M. and D. J. Rue. 1988. The causes and consequences of deforestation among the prehistoric Maya. *Human Ecology* 16 (4): 377–395.
Adeola, Moses Olanre. 1992. Importance of wild animals and their parts in the culture, religious festivals, and traditional medicine of Nigeria. *Environmental Conservation* 19 (2): 125–134.
Aiken, S. Robert and Michael R. Moss. 1975. Man's impact on the tropical rainforest of peninsular Malaysia: A review. *Biological Conservation* 8: 213–230.
Albert, Bruce. 1994. Gold miners and Yanomami Indians in the Brazilian Amazon: The Hashimu massacre. In *Who Pays the Price? The Sociocultural Context of Environmental Crisis*, Barbara Rose Johnston, ed., pp. 47–55. Washington, D.C.: Island Press.
Alcorn, Janis B. 1981. Huastec noncrop resource management: Implications for prehistoric rain forest management. *Human Ecology* 9 (4): 395–417.
———. 1993. Indigenous peoples and conservation. *Conservation Biology* 7 (2): 424–426.
Allegretti, M. H. 1990. Extractive reserves: An alternative for reconciling development and environmental conservation in Amazonia. In *Alternatives to Deforestation: Steps Toward Sustainable Use of the Amazon Rain Forest*, Anthony B. Anderson, ed., pp. 252–264. New York: Columbia University Press.
Allen, Julia C. and Douglas F. Barnes. 1985. The causes of deforestation in developing countries. *Annals of the Association of American Geographers* 75 (2): 163–184.
Alpert, P. 1993. Conserving biodiversity in Cameroon. *Ambio* 22 (1): 44–49.
Alvard, Michael S. 1993. Testing the Ecologically Noble Savage Hypothesis: Conservation and Subsistence Hunting by the Piro of Amazonian Peru. Ph.D. diss., University of New Mexico.
Anderson, Anthony B., ed. 1990. *Alternatives to Deforestation: Steps Toward Sustainable Use of the Amazon Rain Forest.* New York: Columbia University Press.
Appell, George N. 1975. The pernicious effects of development. *Fields Within Fields* 14: 31–45.
Atran, Scott. 1993. Itza Maya tropical agro-forestry. *Current Anthropology* 34 (5): 633–700.
Bahuchet, Serge and Pierre de Maret. 1993. *Situation des populations indigenes des forêts denses humides* (The situation of the indigenous peoples of the tropical rain forests). Parusi, France: Centre National de la Recherche Scientifique.
Bahuchet, Serge, Doyle McKey, and Igor de Garine. 1991. Wild yams revisited: Is independence from agriculture possible for rain forest hunter gatherers? *Human Ecology* 19: 213–244.
Baidya, Kedar N. 1984. Firewood shortage: Ecoclimatic disasters in the Third World. *International Journal of Environmental Studies* 22: 255–272.
Bailey, Robert C. 1990. Exciting opportunities in tropical rain forests: A reply to Townsend. *American Anthropologist* 92 (3): 747–748.

———. 1991. *The Behavioral Ecology of Efe Pygmy Men in the Ituri Forest, Zaire*. Ann Arbor: University of Michigan, Anthropological Papers, Museum of Anthropology, No. 86.

Bailey, Robert C. and Thomas N. Headland. 1991. The tropical rain forest: Is it a productive environment for human foragers? *Human Ecology* 19: 261–285.

Bailey, Robert C., G. Head, M. Jenike, B. Owen, R. Rechtman, and E. Zechenter. 1989. Hunting and gathering in tropical rain forest: Is it possible? *American Anthropologist* 91 (1): 58–82.

Balée, William. 1989. The culture of Amazonian forests. *Advances in Economic Botany* 7: 1–21.

———. 1992. People of the fallow: A historical ecology of foraging in lowland South America. In *Conservation of Neotropical Forests*, Kent H. Redford and C. Padoch, eds., pp. 35–57. New York: Columbia University Press.

———. 1993. Indigenous transformation of Amazonian forests. *L'Homme* 33: 231–254.

———. 1994. *Footprints of the Forest: Ka'apor Ethnobotany—The Historical Ecology of Plant Utilization by an Amazonian People*. New York: Columbia University Press.

Barbier, E. B., J. C. Burgess, and A. Markandya. 1991. The economics of tropical deforestation. *Ambio* 20 (2): 55–58.

Barrau, Jacques. 1982. Plants and men on the threshold of the twenty-first century. *Social Science Information* 21 (1): 127–141.

Beckerman, Stephen. 1979. The abundance of protein in Amazonia: A reply to Gross. *American Anthropologist* 81 (3): 533–560.

Berger, Thomas R. 1991. *A Long and Terrible Shadow: White Values, Native Rights in the Americas*. Seattle: University of Washington Press.

Bernard, H. Russell. 1994. *Research Methods in Anthropology: Qualitative and Quantitative Approaches*. Thousand Oaks, Calif.: Sage.

Berwick, Dennison, 1992. *Savages: The Life and Killing of the Yanomami*. London: Hodder and Stoughton.

Blankespoor, Gilbert W. 1991. Slash and burn shifting agriculture and bird communities in Liberia, West Africa. *Biological Conservation* 57: 41–71.

Blum, Elissa. 1993. Making biodiversity conservation profitable: A case study of the Merck/INBio agreement. *Environment* 35 (4): 16–20, 38–45.

Bodley, John H. 1985. *Anthropology and Contemporary Human Problems*. Menlo Park, Calif.: Cummings.

———. 1990. *Victims of Progress*, 3d ed. Mountain View, Calif.: Mayfield.

———. 1994. *Cultural Anthropology: Tribes, States, and the Global System*. Mountain View, Calif.: Mayfield.

Bodley, John H., ed. 1988. *Tribal Peoples and Development Issues: A Global Overview*. Mountain View, Calif.: Mayfield.

Bonnicksen, Thomas M. 1991. Managing biosocial systems: A framework to organize society environment relationships. *Journal of Forestry* 89 (10): 10–15.

Bowonder, B. 1986. Deforestation in developing countries. *Journal of Environmental Systems* 15 (2): 171–192.

Bowonder, B., S. S. R. Prasad, and N. V. M. Unni. 1987. Deforestation around urban centers in India. *Environmental Conservation* 14: 23–28.

Boza, Mario A. 1993. Conservation in action: Past, present, and future of the national park system of Costa Rica. *Conservation Biology* 7 (2): 239–247.

Bray, David Barton and Dominique Irvine, eds. 1993. Resources and sanctuary: Indigenous peoples, ancestral rights, and the forests of the Americas. *Cultural Survival Quarterly* 17 (1): 12–64.

Brockelman, Warren. 1987. Nature conservation. In *Thailand: Natural Resources Profile*, Anat Arbhabhirama, ed., pp. 90–119. Bangkok: Thailand Development Research.

Brockelman, W. Y. and P. Dearden. 1990. The role of nature trekking in conservation: A case study in Thailand. *Environmental Conservation* 17 (2): 141–148.

Brookfield, Harold C. 1988. The new great age of clearance and beyond. In *People of the Tropical Rain Forest*, Julie Sloan Denslow and Christine Padoch, eds., pp. 205–224. Berkeley: University of California Press.

Brookfield, Harold and Christine Padoch. 1994. Appreciating agrodiversity: A look at the dynamism and diversity of indigenous farming practices. *Environment* 36 (5): 6–11, 37–45.

Brosius, J. Peter. 1991. Foraging in tropical rain forests: The case of the Penan of Sarawak, East Malaysia (Borneo). *Human Ecology* 19: 123–150.

Browder, John O. 1992. The limits of extractivism. *BioScience* 42 (3): 174–182.

Bruijnzeel, L. A. 1991. Hydrological impacts of tropical forest conversion. *Nature and Resources* 27 (2): 36–46.

Budowski, G. 1976. Why save tropical forests? Some arguments for campaigning conservationists. *Amazoniana* 5 (4): 529–538.

Bunker, Stephen G. 1989. The eternal conquest. *North American Conference on Latin America Report on the Americas* 23 (1): 27–35.

Bunyard, Peter. 1985. World climate and tropical deforestation. *Ecologist* 15 (3):125–136.

——. 1989. Guardians of the forest: Indigenous policies in the Colombian Amazon. *Ecologist* 19 (6): 255–258.

Burger, Julian. 1987. *Report from the Frontier: The State of the World's Indigenous Peoples*. Atlantic Highlands, N.J.: Zed Books.

Burkhalter, S. Brian and Robert F. Murphy. 1989. Tappers and sappers: Rubber, gold and money among the Mundurucu. *American Ethnologist* 16 (1): 100–116.

Burnie, David. 1994. Ecotourists to paradise. *New Scientist* 142 (1921): 23–27.

Burwell, Bruce B., William H. Helin, and Steven D. Joyce. 1994. A shared vision: Ghana's collaborative community forestry initiative. *Journal of Forestry* 92 (6): 18–23.

Buschbacher, Robert J. 1986. Tropical deforestation and pasture development. *BioScience* 36 (1): 22–28.

Carneiro da Cunha, Mannela. 1989. Native realpolitik. *North American Conference on Latin America Report on the Americas* 23 (1): 19–22.

Carneiro, Robert L. 1979a. Forest clearance among the Yanomamo: Observations and implications. *Antropologica* 52: 39–76.

———. 1979b. Tree felling with the stone axe: An experiment carried out among the Yanomamo Indians of southern Venezuela. In *Ethnoarchaeology: Implications of Ethnography for Archaeology,* C. Kramer, ed., pp. 21–58. New York: Columbia University Press.

———. 1988. Indians of the Amazonian forest. In *People of the Tropical Rain Forest,* Julie Sloan Denslow and Christine Padoch, eds., pp. 73–86. Berkeley: University of California Press.

Carr, Thomas A., Heather L. Pedersen, and Sunder Ramaswamy. 1993. Rainforest entrepreneurs: Cashing in on conservation. *Environment* 35 (7): 12–38.

Cartwright, John. 1989. Conserving nature, decreasing debt. *Third World Quarterly* 11 (2): 114–127.

Castro, A. H. P. 1992. Social forestry: A cross-cultural analysis. In *Ecosystem Rehabilitation, Vol. 1: Policy Issues,* Mohan K. Wali, ed., pp. 63–78. The Hague: SPB Academic Publishing.

Chernela, Janet M. 1993. *The Wanano Indians of the Brazilian Amazon: A Sense of Space.* Austin: University of Texas Press.

Clark, Colin. 1992. Empirical evidence for the effect of tropical deforestation on climate change. *Environmental Conservation* 19 (1): 39–47.

Clarke, W. C. 1976. Maintenance of agriculture and human habitats within the tropical forest ecosystem. *Human Ecology* 4 (3): 247–259.

Clay, Jason W. 1989. Brazil: Who pays for development *Cultural Survival Quarterly* 13 (1): 1–47.

———. 1991. Cultural survival and conservation: Lessons from the past twenty years. In *Biodiversity: Culture, Conservation, and Ecodevelopment,* Margery L. Oldfield and Janis B. Alcorn, eds., pp. 248–273. Boulder, Colo.: Westview.

———. 1993. Looking back to go forward: Predicting and preventing human rights violations. In *State of the Peoples: A Global Human Rights Report on Societies in Danger,* Marc S. Miller, ed., pp. 64–71. Boston: Beacon Press.

Clay, Jason W., ed. 1988. *Indigenous Peoples and Tropical Forests: Models of Land Use and Management from Latin America.* Report No. 27. Cambridge, Mass.: Cultural Survival.

Clement, Charles R. 1989. A center of crop genetic diversity in Western Amazonia. *BioScience* 39 (9): 624–630.

Cohn, Jeffrey A. 1988. Culture and conservation. *BioScience* 38 (7): 450–453.

Colchester, Marcus. 1991. Guatemala: The clamor for land and the fate of the forests. *Ecologist* 21 (4): 177–185.

———. 1994. The new sultans: Asian loggers move in on Guyana's forests. *Ecologist* 26 (2): 45–52.

Colchester, Marcus and Larry Lohmann, eds. 1993. *The Struggle for Land and the Fate of the Forests.* Atlantic Highlands, N.J.: Zed Books.

Conklin, Harold C. 1957. *Hanunoo Agriculture.* Rome: United Nations.

Coppens, Yves. 1994. East side story: The origin of humankind. *Scientific American* 270 (5): 88–95.

Cory, Stephen. 1993. The rainforest harvest: Who reaps the benefit? *Ecologist* 23 (4): 148–153.

Croll, Elizabeth and David Parkin, eds. 1992. *Bush Base: Forest Farm, Culture, Environment and Development.* New York: Routledge.

Crosby, Alfred W. 1972. *The Colombian Exchange: Biological and Cultural Consequences of 1492.* Westport, Conn.: Greenwood.

———. 1986. *Ecological Imperialism: The Biological Expansion of Europe, 900–1900.* New York: Cambridge University Press.

Cultural Survival. 1987. *Southeast Asian Tribal Groups and Ethnic Minorities.* Report No. 22. Cambridge, Mass.: Cultural Survival.

Dei, George S. 1990. Deforestation in a Ghanaian rural community. *Anthropologica* 32: 3–27.

Denevan, William M. 1992a. The pristine myth: The landscape of the Americas in 1492. *Annals of the Association of American Geographers* 82 (3): 369–385.

———. 1992b. Stone versus metal axes: The ambiguity of shifting cultivation in prehistoric Amazonia. *Journal of the Steward Anthropological Society* 20: 153–165.

Denevan, William M. and Christine Padoch. 1988. Swidden fallow agroforestry in the Peruvian Amazon. *Advances in Economic Botany* 5: 1–107.

Denslow, Julie Sloan and Christine Padoch, eds. 1988. *People of the Tropical Rain Forest.* Berkeley: University of California Press.

Detwiler, R. P. and C. A. S. Hall. 1988. Tropical forests and the global carbon cycle. *Science* 239: 42–47.

Dewees, Peter A. 1989. The woodfuel crisis reconsidered: Observations on the dynamics of abundance and scarcity. *World Development* 17 (8): 1159–1172.

Diamond, Jared. 1992. The golden age that never was. In his *The Third Chimpanzee,* pp. 317–338. New York: HarperCollins.

Dickenson, J. P. 1989. Development in Brazilian Amazonia: Background to new frontiers. *Revista Geografica* 109: 141–155.

Dinerstein, Eric and Eric D. Wikramanayake. 1993. Beyond "hotspots": How to prioritize investments to conserve biodiversity in the Indo-Pacific region. *Conservation Biology* 7 (1): 53–65.

Dove, Michael R. 1983. Theories of swidden agriculture and the political economy of ignorance. *Agroforestry Systems* 1: 85–99.

———. 1992. Foresters' beliefs about farmers: A priority for social science research in social forestry. *Agroforestry Systems* 17: 13–41.

———. 1993. A revisionist view of tropical deforestation and development. *Environmental Conservation* 20 (1): 17–24, 55–56.

Dovers, Stephen R. and John W. Handmer. 1993. Contradictions in sustainability. *Environmental Conservation* 20 (3): 217–222.

Downing, Theodore E. and Gilbert Kushner, eds. 1988. *Human Rights and Anthropology.* Cambridge, Mass.: Cultural Survival.

Downing, Theodore E., Susanna B. Hecht, Henry A. Pearson, and Carmen Garcia-Downing, eds. 1992. *Development or Destruction: The Conversion of Tropical Forest to Pasture in Latin America.* Boulder, Colo.: Westview.

Dufour, Darna L. 1990. Use of tropical rainforests by native Amazonians. *BioScience* 40 (9): 652–659.

Dunn, F. L. 1975. *Rain Forest Collectors and Traders: A Study of Resource Utilization in Modern and Ancient Malaya.* Kuala Lumpur: Monographs of the Malaysian Branch of the Royal Asiatic Society, No. 5.

Durning, Alan T. 1989. Cradles of life. *World Watch* 2 (3): 30–43.

———. 1993. Supporting indigenous peoples. In *State of the World 1993.* Lester R. Brown, ed., pp. 80–100. New York: W. W. Norton.

Dwyer, Peter. 1990. *Pigs That Ate the Garden: A Human Ecology from Papua, New Guinea.* Ann Arbor: University of Michigan Press.

Dwyer, Peter D. and Monica Minnegal. 1991. Hunting in lowland tropical rain forest: Toward a model of nonagricultural subsistence. *Human Ecology* 19: 187–212.

Eckholm, D., G. Foley, and G. Bernard. 1984. *Fuelwood: The Energy Crisis That Won't Go Away.* London: Earthscan.

Ecologist. 1985. To eat or develop? That is the question. *Ecologist* 15 (5/6): 202–305.

———. 1986. Can we trust the World Bank? *Ecologist* 16 (2/3): 57–128.

———. 1987. Special issue: Save the forests: Save the planet. *Ecologist* 17(4/5): 129–204.

Edgerton, Robert B. 1992. *Sick Societies: Challenging the Myth of Primitive Harmony.* New York: Free Press.

Eggert, Manfred K. H. 1993. Central Africa and the archaeology of the equatorial rainforest: Reflections on some major topics. In *The Archaeology of Africa: Food, Metals, and Towns,* Thurstan Shaw, Paul Sinclair, Bassey Andah, and Alex Okpoko, eds., pp. 289–329. New York: Routledge.

Ehrlich, Paul R. and Anne H. Ehrlich. 1991. *The Population Explosion.* New York: Simon & Schuster.

———. 1992. The value of biodiversity. *Ambio* 21 (3): 219–226.

Ehrlich, Paul R. and Edward O. Wilson. 1991. Biodiversity studies: Science and policy. *Science* 253 (5021): 758–762.

Ekachai, Sanitsuda. 1990. *Behind the Smile: Voices of Thailand.* Bangkok: Thai Development Support Committee.

Ekanade, Olusegun. 1987. Small-scale cocoa farmers and environmental change in the tropical rain forest regions of southwestern Nigeria. *Journal of Environmental Management* 25: 61–70.

Elfring, Chris. 1984. Can technology save tropical forests? *BioScience* 34: 350–352.

Eltringham, S. K. 1984. *Wildlife Resources and Economic Development.* New York: Wiley.

Endicott, Kirk and Peter Bellwood. 1991. The possibility of independent foraging in the rain forest of peninsular Malaysia. *Human Ecology* 19: 151–186.

Farnworth, Edward G. and Frank B. Golley, eds. 1974. *Fragile Ecosystems: Evaluation of Research and Applications in the Neotropics.* New York: Springer-Verlag.

Farnworth, Edward G., Thomas H. Tidrick, Webb M. Smathers, and Carl F. Jordan. 1983. A synthesis of ecological and economic theory toward more complex valuation of tropical moist forests. *International Journal of Environmental Studies* 21: 11–28.

Fearnside, Philip M. 1986. Settlement in Rondonia and the token role of science and technology in Brazil's Amazonian development planning. *Interciencia* 11 (5): 229–236.

———. 1989a. Brazil's Balbina Dam: Environment versus the legacy of the pharaohs in Amazonia. *Environmental Management* 13 (4): 401–423.

———. 1989b. The charcoal of Carajas: A threat to the forests of Brazil's eastern Amazon region. *Ambio* 28 (2): 141–143.

———. 1989c. Extractive reserves in Brazilian Amazonia. *BioScience* 39 (6): 387–393.

Fimbel, Cheryl. 1994. The relative use of abandoned farm clearings and old forest habitats by primates and a forest antelope at Tiwai, Sierra Leone, West Africa. *Biological Conservation* 70: 277–286.

Fisher, William H. 1994. Megadevelopment, environmentalism, and resistance: The institutional context of Kayapó indigenous politics in central Brazil. *Human Organization* 53 (3): 220–232.

Flenley, J. R. 1979. *The Equatorial Rain Forest: A Geological History*. London: Butterworth.

Fossey, Dian. 1983. *Gorillas in the Mist*. Boston: Houghton Mifflin.

Gadgil, Madhav, Fikret Berkes, and Carl Folke. 1993. Indigenous knowledge for biodiversity conservation. *Ambio* 22 (2–3): 151–156.

Gentry, A. H. and J. Lopez-Parodi. 1980. Deforestation and increased flooding of the Upper Amazon. *Science* 210: 1354–1356.

Giaimo, Michael S. 1988. Deforestation in Brazil: Domestic political imperative global ecological disaster. *Environmental Law* 18 (3): 537–570.

Gillis, Malcolm. 1988. The logging industry in tropical Asia. In *People of the Tropical Rain Forest*, Julie Sloan Denslow and Christine Padoch, eds., pp. 177–184. Berkeley: University of California Press.

Glover, Ian, ed. 1992. Special issue on the humid tropics. *World Archaeology* 24 (1): 1–165.

Godfrey, Brian J. 1990. Boom towns of the Amazon. *Geographic Review* 80 (2): 103–117.

Godoy, Ricardo. 1994. The effects of rural education on the use of the tropical rain forest by the Sumu Indians of Nicaragua: Possible pathways, qualitative findings, and policy options. *Human Organization* 53 (3): 233–244.

Godoy, Ricardo and Ruben Lubowski. 1992. Guidelines for the economic valuation of nontimber tropical-forest products. *Current Anthropology* 33 (4): 423–433.

Gómez-Pompa, Arturo and F. W. Burley. 1991. The management of natural tropical forests. In *Rain Forest Regeneration and Management*, A. Gómez-Pompa, T. C. Whitmore, and M. Hadley, eds., pp. 3–18. Park Ridge, N.J.: Parthenon.

Gómez-Pompa, Arturo and Andrea Kaus. 1992. Taming the wilderness myth. *BioScience* 42 (4): 271–279.

Gómez-Pompa, A., S. Flores, and V. Sosa. 1987. The "Petkot": A man-made tropical forest of the Maya. *Interciencia* 12 (1): 10–15.

Gómez-Pompa, Arturo, T. C. Whitmore, and M. Hadley, eds. 1991. *Rain Forest Regeneration and Management*. Park Ridge, N.J.: Parthenon.

Goodland, Robert J. A., Anastacio Juras, and Rajendra Pachauri. 1993. Can hydro-reservoirs in tropical moist forests be environmentally sustainable? *Environmental Conservation* 20 (2): 122–130.

Goodland, Robert J. A., Herman E. Daly, and Salah El Serafy. 1993. The urgent need for rapid transition to global environmental sustainability. *Environmental Conservation* 20 (4): 297–309.

Goreau, T. J. and W. Z. de Mello. 1988. Tropical deforestation: Some effects on atmospheric chemistry. *Ambio* 27 (4): 275–281.

Gradwohl, Judith and Russell Greenberg. 1988. *Saving the Tropical Forests.* Washington, D.C.: Island Press.

Graeve, Bernard von. 1989. *The Pacaa Nova: Clash of Cultures on the Brazilian Frontier.* Lewiston, Maine: Broadview Press.

Gray, Andrew. 1991. *Between the Spice of Life and the Melting Pot: Biodiversity Conservation and Impact on Indigenous Peoples.* Copenhagen: International Work Group for Indigenous Affairs, Document No. 70.

Greer, Jed. 1993. The price of gold: Environmental costs of the new gold rush. *Ecologist* 23 (3): 91–96.

Gribel, Rogerio. 1990. The Balbina disaster: The need to ask why? *Ecologist* 20 (4): 133–135.

Gullison, Raymond E. and Elizabeth C. Losos. 1993. The role of foreign debt in deforestation in Latin America. *Conservation Biology* 7 (1): 140–147.

Gunn, Alastair S. 1994. Environmental ethics and tropical rain forests: Should greens have standing? *Environmental Ethics* 16: 21–40.

Haas, Peter M., Marc A. Levy, and Edward A. Parson. 1992. Appraising the Earth Summit: How should we judge UNCED's success? *Environment* 34 (8): 6–11, 26–33.

Haigh, Martin J. 1988. Understanding "Chipko": The Himalayan people's movement for forest conservation. *International Journal of Environmental Studies* 31: 99–110.

Hames, Raymond. 1991. Wildlife conservation in tribal societies. In *Biodiversity: Culture, Conservation, and Ecodevelopment,* Margery L. Oldfield and Janis B. Alcorn, eds., pp. 172–199. Boulder, Colo.: Westview.

Hannah, L. 1990. Rain forests and geothermal energy in Hawaii: Environmental concerns expose flawed state planning process. *Environmental Conservation* 17 (3): 239–244.

Hannah, Lee, David Lohse, Charles Hutchinson, John L. Carr, and Ali Lankerani. 1994. A preliminary inventory of human disturbance of world ecosystems. *Ambio* 23 (4–5): 246–250.

Hayano, David M. 1990. *Road Through the Rain Forest: Living Anthropology in Highland Papua, New Guinea.* Prospect Heights, Ill.: Waveland Press.

Hayes, Randall. 1990. Activism: You make the difference. In *Lessons of the Rainforest,* Suzanne Head and Robert Heinzman, eds., pp. 219–233. San Francisco: Sierra Club Books.

Head, Suzanne. 1990. The consumer connection: Psychology and politics. In *Lessons of the Rainforest,* Suzanne Head and Robert Heinzman, eds., pp. 156–167. San Francisco: Sierra Club Books.

Head, Suzanne and Robert Heinzman, eds. 1990. *Lessons of the Rainforest*. San Francisco: Sierra Club Books.

Headland, Thomas N. 1987. The wild yam question: How well could independent hunter gatherers live in a tropical rain forest ecosystem? *Human Ecology* 15 (4): 463–491.

——. 1988. Ecosystemic change in a Philippine tropical rainforest and its effect on a Negrito foraging society. *Tropical Ecology* 29: 121–135.

——. 1993. Westernization, Deculturation, or Extinction Among Agta Negritos? The Philippine Population Explosion and Its Effect on a Rainforest Hunting and Gathering Society. Paper read at the Seventh International Conference on Hunting and Gathering Societies, August 17–23, Moscow.

——. 1994. Ecological Revisionism: Anthropological "Myths" and the Historical Ecology of "Truth." Paper read at the Conference on Historical Ecology, June 9–11, Tulane University, New Orleans.

Headland, Thomas N. and Robert C. Bailey. 1991. Introduction: Have hunter gatherers ever lived in tropical rain forest independently of agriculture? *Human Ecology* 19: 115–122.

Headland, Thomas N. and Lawrence A. Reid. 1989. Hunter gatherers and their neighbors from prehistory to the present. *Current Anthropology* 30 (1): 43–66.

——. 1991. Holocene foragers and interethnic trade: A critique of the myth of isolated independent hunter gatherers. In *Between Bands and States*, Susan A. Gregg, ed., pp. 333–340. Carbondale: Southern Illinois University.

Hecht, Susanna B. 1982. Agroforestry in the Amazon basin: Practice, theory, and limits of a promising land use. In *Amazonia: Agriculture and Land Use Research*, Susanna B. Hecht, ed., pp. 331–371. Cali, Colombia: Centro Internacional de Agricultura Tropical.

——. 1989. The sacred cow in green hell. *Ecologist* 19 (6): 229–234.

——. 1993. The logic of livestock and deforestation in Amazonia. *BioScience* 43 (10): 687–695.

Hecht, Susanna B. and Alexander Cockburn. 1989. *The Fate of the Forest: Developers, Destroyers, and Defenders of the Amazon*. New York: Verso.

Hecht, Susanna B and D. Posey. 1989. The sacred cow. *North American Conference on Latin America Report on the Americas* 23 (1): 23–26.

Heizer, Robert. 1955. Primitive man as an ecological factor. *Kroeber Anthropological Society Papers* 13: 1–31.

Hendee, John C. and Randall C. Pitstick. 1994. Growth and change in U.S. forest-related environmental groups. *Journal of Forestry* 92 (6): 24–31.

Hildyard, N. 1989. Adios Amazonia: A report from the Altamira gathering. *Ecologist* 19 (2): 53–62.

Hires, George A. and Thomas N. Headland. 1977. A sketch of western Bukidnon Manobo farming practices, past and present. *Philippine Quarterly of Culture and Society* 5: 65–75.

Hoffman, Carl. 1986. *The Punan: Hunters and Gatherers of Borneo*. Ann Arbor, Mich.: UMI Research Press.

Holden, Constance. 1985. Hawaiian rainforest being felled. *Science* 228: 1073–1074.

Holdgate, Martin. 1993. Using wildlife sustainably. *People and the Planet* 2 (3): 26–27.

Holloway, Marguerite. 1993. Sustaining the Amazon. *Scientific American* 269 (1): 90–99.

Homer-Dixon, Thomas F., Jeffrey H. Boutwell, and George W. Rathjens. 1993. Environmental change and violent conflict. *Scientific American* 268 (2): 38–45.

Horta, K. 1991. The last big rush for green gold. *Ecologist* 21 (3): 142–147.

Hurst, Philip. 1987. Forest destruction in Southeast Asia. *Ecologist* 17 (4/5): 170–174.

———. 1990. *Rainforest Politics: Ecological Destruction in Southeast Asia.* Atlantic Highlands, N.J.: Zed Books.

Huston, Michael A. 1994. *Biological Diversity: The Coexistence of Species on Changing Landscapes.* New York: Cambridge University Press.

Hutterer, Karl L. 1988. The prehistory of the Asian rain forests. In *People of the Tropical Rain Forest,* Julie Sloan Denslow and Christine Padoch, eds., pp. 63–72. Berkeley: University of California Press.

Hutts, Richard L. 1988. Is tropical deforestation responsible for the reported declines in the neotropical migrant populations? *American Birds* 42 (3): 375–379.

Hyman, Randall. 1988. Rise of the rubber tappers. *International Wildlife* 18 (5): 24–28.

Hyndman, David. 1994. Conservation through self-determination: Promoting the interdependence of cultural and biological diversity. *Human Organization* 53 (3): 296–302.

Independent Commission on International Humanitarian Issues (ICIHI). 1986. *The Vanishing Forest: The Human Consequences of Deforestation.* Atlantic Highlands, N.J.: Zed Books.

Institute of Social Analysis (INSAN). 1989. *Logging Against the Natives of Sarawak.* Selangor, Malaysia: Institute of Social Analysis.

Janetos, A. C. and W. T. Hinds. 1990. Global effects of tropical deforestation: Toward an integrated perspective. *Environmental Conservation* 17 (3): 201–212.

Johnston, Barbara Rose, ed. 1994. *Who Pays the Price? The Sociocultural Context of Environmental Crisis.* Washington, D.C.: Island Press.

Johnstone, Bob. 1987. Japan saps the world's rain forests. *New Scientist* 114: 18–22.

Jones, M. D. H. and A. Henderson-Sellers. 1990. History of the greenhouse effect. *Progress in Physical Geography* 14 (1): 1–18.

Jones, Philip D. and Tom M. L. Wigley. 1990. Global warming trends. *Scientific American* 263 (2): 84–91.

Jukofsky, Diane. 1991. The uncertain fate of Panama's forests. *Journal of Forestry* 89 (1): 17–19.

Kalapula, E. S. 1989. Woodfuel situation and deforestation in Zambia. *Ambio* 28 (5): 293–294.

Kartawinata, K., S. Adisoemarto, R. Riswan, and A. P. Vayda. 1981. The impact of man on tropical forest of Indonesia. *Ambio* 10: 115–119.

Katzman, Martin T. and William G. Cale Jr. 1990. Tropical forest preservation using economic incentives. *BioScience* 40 (1): 827–832.

Kaufman, Les and Kenneth Mallory. 1993. Beyond the last extinction: Environmental organizations for biodiversity. In *The Last Extinction*, Les Kaufman and Kenneth Mallory, eds., pp. 215–224. Cambridge, Mass.: MIT Press.

Kemp, Elizabeth. 1993. *The Law of the Mother: Protecting Indigenous Peoples in Protected Areas*. San Francisco: Sierra Club Books.

Klein, R. M. and T. D. Perkins. 1987. Cascades of causes and effects in forest decline. *Ambio* 16 (2–3): 86–93.

Kunstadter, Peter. 1988. Hill people of northern Thailand. In *People of the Tropical Rain Forest*, Julie Sloan Denslow and Christine Padoch, eds., pp. 93–110. Berkeley: University of California Press.

Kurlansky, Mark. 1988. Haiti's environment teeters on the edge. *International Wildlife* 18 (2): 35–38.

Lamb, David. 1991. Combining traditional and commercial uses of rain forest. *Nature and Resources* 27 (2): 3–11.

Lamb, F. B. 1987. The role of anthropology in tropical forest ecosystem resource management and development. *Journal of Developing Areas* 21: 429–458.

Langub, Jayl. 1988. The Penan strategy. In *People of the Tropical Rain Forest*, Julie Sloan Denslow and Christine Padoch, eds., pp. 207–208. Berkeley: University of California Press.

Lanly, Jean-Paul, Karn Deo Singh, and Klaus Janz. 1991. FAO's [Food and Agriculture Organization's] 1990 reassessment of tropical forest cover. *Nature and Resources* 27 (2): 21–26.

Lappe, Frances Moore and Rachel Schurman. 1990. Taking population seriously. In *Lessons of the Rainforest*, Suzanne Head and Robert Heinzman, eds., pp. 131–144. San Francisco: Sierra Club Books.

Ledec, George. 1985. The political economy of tropical deforestation. In *Divesting Nature's Capital*, H. Jeffrey Leonard, ed., pp. 179–226. New York: Holmes and Meier.

Leonard, H. Jeffrey, ed. 1985. *Divesting Nature's Capital*. New York: Holmes and Meier.

LeTacon, F. and J. L. Harley. 1990. Deforestation in the tropics and proposals to arrest it. *Ambio* 19 (8): 372–378.

Lewis, L. A. and W. J. Coffey. 1985. The continuing deforestation of Haiti. *Ambio* 14: 158–160.

Lizot, Jacques. 1985. *Tales of the Yanomami: Daily Life in the Venezuelan Forest*. New York: Cambridge University Press.

Lohmann, Larry. 1989. The Thai logging ban and its consequences. *Ecologist* 19 (2): 76–77.

——. 1993. Thailand: Land, power, and forest conservation. In *The Struggle for*

Land and the Fate of the Forests, Marcus Colchester and Larry Lohmann, eds., pp. 198–227. Atlantic Highlands, N.J.: Zed Books.

Loker, William M. 1993. The human ecology of cattle raising in the Peruvian Amazon: The view from the farm. *Human Organization* 52 (1): 14–24.

Lovejoy, Thomas E. 1983. Tropical deforestation and North American migrant birds. *Bird Conservation* 1: 126–128.

Lovelock, James. 1985. Are we destabilizing world climate the lessons of geophysiology? *Ecologist* 15 (1/2): 52–55.

Lugo, Ariel E. 1991. Tropical forestry research: Past, present, and future. *Journal of Forestry* 89 (3): 10–11, 22.

Lugo, Ariel E., Ralph Schmidt, and Sandra Brown. 1981. Tropical forests in the Caribbean. *Ambio* 10 (6): 318–311.

Lundberg, Lea J. 1984. Comprehensiveness of coverage of tropical deforestation. *Journalism Quarterly* 61: 378–382.

Lutzenberger, Jose. 1985. The World Bank's Polonaroeste project : A social and environmental catastrophe. *Ecologist* 15 (1/2): 69–72.

———. 1987. Brazil's Amazonian alliance. *Ecologist* 17 (4/5): 190–191.

Mackie, Cynthia. 1984. The lessons behind East Kalimantan's forest fires. *Borneo Research Bulletin* 16: 63–74.

Malingreau, J. P., G. Stephens, and L. Fellows. 1985. Remote sensing and forest fires: Kalimantan and North Borneo in 1982–83. *Ambio* 14: 314–321.

Malm, Olaf, Wolfgang C. Pfeiffer, Christina M. M. Souza, and Rudolf Reuther. 1990. Mercury pollution due to gold mining in the Madeira River basin, Brazil. *Ambio* 19 (1): 11–15.

Mannion, Antoinette M. 1992. Sustainable development and biotechnology. *Environmental Conservation* 19 (4): 297–306.

Marshall, George. 1990. The political economy of logging: A case study in corruption. *Ecologist* 20 (5): 174–181.

———. 1991. FAO and forestry. *Ecologist* 21 (2): 66–72.

Martinelli, Luis Antonio, Jose Roberto Ferreira, Bruce Rider Forsberg, and Reynaldo Luiz Victoria. 1989. Gold rush produces mercury contamination in Amazon. *Cultural Survival Quarterly* 13 (1): 32–34.

Mather, A. S. 1987. Global trends in forest resources. *Geography* 72 (1): 1–15.

Maybury-Lewis, David. 1992. *Millennium: Tribal Wisdom and the Modern World.* New York: Viking.

McCloskey, J. Michael and Heather Spalding. 1989. A reconnaissance-level inventory of the amount of wilderness remaining in the world. *Ambio* 18 (4): 221–227.

McDade, Lucinda A., Kamaljit S. Bawa, Henry A. Hespenheide, and Gary S. Hartshorn. 1993. *La Selva: Ecology and Natural History of a Neotropical Rainforest.* Chicago: University of Chicago Press.

McGee, R. Jon. 1990. *Life, Ritual, and Religion Among the Lancandon Maya.* Belmont, Calif.: Wadsworth.

McGlone, S. 1983. Polynesian deforestation of New Zealand: A preliminary synthesis. *Archaeology of Oceania* 18: 11–25.

McLarney, William O. 1993. Indigenous peoples and conservation. *Conservation Biology* 7 (4): 748–749.

McNabb, Ken, Joao Borges, and John Welker. 1994. Jari at twenty-five: An investment in the Amazon. *Journal of Forestry* 92 (2): 21–26.

McNeely, Jeffrey A. 1989. How to pay for conserving biological diversity. *Ambio* 18 (6): 308–313.

——. 1993. Diverse nature, diverse culture. *People and Planet* 2 (3): 11–13.

McNeely, J. A. and R. J. Dobias. 1991. Economic incentives for conserving biological diversity in Thailand. *Ambio* 20 (2): 86–90.

McNeely, J. A. and David Pitt, eds. 1984. *Culture and Conservation: The Human Dimension in Environmental Planning.* London: Croom Helm.

McNeely, Jeffrey A. and Paul Spencer Wachtel. 1988. *Soul of the Tiger: Searching for Nature's Answers in Exotic Southeast Asia.* New York: Doubleday.

McNeely, Jeffrey A., Kenton R. Miller, Walter V. Reid, Russell A. Mittermeier, and Timothy B. Werner. 1990a. *Conserving the World's Biological Diversity.* Washington, D.C.: World Resources Institute.

——. 1990b. Strategies for conserving biodiversity. *Environment* 32 (3): 16–20, 36–40.

Meehl, Gerald A. 1987. The tropics and their role in the global climate system. *Geography Journal* 153 (1): 21–36.

Meggers, Betty J. 1971. *Amazonia: Man and Culture in a Counterfeit Paradise.* Chicago: Aldine Atherton.

——. 1988. The Prehistory of Amazonia. In *People of the Tropical Rain Forest,* Julie Sloan Denslow and Christine Padoch, eds., pp. 53–62. Berkeley: University of California Press.

Merchant, Carolyn. 1992. *Radical Ecology: The Search for a Livable World.* New York: Routledge.

Merchant, Carolyn, ed. 1994. *Ecology: Key Concepts in Critical Theory.* Atlantic Highlands, N.J.: Humanities Press International.

Michaels, Patrick J. 1990. The greenhouse effect and global change: Review and reappraisal. *International Journal of Environmental Studies* 36: 55–71.

Milton, Katharine. 1981. Distributional patterns of tropical plant foods as an evolutionary stimulus to primate mental development. *American Anthropologist* 83: 534–548.

——. 1991. Comparative aspects of diet in Amazonian forest dwellers. *Philosophical Transactions of the Royal Society of London* 334: 253–263.

——. 1992. Civilization and its discontents. *Natural History* 3: 37–42.

——. 1993. Diet and primate evolution. *Scientific American* 269 (2): 86–93.

Mohd, Rusli bin and Jan G. Laarman. 1994. The struggle for influence: U.S. nongovernmental organizations and tropical forests. *Journal of Forestry* 92 (6): 32–36.

Molion, Luiz Carlos B. 1989. The Amazon forests and climatic stability. *Ecologist* 19 (6): 211–213.

Moody, Roger. 1988. *The Indigenous Voice: Visions and Realities.* Atlantic Highlands, N.J.: Zed Books.

Morales, Edmundo. 1989. *Cocaine: White Gold Rush in Peru.* Tucson: University of Arizona Press.

Moran, Emilio F. 1983. Government-directed settlement in the 1970s: An assessment of transamazon highway colonization. In *The Dilemma of Amazonian Development,* Emilio F. Moran, ed., pp. 297–317. Boulder, Colo.: Westview.

——. 1993. *Through Amazonian Eyes: The Human Ecology of Amazonian Populations.* Iowa City: University of Iowa Press.

Morris, Brian. 1986. Deforestation in India and the fate of forest tribes. *Ecologist* 16 (6): 253–257.

Mowat, Farley. 1987. *Woman in the Mists: The Story of Dian Fossey and the Mountain Gorillas of Africa.* New York: Warner Books.

Murdy, Garson N. 1990. Prehistoric agriculture and its effects in the valley of Guatemala. *Forest and Conservation History* 34 (4): 179–190.

Murray, Gerald F. 1987. The domestication of wood in Haiti: A case study in applied evolution. In *Anthropological Praxis: Translating Knowledge into Action,* Robert M. Wulff and Shirley J. Fiske, eds., pp. 223–240. Boulder, Colo.: Westview.

Murray, Martyn G. 1990. Conservation of tropical rain forests: Arguments, beliefs, and convictions. *Biological Conservation* 52: 17–26.

Myers, Norman. 1986. Environmental repercussions of deforestation in the Himalayas. *Journal of World Forest Resource Management* 2: 63–72.

——. 1992. *The Primary Source: Tropical Forests and Our Future.* New York: W. W. Norton.

——. 1993a. Environmental refugees in a globally warmed world. *BioScience* 43 (11): 752–761.

——. 1993b. Tropical forests: The main deforestation fronts. *Environmental Conservation* 20 (1): 9–16.

Myers, Norman and R. Tucker. 1967. Deforestation in Central America: Spanish legacy and North American consumers. *Environmental Review* 11: 55–71.

National Research Council. 1980. *Research Priorities in Tropical Biology.* Washington, D.C.: National Academy of Sciences, National Research Council.

——. 1992. *Conserving Biodiversity: A Research Agenda for Development Agencies.* Washington, D.C.: National Academy Press.

Nations, James D. and Daniel I. Komer. 1982. Indians, immigrants, and beef exports: Deforestation in Central America. *Cultural Survival Quarterly* 6: 8–12.

——. 1983. Rainforests and the hamburger society. *Environment* 25 (3): 12–20.

Nations, James D. and Ronald B. Nigh. 1980. The evolutionary potential of Lacandon Maya sustained yield tropical forest agriculture. *Journal of Anthropological Research* 36 (1): 1–30.

Nelson, J. Gordon. 1992. Assessing biodiversity: A human ecological approach. *Ambio* 21 (3): 212–218.

Newson, Linda A. 1985. Indian population patterns in colonial Spanish America. *Latin American Research Review* 20 (3): 41–74.

Nietschmann, Bernard. 1985. Indonesia, Bangladesh: Disguised invasion of indigenous nations. *Fourth World Journal* 1 (2): 89–126.

——. 1987. Militarization and indigenous people. *Cultural Survival Quarterly* 11 (3): 1–16.

Nyoni, Sithembiso. 1987. Indigenous NGOs: Liberation, self-reliance, and development. *World Development* 15: 51–56.

O'Keefe, P. and R. Hosier. 1983. The Kenyan fuelwood cycle: A summary. *GeoJournal* 7 (1): 25–28.

Oelschlaeger, Max. 1991. *The Idea of Wilderness from Prehistory to the Age of Ecology.* New Haven, Conn.: Yale University Press.

Oldfield, Margery L. and Janis B. Alcorn, eds. 1991. *Biodiversity: Culture, Conservation, and Ecodevelopment.* Boulder, Colo.: Westview.

Osmond, C. B. 1987. Stress physiology and the distribution of plants. *BioScience* 37 (1): 38–48.

Paaby, Pia, David B. Clark, and Hector Gonzalez. 1991. Training rural residents as naturalist guides: Evaluation of a pilot project in Costa Rica. *Conservation Biology* 5 (4): 542–546.

Pace, Richard. 1992. Social conflict and political activism in the Brazilian Amazon: A case study of Gurupa. *American Ethnologist* 19 (4): 710–732.

Padua, Suzana M. 1994. Conservation awareness through an environmental education programme in the Atlantic Forest of Brazil. *Environmental Conservation* 21 (2): 145–151.

Palm, C. A., R. A. Houghton, J. M. Melillo, and D. L. Skole. 1986. Atmospheric carbon dioxide from deforestation in Southeast Asia. *Biotropica* 18 (3): 177–188.

Park, Chris C. 1992. *Tropical Rainforests.* New York: Routledge.

Parker, Eugene. 1993. Fact and fiction in Amazonia: The case of the Apêtê. *American Anthropologist* 95: 715–723.

Parker, J. Kathy. 1992. Hanging question marks on our profession. *Journal of Forestry* 90 (4): 21–24.

Parson, Edward A., Peter M. Haas, and Marc A. Levy. 1992. A summary of the major documents signed at the Earth Summit and the Global Forum. *Environment* 34 (8): 12–15, 34–36.

Partridge, William L., ed. 1984. *Training Manual in Development Anthropology.* Washington, D.C.: American Anthropological Association, Special Publication No. 17.

Pasca, T. M. 1988. The politics of tropical deforestation. *American Forests* 94 (11/12): 21–24.

Patterson, Alan. 1990. Debt for nature swaps and the need for alternatives. *Environment* 32 (10): 4–13, 31–32.

Pellek, Richard. 1990. Combating tropical deforestation in Haiti. *Journal of Forestry* 88 (9): 15–19.

Pelzer, Karl J. 1968. Man's role in changing the landscape of Southeast Asia. *Journal of Asian Studies* 27 (2): 269–279.

Perlin, John. 1991. *A Forest Journey: The Role of Wood in the Development of Civilization.* Cambridge, Mass.: Harvard University Press.

Peters, Charles M., Alwyn H. Gentry, and Robert O. Mendelsohn. 1989. Valuation of an Amazonian rainforest. *Nature* 339: 655–656.

Peterson, Jean T. 1978. Hunter gatherer/farmer exchange. *American Anthropologist* 80: 335–351.

Pimentel, David, Ulrich Stachow, David A. Takacs, Hans W. Brubaker, Amy R. Dumas, John J. Meaney, John A.S. O'Neil, Douglas E. Onsi, and David B. Corzilius. 1992. Conserving biological diversity in agricultural/forestry systems. *BioScience* 42 (5): 354–362.

Pinto, Lucio Flavio. 1989. Calha Norte: The special project for the occupation of the frontiers. *Cultural Survival Quarterly* 13 (1): 40–41.

Plotkin, Mark and Lisa Famolare, eds. 1992. *Sustainable Harvest and Marketing of Rain Forest Products*. Washington, D.C.: Island Press.

Plumwood, Val and Richard Routley. 1982. World rainforest destruction the social factors. *Ecologist* 12 (1): 4–22.

Poelhekke, Fabio G. M. N. 1986. Fences in the jungle: Cattle raising and the economic and social integration of the Amazon region in Brazil. *Revista Geografica* 104: 33–43.

Poffenberger, Mark, ed. 1990. *Keepers of the Forest: Land Management Alternatives in Southeast Asia*. West Hartford, Conn.: Kumarian Press.

Pomeroy, Derek. 1993. Centers of high biodiversity in Africa. *Conservation Biology* 7 (4): 901–907.

Posey, Darrell A. 1989a. From war clubs to words. *North American Conference on Latin America Report on the Americas* 23 (1): 13–18.

Posey, Darrell A. 1989b. Alternatives to forest destruction: Lessons from the Mebengokre Indians. *Ecologist* 19 (6): 241–244.

Posey, Darrell A. and William Balée, eds. 1989. *Resource Management in Amazonia: Indigenous and Folk Strategies*. Advances in Economic Botany, No. 7. New York: New York Botanical Garden.

Posey, Darrell A. and William Leslie Overall, eds. 1988. *Ethnobiology: Implications and Applications*. Belem, Brazil: Museu Goeldi.

Posey, Darrell A., John Frechione, John Eddins, Luiz Francelino da Silva, Debbie Myers, Diane Case, and Peter MacBeath. 1984. Ethnoecology as applied anthropology in Amazonian development. *Human Organization* 43 (2): 95–107.

Post, Wilfred M., Tsung-Hung Peng, William R. Emanuel, Anthony W. King, Virginia H. Dale, and Donald L. DeAngelis. 1990. The global carbon cycle. *American Scientist* 78 (4): 310–326.

Postel, Sandra and Lori Heise. 1988. Reforesting the earth. In *State of the World 1988*, Lester R. Brown, ed., pp. 83–100. New York: W. W. Norton.

Prance, Ghillean T. 1990. Rainforested regions of Latin America. In *Lessons of the Rainforest*, Suzanne Head and Robert Heinzman, eds., pp. 53–65. San Francisco, Calif.: Sierra Club Books.

——. 1993. The Amazon: Paradise lost. In *The Last Extinction*, Les Kaufman and Kenneth Mallory, eds., pp. 69–114. Cambridge, Mass.: MIT Press.

Prance, Ghillean T., ed. 1982. *Biological Diversification in the Tropics*. New York: Columbia University Press.

Project for Ecological Recovery. 1992. *The Future of People and Forests in Thailand After the Logging Ban*. Bangkok: Project for Ecological Recovery.

Pushparajah, E. 1985. Development and induced soil erosion and flash floods in Malaysia. *Ecologist* 15 (1/2): 19–20.

Putz, Francis E. 1989. Tropical forestry education in the United States. *Journal of Forestry* 87 (7): 27–30.

Rai, Navin. 1990. *Living in a Lean-to: Philippine Negrito Foragers in Transition*. Ann Arbor: University of Michigan, Museum of Anthropology, Anthropological Papers, No. 80.

Raloff, Janet. 1988. Unraveling the economics of deforestation. *Science News* 133: 366–367.

Rambo, A. Terry. 1985. *Primitive Polluters: Semang Impact on the Malaysian Tropical Rain Forest Ecosystem*. Ann Arbor: University of Michigan, Museum of Anthropology.

Randall, A. 1991. The value of biodiversity. *Ambio* 20 (2): 64–68.

Randrianarijaona, P. 1983. The erosion of Madagascar. *Ambio* 12: 308–311.

Ranjitsinh, M. K. 1979. Forest destruction in Asia and the South Pacific. *Ambio* 8: 192–201.

Rasoanaivo, P. 1990. Rain forests of Madagascar: Sources of Industrial and Medicinal Plants. *Ambio* 19 (8): 421–424.

Raup, David M. 1988. Diversity crises in the geological past. In *Biodiversity*, E. O. Wilson and Frances M. Peter, eds., pp. 51–57. Washington, D.C.: National Academy Press.

Redford, Kent H. 1991. The ecologically noble savage. *Cultural Survival Quarterly* 15 (1): 46–48.

Redford, Kent H. and Christine Padoch, eds. 1992. *Conservation of Neotropical Forests: Working from Traditional Resource Use*. New York: Columbia University Press.

Redford, Kent H. and John G. Robinson. 1985. Hunting by indigenous peoples and conservation of game species. *Cultural Survival Quarterly* 9 (1): 41–44.

Redford, Kent H. and Allyn Maclean Stearman. 1993. On common ground: Response to Alcorn. *Conservation Biology* 7 (2): 427–428.

Reichel-Dolmatoff, Gerardo. 1971. *Amazonian Cosmos: The Sexual and Religious Symbolism of the Tukano Indians*. Chicago: University of Chicago Press.

——. 1976. Cosmology as ecological analysis: A view from the rainforest. *Man* 11: 307–318.

Reid, Walter V., Sara A. Laird, Carrie A. Meyer, Rodrigo Gamez, Ana Sittenfeld, Daniel H. Janzen, Michael A. Gollin, and Calestous Juma. 1993. *Biodiversity Prospecting: Using Genetic Resources for Sustainable Development*. Washington, D.C.: World Resources Institute.

Repetto, R. 1987. Creating incentives for sustainable forest development. *Ambio* 16 (2–3): 94–99.

——. 1990. Deforestation in the tropics. *Scientific American* 262 (4): 36–42.

——. 1992. Accounting for environmental assets. *Scientific American* 266 (6): 94–101.

Rice, Don Stephen and Prudence M. Rice. 1984. Lessons from the Maya. *Latin American Research Review* 19 (3): 7–34.

Rich, Bruce. 1990. Multilateral development banks and tropical deforestation. In *Lessons from the Rainforest*, Suzanne Head and Robert Heinzman, eds., pp. 118–130. San Francisco: Sierra Club Books.

Robinson, John G. 1993. The limits to caring: Sustainable living and the loss of biodiversity. *Conservation Biology* 7 (1): 20–28.

Robinson, John G. and Kent H. Redford, eds. 1991. *Neotropical Wildlife Use and Conservation*. Chicago: University of Chicago Press.

Roosevelt, Anna. 1980. *Parmana: Prehistoric Maize and Manioc Subsistence Along the Amazon and Orinoco*. New York: Academic Press.

———. 1989. Resource management in Amazonia before the Conquest: Beyond ethnographic projection. *Advances in Economic Botany* 7: 30–62.

———. 1992. Secrets of the forest. *Sciences* 32 (6): 22–28.

Round, Philip D. 1990. Bangkok Bird Club survey of the bird and mammal trade in the Bangkok weekend market. *Natural History Bulletin of the Siam Society* 38 (1): 1–43.

Rubinoff, Ira. 1983. A strategy for preserving tropical rainforests. *Ambio* 12 (5): 255–258.

Rudel, T. K. 1989. Population, development, and tropical deforestation: A cross-national study. *Rural Sociology* 54 (3): 327–338.

Rush, James. 1991. *The Last Tree: Reclaiming the Environment in Tropical Asia*. New York: Asia Society.

Ryan, John C. 1992. Goods from the woods: Managing tropical forests for preservation and profit. *Journal of Forestry* 90 (4): 25–28.

Sakai, Howard F. 1988. Avian response to mechanical clearing of a native rainforest in Hawaii. *Condor* 90: 339–348.

Salafsky, Nick. 1993. Mammalian use of a buffer zone agroforestry system bordering Gunung Palung National Park, West Kalimantan, Indonesia. *Conservation Biology* 7 (4): 928–933.

Salafsky, Nick, Barbara L. Dugelby, and John W. Terborgh. 1993. Can extractive reserves save the rain forest? An ecological and socioeconomic comparison of nontimber forest product extraction systems in Peten, Guatemala, and West Kalimantan, Indonesia. *Conservation Biology* 7 (1): 39–52.

Salati, Eneas. 1990. Amazonia. In *The Earth as Transformed by Human Action*, B. L. Turner II, ed., pp. 479–493. New York: Cambridge University Press.

Santilli, Marcio. 1989. The Calha Norte project: Military guardianship and frontier policy. *Cultural Survival Quarterly* 13 (1): 42–43.

Schefold, Reimar. 1988. The Mentawai equilibrium and the modern world. In *The Real and the Imagined Role of Culture in Development: Case Studies from Indonesia*, Michael R. Dove, ed., pp. 201–215. Honolulu: University of Hawaii Press.

Schmink, Marianne. 1988. Big business in the Amazon. In *People of the Tropical Rain Forest*, Julie Sloan Denslow and Christine Padoch, eds., pp. 163–174. Berkeley: University of California Press.

Schultes, Richard Evans. 1991. Ethnobotany and technology in the Northwest Amazon: Example of partnership. *Environmental Conservation* 18 (3): 264–267.

Schwartz, Tanya. 1989. The Brazilian forest people's movement. *Ecologist* 19 (6): 245–247.

Schwartzman, Stephen. 1989. Extractive reserves: The rubber tappers' strategy for sustainable use of the Amazon rainforest. In *Fragile Lands of Latin America: Strategies for Sustainable Development*, John O. Browder, ed., pp. 150–165. Boulder, Colo.: Westview.

Secrett, C. 1986. The environmental impact of transmigration. *Ecologist* 16: 77–88.

Sedjo, Roger A. 1989. Forests: A tool to moderate global warming. *Environment* 31 (1): 14–20.

Shiva, Vandana. 1990. Biodiversity: A peoples' plan. *Ecologist* 20 (2): 44–47.

Shiva, Vandana and Jayanta Bandyopadhyan. 1987. Chipko: Rekindling India's forest culture. *Ecologist* 17 (1): 26–34.

Shiva, Vandana, Patrick Anderson, Heffa Schücking, Andrew Gray, Larry Lohmann, and David Cooper. 1991. *Biodiversity: Social and Ecological Perspectives*. Atlantic Highlands, N.J.: Zed Books.

Shrader-Frechette, K. S. 1981. *Environmental Ethics*. Pacific Grove, Calif.: Boxwood Press.

Signor, P. W. 1990. The geological history of diversity. *Annual Review of Ecology and Systematics* 21: 509–539.

Silver, Cheryl Simon and Ruth S. DeFries. 1990. *One Earth, One Future: Our Changing Global Environment*. Washington, D.C.: National Academy Press.

Simberloff, Daniel. 1986. Are we on the verge of a mass extinction in tropical rain forests? In *Dynamics of Extinction*, David K. Elliott, ed., pp. 165–180. New York: Wiley.

Simons, Paul. 1988. Belize at the crossroads. *New Scientist* 120: 61–65.

Singh, K. D. 1994. Tropical forest resources: An analysis of the 1990 assessment. *Journal of Forestry* 92 (2): 27–31.

Smith, Nigel J. H. 1981. Colonization lessons from a tropical forest. *Science* 214: 755–761.

Smith, Nigel J. H. and Richard Evans Schultes. 1990. Deforestation and shrinking crop gene pools in Amazonia. *Environmental Conservation* 17 (3): 227–234.

Smith, Nigel J. H., J. T. Williams, and Donald L. Plucknett. 1991. Conserving the tropical cornucopia. *Environment* 33 (6): 7–32.

Solbrig, Otto T. 1991. The origin and function of biodiversity. *Environment* 33 (5): 17–38.

Solo, Pam, ed. 1993. Resource and sanctuary: Indigenous peoples, ancestral rights, and the forests of the Americas. *Cultural Survival Quarterly* 17 (1): 1–63.

Sponsel, Leslie E. 1986. Amazon ecology and adaptation. *Annual Review of Anthropology* 15: 67–97.

——. 1989. Foraging and farming: A necessary complementarity in Amazonia. In *Farmers as Hunters*, Susan Kent, ed., pp. 37–45. New York: Cambridge University Press.

———. 1992a. The environmental history of Amazonia: Natural and human disturbances and the ecological transition. In *Changing Tropical Forests: Historical Perspectives on Today's Challenges in Central and South America,* Harold K. Steen and Richard P. Tucker, eds., pp. 233–251. Durham, N.C.: Forest History Society.

———. 1992b. Information asymmetry and the democratization of anthropology. *Human Organization* 51 (3): 299–301.

———. 1994. The Yanomami holocaust continues. In *Who Pays the Price? Examining the Sociocultural Context of the Environmental Crisis,* Barbara R. Johnston, ed., pp. 37–46. Washington D.C.: Island Press.

———. 1995. Relationships among the world system, indigenous peoples, and ecological anthropology in the endangered Amazon. In *Indigenous Peoples and the Future of Amazonia: An Ecological Anthropology of an Endangered World,* Leslie E. Sponsel, ed., pp. 263–293. Tucson: University of Arizona Press.

Sponsel, Leslie E. and Poranee Natadecha. 1988. Buddhism, ecology, and forests in Thailand: Past, present, and future. In *Changing Tropical Forests: Historical Perspectives on Today's Challenges in Asia, Australasia, and Oceania,* John Dargavel, Kay Dixon, and Noel Semple, eds., pp. 305—325. Canberra, Australia: Center for Resource and Environmental Studies.

Sponsel, Leslie E. and Poranee Natadecha-Sponsel. 1993. The potential contribution of Buddhism in developing an environmental ethic for the conservation of biodiversity. In *Ethics, Religion, and Biodiversity: Relations Between Conservation and Cultural Values,* Lawrence S. Hamilton, ed., pp. 75–97. Cambridge, England: White Horse Press.

Spradley, James P. 1979. *The Ethnographic Interview.* New York: Holt, Rinehart, and Winston.

———. 1980. *Participant Observation.* New York: Holt, Rinehart, and Winston.

Stapp, William B., and N. Polunin. 1991. *Global Environmental Education: Towards a Way of Thinking and Acting. Environmental Conservation* 18(1):13–18.

Stearman, Allyn MacLean. 1989. *Yuqui: Forest Nomads in a Changing World.* New York: Holt, Rinehart, and Winston.

———. 1991. Making a living in the tropical forest: Yuqui foragers in the Bolivian Amazon. *Human Ecology* 19: 245–260.

Stevens, Richard L. 1993. *The Trail: A History of the Ho Chi Minh Trail and the Role of Nature in the War in Viet Nam.* New York: Garland.

Stevenson, Glenn G. 1989. The production, distribution, and consumption of fuelwood in Haiti. *Journal of Developing Areas* 24: 59–76.

Stiles, Daniel. 1992. The ports of East Africa, the Comoros, and Madagascar: Their place in Indian Ocean trade from 1–1500 A.D. *Kenya Past and Present* 24: 27–36.

Stoddart, D. R. 1990. Coral reefs and islands and predicted sea level rise. *Progress in Physical Geography* 14 (4): 521–536.

Stone, Roger D. 1992. *Dreams of Amazonia.* Baltimore: Penguin.

Stowe, Robert C. 1987. United States foreign policy and the conservation of natural resources: The case of tropical deforestation. *Natural Resources Journal* 27 (1): 55–101.

Swinbanks, David. 1987. Sarawak's tropical rainforests exploited by Japan. *Nature* 328: 373–380.

Tangley, Laura. 1988. Fighting deforestation at home. *BioScience* 38 (4): 220–224.

Thapa, G. B. and K. E. Weber. 1990. Actors and factors of deforestation in tropical Asia. *Environmental Conservation* 17 (1): 19–27.

Tierney, Patrick. 1995. *Last Tribes of El Dorado: The Gold Wars in the Amazon Rain Forest*. New York: Viking and Penguin.

Tobias, D. and R. Mendelsohn. 1991. Valuing ecotourism in a tropical forest reserve. *Ambio* 20 (2): 91–93.

Toth, Nicholas, Desmond Clark, and Giancarlo Ligabue. 1992. The last stone ax makers. *Scientific American* 267 (1): 88–93.

Townsend, William H. 1969. Stone and steel tool use in a New Guinea society. *Ethnology* 8: 199–205.

Treacy, J. 1982. Bora Indian agroforestry: An alternative to deforestation. *Cultural Survival Quarterly* 6 (2): 15–16.

Treece, Dave. 1989. The militarization and industrialization of Amazonia: The Calha Norte and Gran Carajas programs. *Ecologist* 19 (6): 225–228.

Turnbull, Colin M. 1961. *The Forest People*. New York: Simon & Schuster.

———. 1983. *The Mbuti Pygmies: Change and Adaptation*. New York: Holt, Rinehart, and Winston.

Uhl, Christopher and Geoffrey Parker. 1986. Our steak in the jungle. *BioScience* 36 (10): 642–649.

Uhl, Christopher, Daniel Nepstad, Robert Buschbacher, Kathleen Clark, Boone Kauffman, and Scott Subler. 1989. Disturbance and regeneration in Amazonia. *Ecologist* 19 (6): 235–240.

United Nations Educational, Scientific, and Cultural Organization (UNESCO). 1991. *Debt for Nature*. Environmental Brief No. 1, pp. 1–16.

Upreti, Gopi. 1994. Environmental conservation and sustainable development require a new development approach. *Environmental Conservation* 21 (1): 18–29.

van Willigen, John and Billie R. DeWalt, eds. 1985. *Training Manual in Policy Ethnography*. Washington, D.C.: American Anthropological Association, Special Publication No. 19.

Vanclay, Jerome K. 1993. Saving the tropical forest: Needs and prognosis. *Ambio* 22 (4): 225–231.

Vansina, Jan. 1990. *Paths in the Rainforests: Toward a History of Political Tradition in Equatorial Africa*. Madison: University of Wisconsin Press.

Vincent, Jeffrey R. 1992. The tropical timber trade and sustainable development. *Science* 256 (June 19): 1651–1655.

Visser, Dana R. and Guillermo A. Mendoza. 1994. Debt-for-nature swaps in Latin America. *Journal of Forestry* 92 (6): 13–16.

Volbeda, Sjoukje. 1986. Pioneer towns in the jungle: Urbanization at an agricultural colonization frontier in the Brazilian Amazon. *Revista Geografica* 104: 115–140.

Wagner, Michael R. and Joseph R. Cobbinah. 1993. Deforestation and sustainability in Ghana: The role of tropical forests. *Journal of Forestry* 91 (6): 35–39.

Wallace, George N. 1993. Wildlands and ecotourism in Latin America: Investing in protected areas. *Journal of Forestry* 91 (2): 37–40.

Walschburger, T. and P. von Hildebrand. 1991. The first twenty-six years of forest regeneration in natural and man-made gaps in the Colombian Amazon. In *Rain Forest Regeneration and Management*, A. Gómez-Pompa, T. C. Whitmore, and M. Hadley, eds., pp. 257–264. London: Parthenon.

Westing, Arthur H. 1992. Environmental refugees: A growing category of displaced persons. *Environmental Conservation* 19 (3): 201–207.

Whelan, Tensie. 1988. The Panama Canal: Will the watershed hold? *Environment* 30 (3): 12–15, 37–40.

Whitehouse, John F. 1991. East Australian rain forests: A case study in resources. *Environmental Conservation* 18 (1): 33–43.

Whitmore, T. C. 1989. Forty years of rain forest ecology: 1948–1988 in perspective. *GeoJournal* 19 (4): 347–360.

———. 1990. *An Introduction to Tropical Forests*. Oxford, England: Clarendon.

Whitmore, T. C. and J. A. Sayer, eds. 1992. *Tropical Deforestation and Species Extinction*. New York: Chapman and Hall.

Whitmore, T. C., J. R. Flenley, and D. R. Harris. 1982. The tropics as the norm in biogeography. *Geographical Journal* 148 (1): 8–21.

Wilkie, David S. 1988. Hunters and farmers of the African forest. In *People of the Tropical Rain Forest*, Julie Sloan Denslow and Christine Padoch, eds., pp. 111–126. Berkeley: University of California Press.

Wilkie, David S. and John T. Finn. 1990. Slash and burn cultivation and mammal abundance in the Ituri Forest, Zaire. *Biotropica* 22 (1): 90–98.

Williams, Michael. 1989. Deforestation: Past and present. *Progress in Human Geography* 13 (2): 176–208.

———. 1990. Forests. In *The Earth as Transformed by Human Action*, B. L. Turner III, ed., pp. 179–201. New York: Cambridge University Press.

Wilson, Edward O. 1989. Threats to biodiversity. *Scientific American* 261 (3): 108–117.

———. 1992. *The Diversity of Life*. Cambridge, Mass.: Belknap Press.

Wilson, Edward O., ed. 1988. *Biodiversity*. Washington, D.C.: National Academy Press.

Wilson, Wendell L. and Andrew D. Johns. 1982. Diversity and abundance of selected animal species in undisturbed forest, selectively logged forest, and plantations in East Kalimantan, Indonesia. *Biological Conservation* 24: 205–218.

Winterbottom, Robert. 1990. *Taking Stock: The Tropical Forestry Action Plan After Five Years*. Washington, D.C.: World Resources Institute.

Winterbottom, R. and P. T. Hazelwood. 1987. Agroforestry and sustainable development: Making the connection. *Ambio* 16 (2–3): 100–110.

Witte, John. 1992. Deforestation in Zaire. *Ecologist* 22 (2): 58–64.

Wolf, Eric R. 1982. *Europe and the People Without History*. Berkeley: University of California Press.

Wood, Peter H. 1992. Re-viewing the map: America's empty wilderness. *Cultural Survival Quarterly* 16 (3): 59–62.

World Resources Institute, World Conservation Union, United Nations Environment Program, Food and Agriculture Organization, and United Nations Educational, Scientific, and Cultural Organization. 1992. *Global Biodiversity Strategy: Guidelines for Action to Save, Study, and Use Earth's Biotic Wealth Sustainably and Equitably*. Washington, D.C.: World Resources Institute.

Wright, Robin M. 1988. Anthropological assumptions of indigenous advocacy. *Annual Review of Anthropology* 17: 365–390.

Yoon, Carol Kaesuk. 1993. Rain forests seen as shaped by human hand. *New York Times*, July 27, pp. C1, C10.

Yuill, Thomas M. 1983. Disease-causing organisms: Components of tropical forest ecosystems. In *Tropical Rain Forest Ecosystems*, Frank B. Golley, ed., pp. 315–324. New York: Elsevier.

II

Insights from Prehistory

Part 2, Insights from Prehistory, draws on two historically important regions of indigenous populations, the Maya of Central America as discussed in chapter 2 by Elliot Abrams, AnnCorinne Freter, David Rue, and John Wingard, and the Polynesians of part of the Pacific islands, as reviewed in chapter 3 by Brien Meilleur. One of the most important points of these two case studies is that anthropogenic deforestation is nothing new—it transpired in prehistory, although the scale and combination of factors involved may have been quite different because of the variations in place and time. That anthropogenic deforestation occurred in prehistory does not depreciate the sustainable resource use and management practices of sectors of the Mayan, Polynesian, and other indigenous populations at some places and times. They have positive as well as negative lessons to teach other societies about environmental conservation and economic development in tropical forests in the modern era.

Another message from these two articles is that in many areas population pressure on the land and resource base may have been an important factor contributing to deforestation in the past as well as in the present. Numerous demographic studies suggest that population pressure will increase in many areas of tropical forest in coming decades before it can be relieved through family planning and changes in the age structure.

The Mayan case provides yet another important lesson—deforestation may lead to the disintegration and even the eventual collapse of a society. However, in the Copán subregion of the Maya, it is noteworthy that the forests did eventually recover after the area was essentially abandoned, although the natural reforestation occurred over five centuries. This indicates that the ancient Mayan ecosystems were not degraded beyond their natural capacity for regeneration, a condition that is unlikely to be true of areas that have undergone rapid and massive deforestation more recently.

The case study of the Maya is also of special interest because of the rich archaeological record that has been recovered from numerous sites during long-term team research. That record has been elucidated through computer simulation models and complementary evidence from special methods, such as the analysis of fuelwood consumption and the reconstruction of vegetation history through pollen analysis.

Polynesia is of particular interest, because the ecology of islands is rather special, given their isolation, circumscription, and limited land and resources. As a result islands are not only extremely fragile and vulnerable ecosystems; because of keen competition over resources and adaptive radiation they often have a high frequency of endemic species. Thus, although the size of forests on islands may be limited compared to continental areas, their biodiversity is important because it is unique, resulting from relatively high endemicity. And, although the ancient Polynesians did contribute to deforestation, species extinction, biodiversity reduction, and so on, Polynesia's environmental degradation has been greatly magnified with Western contact, colonization, and "development." For instance, modern Hawaii has gained the infamous distinction of being the species extinction capital of the United States.

Because many islands were relatively isolated, they provide natural laboratories or experiments for studying processes of parallel or convergent evolution and adaptation. Making a controlled comparison of a sample of islands in the Pacific, Caribbean, and Indian Ocean could prove insightful. Islands may serve as convenient heuristic models, because they are microcosms of human ecology.

2

The Role of Deforestation in the Collapse of the Late Classic Copán Maya State

Elliot M. Abrams, AnnCorinne Freter,
David J. Rue, and John D. Wingard

The current extent of global deforestation is unparalleled in the history of human existence, measured in geographic scope, intensity, rate of expansion, and human and environmental costs. Perhaps as a consequence of this unprecedented scale and impact, and in conjunction with our biases concerning nonindustrialized cultures, contemporary observers often find it easy to view deforestation as a recent and seemingly historic event that could only have occurred in the modern industrial age.

However, deforestation is not a new cultural-ecological process and is not restricted causally to expanding industrial states. Contrary to the notion that the ancients lived in constant Zenlike harmony with nature, deforestation affected many culturally and geographically diverse preindustrial societies. For example, deforestation has been suggested as a factor contributing to the failure of various ancient state systems as culturally distinct as Teotihuacán in central Mexico (Sanders, Parsons, and Santley 1979:406–407, 409) and the Harrapan state in the Indus River valley (Fairservis 1975:304). Given its focus on understanding long-term diachronic social processes from a comparative perspective, the unique contribution of anthropological archaeology is data and thus insights into the process, consequences, and lasting effects of deforestation.

The Mayan civilization represents a unique opportunity to study the independent rise and fall of a prehispanic state system, because the Mayan state developed and collapsed independent of outside influences or politi-

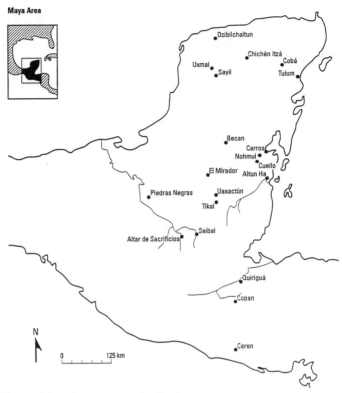

Figure 2.1   The Maya Lowlands

cal pressures. The prehispanic Maya occupied what is today southern
Mexico, Guatemala, Belize, and Honduras (figure 2.1). Between ca. A.D. 1
and A.D. 800 they developed one of the most complex societies in the New
World. By A.D. 800, however, the collapse of this state system is apparent
in the archaeological record. The Maya erected few historical monuments
(stelae) after this time, most construction within major Mayan centers
ceased, and the population within the southern lowland Mayan region
rapidly decreased.

Many factors have been cited as being largely responsible for the collapse of Mayan civilization. Some major factors include warfare (Lemonick
1993), internal rebellion (Hamblin and Pitcher 1980), diseases (Shimkin
1973), trade disruption (Rathje 1973), and stochastic natural disasters such
as earthquakes (MacKie 1961). A key to evaluating these factors is recog-

nizing that they are not mutually exclusive. For example, diseases that seriously threaten the security of a small kingdom may trigger warfare, which in turn may lower the trading capabilities of that kingdom. Nonetheless most factors may be linked to, or perhaps triggered by, some failure of the agricultural economy on which the Classic Maya depended for their food. Thus all sociopolitical factors, as well as ideological factors, rely ultimately on the success or failure of the economy, albeit in a complex interactive manner.

The idea that deforestation may have played a role in the Classic Maya collapse is certainly not new (Cooke 1931; Sanders 1973). However, this specific cultural-ecological model was not tested and confirmed against the empirical archaeological data until the late 1970s (Deevey et al. 1979). Since then mounting evidence points to deforestation of vital arboreal species as helping to undermine the economic infrastructure of the Late Classic Maya state in the southern lowlands (e.g., Wiseman 1985; Rue 1986, 1987; Abrams and Rue 1988). By focusing on deforestation we in no way wish to suggest that deforestation was prime mover in some unilineal cause-and-effect model. Quite to the contrary, one of the more important conclusions drawn in the 1970s concerning the Classic Maya collapse was that a systemic interactive model incorporating a wide range of variables, including those cited, must be considered in the analysis of this complex phenomenon (Willey and Shimkin 1973). On a more theoretical level is the cogent argument that a cultural materialist approach that includes cultural ecology is a de facto mandate for a systemic analysis of causality; the critical distinction between this and other approaches is that those variables linked more directly to the material conditions of life receive greater value or weight within the equation of collapse (Price 1982:710).

In this article we examine the archaeological evidence that links deforestation with the failure, or collapse, of one of the Classic Maya states—that of Copán, Honduras (see figure 2.1). From roughly A.D. 400 to A.D. 850 this specific Maya polity grew in population size and institutional complexity, achieving many hallmarks of early civilizations, including monumental architecture, elaborate tombs, and a sophisticated calendrical and writing system. After ca. A.D. 850, however, the population within this valley steadily declined until A.D. 1250, at which time the valley was nearly devoid of human occupation. These archaeological data describe the rise and fall of a closed state system—in other words, a collapse caused by endogenous rather than exogenous factors.

We are fortunate to have an extremely rich archaeological data base from Copán, the result of a large-scale archaeological project, the Proyecto Arqueologico Copán (PAC) conducted from 1975 to 1988. During this project 1,425 archaeological sites containing 4,507 structures were located and

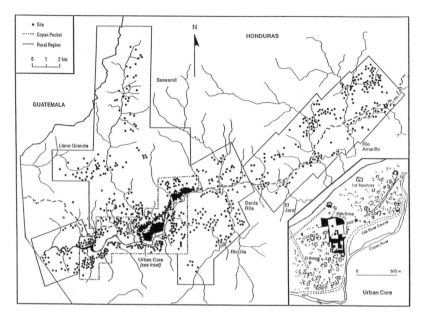

Figure 2.2   Sites Within the Copán Valley *Source:* Freter (1994)

mapped over an area of about 135 square kilometers (figure 2.2). This surveyed area includes virtually all the Copán Valley in Honduras suitable for
prehistoric habitation. Since 1980 sixteen more archaeological sites have been
extensively excavated and 117 structures exposed and restored. In conjunction with these excavations detailed settlement (Freter 1988, 1992, 1994;
Webster and Freter 1990a), pollen (Rue 1986, 1987), agricultural (Rue 1986;
Freter 1988; Wingard 1992), and energy (Abrams and Rue 1988) analyses
have been conducted. We believe that the varied yet convergent information
from a single archaeological site clarifies our understanding of the deforestation process and its complex interactive impact on human populations.
Specifically, we show that deforestation played a significant role in the failure of the Copán political system. The logical implication of our argument is
that the process of deforestation should affect societies in analogous cultural
and ecological settings, recognizing that the Maya collapse is not an isolated
phenomenon but one that bears similarities to other societal declines.

Copán

Copán, located in west central Honduras, was one of the largest Classic
period political centers built by the Maya and was the largest such site in the

southeastern region of the southern Maya lowlands. The entire prehispanic Copán population resided within the Copán Valley, defined by the drainage system of the Copán River that covers an area of approximately 400 square kilometers within Honduras (figures 2.3 and 2.4). This river valley is constricted by mountainous terrain and in many ways represents a classic case of a circumscribed environment instrumental in increasing cultural and institutional complexity (Carneiro 1970). In addition, the Copán Valley is relatively isolated from other valleys; thus its population had to adapt over time to the incentives and limitations of this specific environment.

Four important smaller tributaries flow into the Copán River; expanses or pockets of alluvial bottomlands have formed at their confluences. The largest is the Copán Pocket, which is about 600 meters above sea level and measures 12 kilometers long and 2 to 4 kilometers wide (figure 2.4). These alluvial pockets comprise the bulk of the prime agricultural land in the valley, and although microenvironmental variations exist between them, the general structure of their physical environments in terms of ecological zonation is similar. These four alluvial pockets, and the adjacent alluvial soils in the Sesesmil Valley, have had a profound effect on the settlement and land-use patterns of the ancient Copánecos and for the contemporary Honduran population.

The alluvial bottomlands are surrounded by foothills that are covered with mixed tropical deciduous trees and brush (Turner et al. 1983). The upland forest zone rises above the foothills beginning at approximately 800 meters above sea level. This upland forest contains a wide range and number of arboreal species but is dominated by *Pinus oocarpa Schiede* (Turner et al. 1983:137). Based on the palynological data from Copán, the dominance of this pine species extends to the Classic Maya occupation of the valley;it plays a major role in our reconstructions.

Copán Settlement History and Demographic Reconstruction

It has long been recognized that the archaeological reconstruction of diachronic patterns of resource exploitation is the basis for assessing the relative stability, growth, or decline of a political economy, because the salient features of the economy are revealed through settlement patterns and demographic reconstructions. Because the primary thesis here is that the collapse of the Copán Maya state was in part a consequence of the failure of the agricultural component of the economy, which was systemically linked to increasing deforestation, we must describe the changing demographics of the settlement during these centuries.

Systematic settlement surveys of the Copán Valley were conducted from 1975 to 1988 (Leventhal 1979; Fash 1983; Webster 1985; Freter 1988, 1992, 1994; Webster and Freter 1990a, 1990b). This research identified and

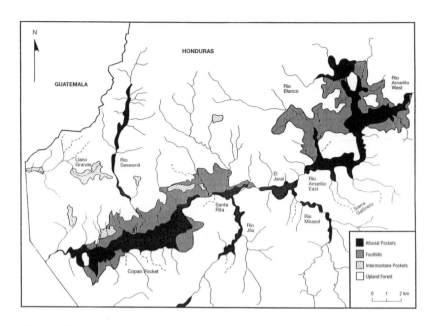

Figure 2.3   Environmental Zones of the Copán Valley

Figure 2.4   The Copán Pocket, Showing Environmental Zones and Degree of Deforestation

mapped 1,425 sites. In addition, test excavations were performed on a 13% stratified random sample of the sites within the Copán Pocket and a 30% stratified random sample of the sites in the rural region (Freter 1988, 1992, 1994; Webster and Freter 1990b). In conjunction with these surveys and test excavations 2,090 obsidian hydration dates were processed from 200 of the sites tested, representing a 15% stratified random sample of all sites in the Copán Valley (Freter 1988, 1992). These combined efforts have created one of the richest, most chronologically detailed settlement data bases within the lowland Maya region.

Based on these data there appear to be three zones of prehistoric settlement in the valley. First is the Urban Core (figure 2.2, inset), which, although small in area (0.75 square kilometers), is the most densely occupied, containing 1,425 structures per square kilometer. This area includes the Copán Main Group and its two flanking barrios—El Bosque and Las Sepulturas. The second settlement zone is the Copán Pocket, defined as the 24-square-kilometer area surrounding the Urban Core but exclusive of it (figure 2.2, dotted line). This settlement zone averages 102 structures per square kilometer, most located on the alluvial bottomlands or the lower foothills adjacent to them. The third settlement zone is the Rural Region, consisting of the surveyed area outside the Copán Pocket (figure 2.2, solid line). This zone has the largest geographic area, about 111 square kilometers, but contains the lowest settlement densities, only 10 structures per square kilometer.

In conjunction with our detailed chronological control of these sites through obsidian hydration dating, we can reconstruct the demographic changes that occurred in the Copán Valley from the rise of the Copán polity through the period of political collapse and demographic decline. For the purposes of this essay we have divided the demographic history of the Copán Valley into three broad time periods that represent the rise (A.D. 400–700), height (A.D. 700–850), and decline (A.D. 850–1300) of the Copán polity. When necessary we subdivide these broad time periods into phases.

This demographic reconstruction is based on the methodology outlined in the work of AnnCorinne Freter (1988, 1992) and David Webster and Freter (1990b). The population estimates we present here are based on four people per residential room occupied during that time phase. Adjustments were made to account for nonresidential room functions, nonresidential sites, hidden structures, abandoned sites, and unsurveyed areas within the valley. We should also note that these are conservative population estimates that assume relatively small family sizes and relatively low population growth rates. As will become apparent in the demographic reconstruction that follows, during the growth and height of the Copán polity the actual population probably was higher than our estimates, because it was growing rapidly during those periods. Our choice of conservative population estimates here is to demonstrate that, even with modest population growth assumptions, the population within the Copán Valley would have generated significant stresses on the natural environment.

The Rise of the Copán Polity: A.D. 400 to 700

The period A.D. 400–700 represents the rise and development of the Copán polity. Unfortunately, it is the least understood time phase demographically, because the population during this period is underrepresented archaeologically as a result of differential preservation and sampling. Clear evidence points to the destruction of Middle Classic Acbi–phase structures (A.D. 400–700) by Classic period Coner-phase (A.D. 700–900) inhabitants, who either looted the earlier buildings for reusable materials or built over their remains. As a result the number of extant structures occupied during the Middle Classic period underrepresents the actual population of the valley. In addition, a recent archaeological survey on the Guatemalan side of the Copán Valley (Carson Murdy, personal communication 1990) demonstrates that a sizable Acbi-phase population lived just outside the survey area. Because of these problems the population estimates for this time period must be viewed as tentative.

Based on our current data, however, it appears that from A.D. 400 to 550 the population within the Copán Valley was quite low and was concentrated in and around the alluvial bottomlands of the Urban Core, where at least twelve hundred people lived (table 2.1). The Copán Pocket and Rural Region contained only about six hundred people, most living along alluvial bottomlands or low foothills in isolated hamlets.

Table 2.1
Copán Valley Population Estimates

Broad periods	Time phase A.D.	Urban Core	Copán Pocket	Rural region	Total valley
Rise	400–550	1,200	500	100	1,800
	550–700	1,800	2,800	1,000	5,600
Height	700–850	6,000	10,000	4,000	20,000
	850–1000	3,000	5,500	5,500	14,000
Collapse	1000–1150	1,500	3,000	2,500	7,000
	1150–1300	600	1,200	500	2,300
	post–1300	0	0	0	0

SOURCE: Data compiled from Freter (1988, 1992), with methodology from Webster and Freter (1990a)

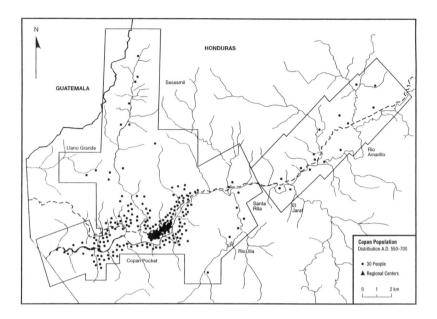

Figure 2.5   Mayan Population Distribution, A.D. 550–700   *Source:* Freter (1992)

From A.D. 550 to 700 the Copán Valley population grew rapidly and continued to use the same basic geographic areas and ecological zones as in the previous period (figure 2.5; table 2.1). This rapid population growth—from at least 1,800 people in the entire valley to at least 5,600 people over 150 years—can be attributed to several factors. Murdy's recent rural settlement survey of the Guatemalan side of the Copán Valley indicates that a relatively large population lived there from A.D. 400 to 600. Some of these people probably migrated toward the Honduran side of the Copán Valley, attracted by the then emerging Copán polity. Second, the population within the Copán Valley at this time appears to have been a founding population, with large expanses of available land and as yet no stresses on the valley's resource base. Under such conditions agrarian populations tend to have higher internal population growth rates (Redfield and Villa Rojas 1962).

The Height of the Copán Polity: A.D. 700 to 850

From A.D. 700 to 850 the Copán Valley reached its height in terms of sociopolitical complexity, with a well-established royal lineage, an elite-

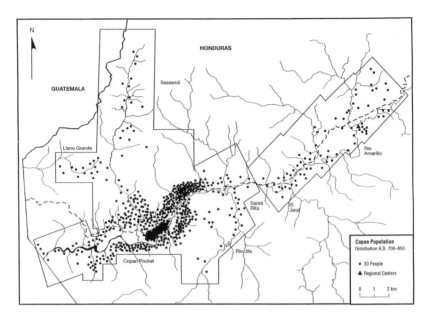

Figure 2.6   Mayan Population Distribution, A.D. 700–850   *Source:* Freter (1992)

commoner stratified class system, craft specialists, and a complex writing system. The population during this period again increased very rapidly— from 5,600 people to a conservative estimate of 20,000 people toward the end of the period (table 2.1; Freter 1992). This population was concentrated primarily in the Urban Core and Copán Pocket settlement zones, where at least 16,000 people resided (figure 2.6). However, the Rural Region also experienced a fourfold increase in absolute numbers— from 1,000 to 4,000 people—and an expansion into the foothill ecological zone.

The Decline of the Copán Polity: A.D. 850 to 1300

After A.D. 850 there was a gradual depopulation and decentralization of the Copán Valley, a trend that continued until the valley was totally abandoned around A.D. 1300 (table 2.1). This depopulation and outward rural settlement migration represents the onset of the Classic collapse of the Copán state. It was followed by a 250-year period of rural community settlement formation, complete with rural ceremonial centers, which replaced the administrative and ideological functions that the Main Group had provided (figure 2.7). These newly founded rural communities were generally

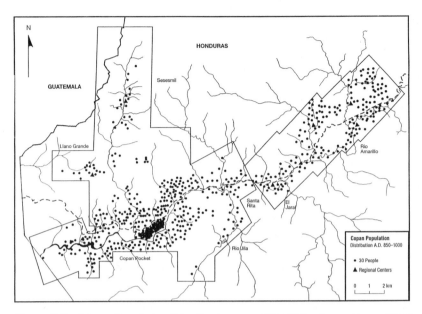

Figure 2.7   Mayan Population Distribution, A.D. 850–1000   *Source:* Freter (1992)

located on the best alluvial soils outside the Copán Pocket, in El Jaral and Rio Amarillo or their adjacent foothills. Few rural sites, most of which were nonresidential, were located in the upland forest zone.

Interestingly, these rural communities survived for more than 300 years in these smaller alluvial pockets before also being gradually abandoned by A.D. 1300 (figures 2.8 and 2.9). As we discuss later, the occupants of these smaller alluvial pockets—the descendants of the Classic collapse population—probably exploited their environment in basically the same fashion as their ancestors in the Copán Pocket, eventually producing the same environmental degradation that contributed to the collapse of the original political system 350 years earlier. Thus it appears that a series of "collapses," rather than a single event, may have occurred within this valley system, although the original event would have had a greater effect than the others on the sociopolitical system.

Agricultural Land-Use Patterns

Using this demographic information, John Wingard (1992) did a computer simulation of the land-use patterns within the Copán Valley to determine

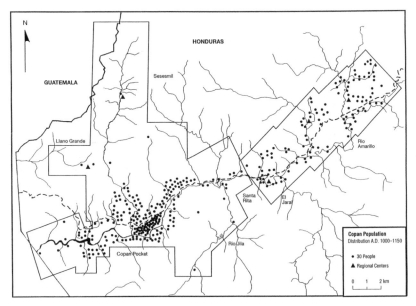

Figure 2.8   Mayan Population Distribution, A.D. 1000–1150
Source: Freter (1992)

Figure 2.9   Mayan Population Distribution, Post—A.D. 1150
Source: Freter (1992)

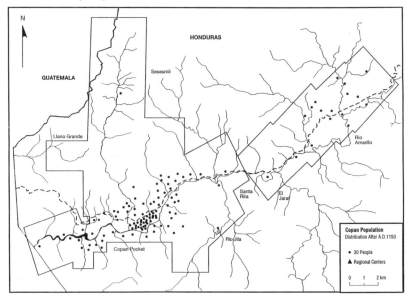

the degree, extent, and effect that the agricultural practices of this population could have had on their natural environment. The primary factors limiting the long-term productivity of soils are nutrient depletion and erosion. To examine these processes at Copán we used the Erosion-Productivity Impact Calculator (EPIC), a comprehensive model developed by the U.S. Department of Agriculture's Agricultural Research Service and the Texas Agricultural Experiment Station, to determine the relationship between soil erosion and soil productivity. This computer program uses soil, weather, and management data supplied by the user and combines these with its internal physiologically based plant growth and soil erosion models. The program produces information on crop yields, soil loss, and nutrient availability.

For the purposes of modeling the ancient Copán Maya agricultural system we collected and analyzed 180 soil samples representing all the different microenvironmental characteristics of the Copán Valley. The dominant soil groups were comprised of inceptisols and entisols, with mollisols, ultisols, and histols also present. The agricultural potential of the Copán soils appears to be comparable to that in other lowland Maya areas. The local Copán farmers reported mean maize yields of just over 1,000 kilograms per hectare, also comparable to other areas within the southern lowlands.

To simulate the Copán Maya agricultural system we ran soil profile data representing eight agriculturally dominant subgroups with as many as six slope classes, using three weather regimes representing the lower, middle, and upper sections of the valley. We used four management strategies. Because EPIC allows a maximum management cycle of only ten years, the longest cycle we used was one with an eight-year fallow period followed by two years of cropping. However, because of the long-term sustainability of even such a short cycle, this is an adequate proxy for longer-term management strategies. The other strategies we incorporated in the model were five years of fallow followed by three years of cropping; three years of fallow followed by three years of cropping; and continuous cropping. In all we conducted 324 runs, each representing a 100-year time span. The results yielded information on the long-term productivity of the Copán Valley soils under prehispanic agricultural techniques and demographic conditions. The four strategies were also designed to maximize the use of the most productive soils in the valley—the alluvial and foothill soils—using the most sustainable management strategy that would provide adequate production for the population.

Starting at A.D. 1 the simulation yields five distinct periods of agricultural development. First is the initial phase of extensive agriculture, during which population and agricultural expansion occurs. A long-term fallow system starting in the Copán Pocket would use the alluvial and foothill soils, while the population grows to about forty-eight hundred people by the middle of the sixth century.

The second period of agricultural development is the first phase of intensification. The management strategy changes from a long-term to a short-term fallow system. By A.D. 600 all areas of alluvium and foothills are under continuous cultivation, with a population of approximately fifty-three hundred people.

The third period is the second phase of extensive agriculture. Already cultivating the alluvial and foothill soils on a continuous basis, the people now must expand their agricultural area to the upper hillsides, using a long-term fallow strategy to meet production requirements. The process starts in the Copán Pocket, still the primary focus of the population. This period lasts more than two centuries, until early in the ninth century, by which time the population has grown from fifty-three hundred to seventeen thousand people.

The fourth period is the second phase of intensification, which corresponds to the climax of the Classic civilization at Copán and the highest population levels. It is also the shortest period, lasting only ten to twenty years. Because the upper hillsides and upland forest cannot support the rapidly growing population, the people clear them and quickly bring them into continuous cultivation, only to see the dramatic increase in production wiped out within a decade by devastating soil erosion. This strategy supported the highest level of aggregate population but was clearly the least sustainable, with the resultant deforestation having severe long-term consequences on the Copán Valley environment. The fifth and final period of production returns to the previous levels, as the valley population decreases and decentralizes. This final phase lasts until the valley eventually is abandoned.

One critical implication of this simulation is that the intensification of agriculture through the shortening of fallow periods within the circumscribed Copán Valley predictably led to extreme soil erosion. In fact, the EPIC simulation indicates that soil erosion—the actual loss of soil—was the inevitable consequence of agricultural intensification and represents the greatest single factor in subsequently lower agricultural productivity. This model further suggests that soil erosion may have occurred in a rapid and severe fashion in a matter of one or two decades at the onset of the ninth century.

Various archaeological data from this region support this simulation. Soil loss as a direct consequence of the removal of trees has been confirmed through controlled experiments in central Honduras (Hudson et al. 1983). The heavy substratum of clay soil in the Copán Valley would have provided an effective medium for accelerated slippage of topsoil. Late Classic architecture buried beneath as much as 3 meters of soil has been excavated in the Copán Pocket (Webster, Sanders, and van Rossum 1992). Finally, Gerald Olson (1975) reported that Mayan artifacts in the Rio Amarillo Pocket were buried beneath 42 centimeters of soil, the result of sheet ero-

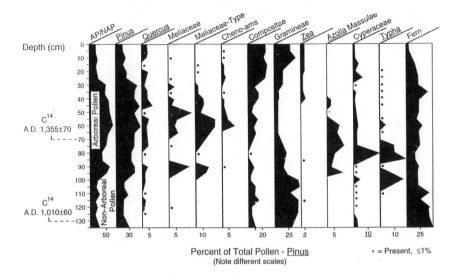

Figure 2.10   Petapilla Pollen Diagram   *Source:* Rue (1986)

sion caused by mismanagement of forested slopes. Because this area was not occupied in any significant numbers until about A.D. 1000 (Freter 1988, 1992, 1994), we can infer that deforestation and subsequent soil erosion were neither localized to the Copán Pocket nor restricted to the end of the Classic period.

Palynological Evidence of Deforestation at Copán

Palynological data also collected from the Copán Pocket directly confirm that deforestation in fact occurred coeval with the collapse of the Copán state. These data were recovered from a small bog in the Petapilla foothill zone of the Copán Pocket (Rue 1986, 1987). The percentage curves for those significant taxa are presented in figure 2.10. A corrected radiocarbon date of A.D. 1010, plus or minus sixty years, is associated with the depth of 125 to 130 centimeters from the core sample, and a second corrected radiocarbon date of A.D. 1355, plus or minus seventy years, is assigned to the depth of 65 to 70 centimeters. Thus these data do not extend to a period predating the initial failure of the Copán state but rather capture some midpoint in the process of collapse.

In the pollen profile the arboreal pollen—nonarboreal pollen and pine (*Pinus* spp.) percentages are based on the total pollen, whereas the percentages for the remaining taxa are based on total pollen minus *Pinus*; given the high percentage of *Pinus*, subtracting it made the nondominant species easier to observe. *Pinus* includes all pine but is dominated by *P. oocarpa*, representing 90% of the pine pollen. The tropical deciduous species in the mahogany family are represented by the *Meliaceae* and *Meliaceae*-type pollen curve; colonizing grasses, reflecting an early stage in ecological succession, are represented by the *Gramineae* (for a more complete description of methodology and analysis, see Rue 1986:89–93).

At a depth of 100 to 135 centimeters pine percentage is relatively low in the profile, whereas grasses and ferns are dominant. This would indicate that the upland forest zone ca. A.D. 1000 was being deforested to some extent and replaced by those nonarboreal species. At the lowest level in the profile tropical deciduous species are virtually nonexistent, indicating that the foothill zone was largely deforested during the Classic period. This confirms a similar conclusion drawn by Frederick Wiseman based on previous palynological research at Copán (Turner et al. 1983) and is further supported by the settlement data from this ecozone.

The 100 centimeter mark corresponds to approximately the beginning of the thirteenth century. As previously discussed, population in the Copán Valley is gradually approaching zero by this time. Congruent with this population abandonment is a regeneration of the upland pine and foothill deciduous species, which correlates with a decline in grasses and maize.

Use Rates of *P. oocarpa*

We can take the analysis of deforestation a step further by considering the human demands for the wood itself. From the data presented thus far we know that the Maya had heavily settled and used the foothill zone of the Copán Pocket by A.D. 600. They were increasingly cultivating this area for agricultural purposes, expanding their forest clearing to the steeper slopes of the uplands after A.D. 600. With the consequent elimination of deciduous species the extensive upland pine forest must then have become the primary, if not the exclusive, reservoir for wood and other forest resources.

To assess the magnitude of differential demands placed on the upland forest by the Classic Maya, Elliot Abrams and David Rue carried out an energy analysis (1988). Aware that a large number of demands—some known, others unknown—were placed upon the pine forest, Abrams and Rue isolated and quantified what they considered the major demands. Thus they were able to calculate the consumption rates for (1) domestic fuelwood for cooking and heating, (2) fuelwood for economic production of lime-based plaster, a major artifact consumed in architectural construction, and (3) wood used in constructing the frame of wattle-and-daub houses occupied by the commoners of Classic Maya society. Conjoining data relating the general characteristics of *P. oocarpa* with their estimates of annual consumption rates for each of the three demands, Abrams and Rue demonstrated that, regardless of the specific population estimates and consumption rates selected (within reasonable order of magnitude parameters), the demand for domestic fuelwood outstripped the other demands for wood by several hundredfold (1988:391). Clearly, the Maya recognized as physical imperatives the continuous and ubiquitous acts of cooking their food and heating their houses and did not see these as cultural values that they could simply discard or modify over time.

Abrams and Rue expanded this analysis of fuelwood demand to provide an estimate of the areal extent of deforestation as it affected the upland forest. Focusing solely on the annual demand for domestic fuelwood by the Late Classic population, and accounting for the regrowth rate of upland pine, they estimated that approximately 23 square kilometers of upland forest would have been completely devoid of pine cover by A.D. 800 (Abrams and Rue 1988:391). This figure is considered a minimal estimate, because Abrams and Rue used only one of many demands on the upland forest in their specific analysis of energy usage.

We can use these data to reconstruct the consequences of decisions made by the Maya that led to deforestation. As the population size and density increased within the Copán Pocket, food and housing needs demanded that the Copán Maya extend their agricultural and settlement areas into the foothill zone, thus effectively eliminating that ecozone as a source of wood. With continued demographic increase generated perhaps by the complex interplay of economics, politics, and ideology, the Maya began exploiting the upland pine forest at a rate far greater than that of regrowth; the Maya's primary and overwhelming demands were for agricultural land and domestic fuelwood. The initial deforestation of the foothills, coupled with later deforestation of the contiguous upland slopes, exacerbated the rate of soil erosion, which greatly lowered the agricultural viability of the foothill region. The alluvial bottomlands apparently never lost their agricultural viability, although we could argue that the erosion of poorer soils from the foothill and upland zones may have reduced the fertility of bottomland soil. Ultimately, the process of urbanization in a small and relatively circumscribed wet tropical river valley placed far too much stress on the environment to successfully sustain the growing nucleated Copán state. Gradual, and then rather pronounced, soil loss within this agrarian system produced the complex interaction of population decline, out-migration, ideological and psychological stress, and political decentralization that summarily is termed the *Maya collapse.*

However, the collapse of the Copán state was not restricted to the Copán

Pocket, although that failure is archaeologically most conspicuous. The Copán Pocket population in part seems to have migrated first into the adjacent Sesesmil tributary, soon followed by more pronounced out-migration from the Copán Pocket to all three eastern pockets. After several generations of settlement and use the Maya abandoned each of these relatively small rural pockets. Thus several collapses appear to have affected the population in the Copán Valley over time; the first and largest occurred in the Copán Pocket, and it was followed by smaller collapses in the adjacent rural areas. If this model of multiple collapses is correct, we need to look at and compare the secondary failures with the "Classic" collapse in the Copán Pocket, particularly with reference to deforestation.

Unfortunately, we lack the comparable quality and variety of data we need to test and measure the condition of deforestation in these outlying regions. Nonetheless some data suggest that deforestation may have occurred in these areas after the Classic collapse. As we already have indicated, soil erosion was in fact significant in the Rio Amarillo Pocket after A.D. 1000, which was several generations after initial settlement (Olson 1975). The energy analysis shows that the overwhelming demand on the forest was for domestic fuelwood consumption—such a daily demand could hardly be overlooked by any population. Based on these admittedly few data, we can hypothesize that the emigrant population and their descendants must have been involved in a pattern of colonization and resource exploitation similar to that which had earlier affected the Copán Pocket.

One interesting implication of these long-term data is that the concept of some ecological omniscience among preindustrial nonstate cultures must be tempered by a recognition of the institutional and demographic context for decision making. For several millennia the ancient Maya were able to colonize and successfully adapt to a wet tropical environment, and undoubtedly they possessed valuable ecological and economic knowledge of the rain forest. As is true of the contemporary Maya (Rice and Rice 1984), the prehispanic Maya fully understood the basic principles of soil movement, and they presumably encoded and transmitted this compilation of ecological information through the generations via formal ideology and oral traditions. Nevertheless the Classic Maya cultural system of adaptation failed. We can infer from this that the context of economic and demographic stress, regardless of the complex causal factors producing those conditions, ultimately places tremendous pressure on society as a whole to incrementally degrade its environment to meet the compelling short-term needs of food and essential fuel.

The preponderance of evidence points to forest mismanagement as a factor contributing to systemic failure, despite the fact that the Maya were extremely knowledgeable of the structure and operation of nature. In the context of overpopulation not only the Classic Maya were "forced" in a sense to mismanage their forest environment; apparently their descendants, residing in smaller pockets, repeated that pattern of decision making, which led to smaller-scale collapses. From this case it is apparent that individuals in general are compelled to set aside their traditional philosophies of ecological balance in order to meet critical material needs.

As we mentioned in the introduction, understanding past processes may provide insights into how we conceptualize and respond to current human conditions. Human and environmental relations in the Copán Valley today are a case in point, for the parallels between contemporary cultural-ecological conditions and those that existed during the Late Classic period are considerable. The population within the Copán Pocket today has reached if not exceeded those levels established for the Classic period. Similarly, the alluvial bottomlands of highest agricultural productivity today are continuously cultivated in an intensive double-cropping system in tobacco and maize or beans or have been converted to pasture for cattle. The foothill zone already is heavily populated and deforested; as a result the upland forests are heavily exploited for agricultural land, fuelwood, and lumber.

The Copán Valley, as we have discussed, was completely abandoned by ca. A.D. 1300. It was not until the mid-nineteenth century that migrants from Guatemala recolonized it. Apparently five hundred years or so were necessary for ecological succession to regenerate natural resources in the valley. An examination of the Petapilla pollen profile (figure 2.10), beginning at a depth of 25 centimeters, illustrates the consequences of human recolonization of the valley. Since the mid-nineteenth century pine has declined severely and is today at a level of overexploitation comparable to that of the Late Classic—the time of the major collapse. Tropical deciduous species similarly have declined at the expense of increased maize production and grass cover for cattle pasture. The general cultural-ecological relations that appear to have affected the Classic Maya are conspicuous in contemporary Copán, but the consequences of continued unchecked growth and forest mismanagement are starkly known.

Although this essay has focused on deforestation in the Copán River valley, this specific tropical ecosystem is hardly unique globally. On the contrary, it typifies many tropical riverine systems that are experiencing similar cultural-ecological stresses. The critical role of the archaeological record is to provide the present with a picture of the past so that contemporary societies can preview their future. This comparative view of the long-term diachronic processes available only to archaeology highlights the importance of the study, as well as the preservation, of the archaeological record.

Acknowledgments We thank the Instituto Hondureno de Antropologia e Historia for permission to work at Copan. Figures 2.5 through 2.9 are reprinted from Freter 1992 with the permission of Cambridge University Press.

References Cited

Abrams, Elliot and David Rue. 1988. The causes and consequences of deforestation among the prehistoric Maya. *Human Ecology* 16 (4): 377–395.
Carneiro, Robert. 1970. A theory of the origin of the state. *Science* 169: 733–738.
Cooke, C. W. 1931. Why the Maya cities of the Petén District, Guatemala, were abandoned. *Journal of the Washington Academy of Sciences* 21 (13): 283–287.
Deevey, E. S., D. Rice, P. Rice, H. Vaughan, M. Brenner, and M. Flannery. 1979. Mayan urbanism: Impact on a tropical karst environment. *Science* 206: 298–306.
Fairservis, Walter Jr. 1975. *The Roots of Ancient India.* Chicago: University of Chicago.
Fash, William. 1983. Maya State Formation: A Case Study and Its Implications. Ph.D. diss., Harvard University, Department of Anthropology.
Freter, AnnCorinne. 1988. The Classic Maya Collapse at Copán, Honduras: A Regional Settlement Perspective. Ph.D. diss., Pennsylvania State University, Anthropology Department.
———. 1992. Chronological research at Copán: Methods and implications. *Ancient Mesoamerica* 3 (1): 117–134.
———. 1994. An analysis of rural settlement and its implications on the Classic Maya collapse at Copán, Honduras. In *Village Communities in Early Complex Societies,* G. Schwartz and S. Falconer, eds., pp. 160–176. Washington, D.C.: Smithsonian Series in Archaeology.
Hamblin, Robert and Brian Pitcher. 1980. The Classic Maya collapse: Testing class conflict hypotheses. *American Antiquity* 45 (2): 246–267.
Hudson, J., M. Kellman, K. Sanmugadas, and C. Alvarado. 1983. Prescribed burning of *Pinus oocarpa* in Honduras. Effects on surface runoff and sediment loss. *Forest Ecology and Management* 5: 269–281.
Lemonick, Michael D. 1993. Lost secrets of the Maya. *Time,* August 9, pp. 44–50.
Leventhal, Richard. 1979. Settlement Patterns at Copán, Honduras. Ph.D. diss., Harvard University, Anthropology Department.
MacKie, Euan W. 1961. New light on the end of Classic Maya at Benque Viejo, British Honduras. *American Antiquity* 27 (2): 216–224.
Olson, Gerald. 1975. Study of the soils in Valle de Naco (near San Pedro Sula) and La Canteada (near Copán, Honduras): Implications to the Maya mounds and other ruins. *Cornell Agronomy Mimeo,* Vol. 19.
Price, Barbara. 1982. Cultural materialism: A theoretical review. *American Antiquity* 47: 709–741.
Rathje, William. 1973. Classic Maya development and denouement: A research design. In *The Classic Maya Collapse,* T. Patrick Culbert, ed., pp. 405–454. Albuquerque: University of New Mexico Press.
Redfield, Robert and Alfonso Villa Rojas. 1962. *Chan Kom: A Maya Village.* Chicago: University of Chicago Press.
Rice, Don and Prudence Rice. 1984. Lessons from the Maya. *Latin American Research Review* 19 (3): 7–34.

Rue, David. 1986. A Palynological Analysis of Prehispanic Human Impact in the Copán Valley, Honduras. Ph.D. diss., Pennsylvania State University, Anthropology Department.

———. 1987. Early agriculture and early postclassic occupation in western Honduras. *Nature* 326 (6110): 285–286.

Sanders, William T. 1973. The cultural ecology of the lowland Maya: A reevaluation. In *The Classic Maya Collapse*, T. Patrick Culbert, ed., pp. 325–365. Albuquerque: University of New Mexico Press.

Sanders, William T., Jeffery Parsons, and Robert Santley. 1979. *The Basin of Mexico: Ecological Processes in the Evolution of a Civilization*. New York: Academic Press.

Shimkin, Demitri. 1973. Models for the downfall: Some ecological and culture historical considerations. In *The Classic Maya Collapse*, T. Patrick Culbert, ed., pp. 117–143. Albuquerque: University of New Mexico Press.

Turner, B. L. II, W. Johnson, G. Mahood, F. Wiseman, B. Turner, and J. Poole. 1983. Habitat y agricultura en la region de Copán (Habitat and agriculture in the region of Copán). In *Introduccion a la arqueologia de Copán, Honduras* (Introduction to the archeology of Copán, Honduras), Claude Baudez, ed., Vol. 1., pp. 35–142. Tegucigalpa: Instituto Hondureno de Antropologia e Historia.

Webster, David. 1985. Recent settlement survey in the Copán Valley, Honduras. *Journal of New World Archaeology* 5: 39–51.

Webster, David and AnnCorinne Freter. 1990a. Settlement history and the classic collapse at Copán: A redefined chronological perspective. *Latin American Antiquity* 1: 66–85.

———. 1990b. The demography of Late Classic Copán. In *Prehistoric Population History in the Maya Lowlands*, T. Patrick Culbert and Don Rice, eds., pp. 37–61. Albuquerque: University of New Mexico Press.

Webster, David, William T. Sanders, and Peter van Rossum. 1992. A simulation of Copán population history and its implications. *Ancient Mesoamerica* 3 (1): 185–197.

Willey, Gordon and Dimitri Shimkin. 1973. The Maya collapse: A summary view. In *The Classic Maya Collapse*, T. Patrick Culbert, ed., pp. 457–501. Albuquerque: University of New Mexico Press.

Wingard, John. 1992. The Role of Soils in the Development and Collapse of Classic Maya Civilization at Copán, Honduras. Ph.D. diss., Pennsylvania State University, Anthropology Department.

Wiseman, Frederick. 1985. Agriculture and vegetation dynamics of the Maya collapse in central Petén, Guatemala. In *Prehistoric Lowland Maya Environment and Subsistence Economy*, Mary Pohl, ed., pp. 63–71. Cambridge, Mass.: Papers of the Peabody Museum of Archaeology and Ethnology.

3

Forests and Polynesian Adaptations

Brien A. Meilleur

A vast culture area in the central Pacific Ocean, Polynesia is composed of well over three hundred inhabited or once-inhabited islands (Douglas 1969:347; Kirch 1984:19). The four or five major island types range from diminutive coral atolls to high volcanic islands and one that is nearly a continent. Their climatic range spans the equatorial heat and humidity of Samoa and Tonga, the perpetual snows of mountainous New Zealand, and the subantarctic conditions of the Chatham Islands.

Samoa and Tonga to the west, Hawai'i to the northeast, New Zealand and Easter Island to the southwest and the southeast, respectively, with some thirty-five archipelagoes between, together comprise one of the most biologically diverse culture areas on earth (Fosberg 1953; Klee 1980:245–253; Steadman 1989:196). These islands were entirely covered or nearly so by a wide range of natural forest communities when Polynesians arrived between about 1500 B.C. and 1200 A.D. (Cuddihy and Stone 1990:13; Decker 1971:82; Merrill 1940:637; Olson and James 1984:776; Zimmerman 1965:57). Although the specifics vary in time and space, Polynesian attitudes toward the forest and forest use have been on the whole remarkably similar.

Gathering and Hunting

Harvesting wild plant and animal products from forests has been recognized as a significant part of life everywhere in Polynesia throughout the prehistoric and historic periods. Gathering plants, especially for human ornamentation, medicine, and crafts, is still widely practiced. The inhabitants of several central and western Polynesian islands continue to hunt forest birds and bats. However, the importance of hunting, the types of harvest, and the species involved vary among island groups and during time periods within the same groups. This is the result of at least five factors: the diversity of climates and physical conditions as they affected classical Polynesian agriculture, the need to emphasize different subsistence strategies at different periods of island occupation, the differential availability and productivity of marine resources, the specific societal interest shown toward certain forest species or groups of species, and of course the range of natural and introduced terrestrial flora and fauna within the culture area. With respect to forest-related gathering and hunting, anthropologists recognize significant variation between islands in tropical versus temperate Polynesia, between large high islands and atolls, and between the subsistence strategies at the time of colonization versus those of later periods.

On the tropical and subtropical islands, colonization-phase subsistence strategies differed in emphasis from those of later periods (Kirch 1973:38, 1982a:2, 1984:84; Kirch et al. 1992). In contrast, overall subsistence strategies on temperate islands were more stable. Harvesting of marine resources, nesting seabird and flightless land bird populations, and undoubtedly some wild forest plants dominated the earliest economic activities on newly reached islands of whatever type (Emory and Sinoto 1961:17; Kirch 1973:37, 1984:84). As Polynesian-introduced root- and tree-crop horticulture and animal husbandry took hold, eventually joining marine fishing as the dominant production techniques, Polynesians' gathering strategies changed, especially on the tropical and subtropical high islands. Here, complex social, religious, and political hierarchies evolved in later periods, and the trend was away from subsistence gathering and hunting primarily for food and toward appropriation of a wider range of forest products. Growing familiarity with the new flora and fauna also would have contributed to more diverse forest use.

In temperate Polynesia, and especially in the cooler uplands of North Island and on South Island, New Zealand, where classical Polynesian horticulture was much less suitable, gathering and hunting were substantially more important to subsistence throughout the prehistoric and historic periods (Anderson 1983; Best 1942:2, 312; Cumberland 1961:144; Firth 1929:49–52). At the subantarctic extreme of the Polynesian adaptation the Moriori of the Chatham Islands did not practice agriculture in any form, depending instead on coastal resources such as seals and seabirds (Sutton 1980; Sutton and Marshall 1980). Because of the depauperate nature of the indigenous floras and faunas (Fosberg 1953; Merrill 1981:207–208), the easily depleted seabird and limited land bird populations, and very different physical constraints on horticulture, the inhabitants of the equatorial atolls turned predominantly to marine fisheries for subsistence (Barrau 1957; Emory 1975:2, 7–8).

In the absence of terrestrial mammals hunting and collecting nonmarine animals in pre-European Polynesia focused on sea and forest birds. A substantial literature, developed since the 1970s, describes the often spectacular effects of various prehistoric Polynesian activities and events on birds. Such documentation is now being increasingly linked to the archaeological, ethnographic, and ethnohistoric records of wild bird use in Polynesia (for Anuta see Yen 1973:117; for Anuta and Tikopia see Firth 1975:60 and Steadman, Pahlavan, and Kirch 1990:146–147; for Hawai'i see Emerson 1894, Lyons 1903:25, Malo 1951:37–40; Kamakau 1976:108, 124; Kepelino 1932:88, 148, 150, 160; Manning 1981; Samwell in Beaglehole 1967:1167, 1173, 1235; for the Marquesas, Handy 1923:180–181; for New Zealand, Best 1942:137, 253, 288, 312, 345, 357, 386, 406, 410, 412; Downes 1928; Ranapiri 1895; for the Cook Islands, Utanga 1989:102; for Tahiti, Ferdon 1981:52, 63; for Tonga, Ferdon 1987:180–184; for the Tuamotus, Emory 1975:225–227). Together they describe a coherent and strongly entrepreneurial Polynesian attitude toward birds.

Contrary to what was believed at one time, aboriginal human activities seriously depleted bird species and avian populations throughout Polynesia (Amadon 1950:210; Berger 1972:5). Storrs Olson and Helen James (1982:633, 1984:770), for example, estimate that more than 50% of the species of endemic Hawaiian land birds became extinct prehistorically. To this figure must be added the many local extirpations and range reductions of extant species, which have been less thoroughly investigated (Steadman 1989:183). However, it is often difficult to know exactly which of three causes—direct human predation, human-induced habitat destruction, or human-introduced organisms—was ultimately responsible. Undoubtedly, the three often worked together to reduce avian numbers and species.

Human predation on birds began immediately upon humans' arrival at a given island (Steadman, Pahlavan, and Kirch 1990:147). Humans initially preyed upon what must often have been teaming colonies of nesting seabirds. Bird bones commonly dominate the earliest strata of prehistoric food middens, only to drop off later, presumably as avian numbers are reduced, as nesting habitat is disturbed or destroyed, and as agricultural production rises (Anderson 1983, 1984:734; Kirch 1973:37; Kirch and Yen 1982:284; Kirch et al. 1992:173–174; Steadman 1989:181; Steadman, Pahlavan, and Kirch 1990:147–148). Both sea and forest birds are found in middens, although forest birds are generally found in much smaller numbers (Kirch and Yen 1982:282; Sutton and Marshall 1980:30). This probably reflects less important initial densities in forest species, coupled with the ease with which nesting seabirds and their young could be taken. The opposite was true in New Zealand: huge midden sites of many hectares have been found to contain extinct moa bones (Cassels 1984:743, 748;

Cumberland 1963:190); the approximately fifteen endemic species of moa are now believed to have been brushland or forest dwellers (Anderson 1984:728; Trotter and McCulloch 1984:710, 723).

Many other endemic forest birds were either extinguished or extirpated throughout Polynesia. Direct human predation is implicated in part, especially among the larger forest birds, some of which were flightless like the moa. Such was the case with geese, ibis, raptors, rails and crows, and a large honey eater in Hawai'i (James and Olson 1991; Olson and James 1984:770, 775, 1991), a flightless megapode and a rail species in Tikopia (Kirch and Yen 1982:283–284), several species, some flightless, of swan, eagle, goose, rail, crow, and parrot in New Zealand, a wide range of large and small species—volant and flightless—from several environments in the Chatham Islands (Cassels 1984:750–751), and many other taxa of parrot, pigeon, dove, and flightless rail throughout the rest of Polynesia (Cassels 1984:756; Steadman 1989). At least one forest-dwelling bat species may have been extinguished in Hawai'i (James et al. 1987:2353). The demise of these species and many others is attributed to a combination of human-induced causes, including direct predation.

Although this archaeological and ethnographic evidence demonstrates oftentimes heavy direct human predation on forest birds as items of food and material culture, except for the occasional finding of a canoe part or other large piece of carved wood in a well-dated archaeological context, we have little direct evidence or detail regarding the gathering of wild forest plants in Polynesian prehistory (Sinoto 1979, 1988). Even when found, the species of plants used prehistorically by Polynesians have not been botanically determined with much regularity (but see Orliac 1990). Although charcoal is commonly found in prehistoric earth ovens and indicates the importance of firewood gathering, again, surprisingly few efforts have been made to identify such remains to the species level (but see Murakami 1983a, 1983b, 1989). Rather, the widely held and certainly correct assumption that forest plants were important in Polynesian prehistory is drawn mostly from three other types of evidence: by reference and comparison to archaeologically derived descriptions of prehistoric use of forest animals, by extrapolation from what is known about contemporary gathering and from the ethnographic and ethnohistoric records on wild plant use, and through consultation of native language texts, often mythological, that allude to wild plant use and that clearly originated before the arrival of Europeans (Alpers 1987; Beckwith 1972; Handy 1923:322–341; Henry 1928:336; Langridge and Terrell 1988; Métraux 1940:55–88, 362–389).

The literature on Polynesian ethnobotany is extensive, although variable by region; it amply demonstrates the great importance of a wide range of indigenous forest species in ocean voyaging, medicine, warfare, secular and religious material culture, building construction, body ornamentation, secondary foods, and many other features of traditional life (Abbott 1992;

Brooker, Cambie, and Cooper 1987, 1989; Chock 1968; Cox and Banack 1991; Croft and Tu'ipulotu 1980; Emory 1947; Kaaiakamanu and Akina 1922; Krauss 1975, 1993; Maclet and Barrau 1959; Nagata 1971; Office of Comprehensive Health Planning 1974; Pétard 1986; Uhe 1974; Walls 1988; Whistler 1985, 1988; etc.). However, because of a general absence of quantitative data in this literature it is difficult to accurately estimate the effects that Polynesian plant gathering had on forests.

Ethnohistoric accounts dating from the period of first Euroamerican interaction with Polynesians, when combined with recent assessments of tree cover during prehistory, paint a picture of significant forest removal throughout pre-European Polynesia. Such habitat modification is commonly associated with Polynesian use of fire to clear land for swidden plots and permanent agriculture. But as human populations grew throughout Polynesia—and recent estimates indicate that densities of 250 people per square kilometer and sometimes more were not uncommon (Firth 1975:39; Kirch 1982b:4; Kirch and Yen 1982:56; Stannard 1989:40; Yen 1973:122)—harvesting of large trees for canoes and for other activities such as statue moving on Easter Island (McCoy 1979:160; Mulloy 1970) and the ubiquitous gathering of firewood and other native economic plants also must have played some part in the retreat of the natural forest (Cuddihy and Stone 1990:29, 34–35; Handy and Pukui 1972:224–225; Menzies 1920:81, 156; Rock 1913:100).

Removal of Natural Forest

Prehistoric Polynesians on all islands removed major portions of aboriginal forest, often within relatively short periods of time. Although M. McGlone (1983:11) estimates, for example, that as much as 50% of the some 15 million original hectares of New Zealand forest were removed within one thousand years, he believes that Polynesians actually cleared most of the forest in less than half that time. Although impressive, this destruction is much less severe than the more than 300 million hectares of forests removed or seriously degraded by Euroamericans in the coterminous United States within about three hundred years of their arrival in the New World (Williams 1989:3–4).

Eighteenth-century European visitors to Polynesia noted that native forests on the larger islands were often found well back from the coast (for Kaua'i, Clerke in Beaglehole 1967:1322, Menzies 1920:38; for Moloka'i, Samwell in Beaglehole 1967:1151; for Hawai'i, MacCaughey 1918:392, citing Ledyard, Menzies 1920:51; for Tahiti, Ferdon 1981:179–180). On several smaller, and especially drier islands, they noted that nearly the entire forest had been cleared (for Easter Island, Flenley 1979, McCoy 1979:160, Métraux 1940:12–14; for the Hawaiian island of Kaho'olawe, Samwell in

Figure 3.1   Taro (*Colocasia esculenta*, Hawaiian: *kalo*) planted at an elevation of 1,200 meters in south Kona, Hawai'i, within a cleared forest of *'ohi'a lehua* (*Metrosideros polymorpha*).

Beaglehole 1967:1219; for Lana'i, Samwell in Beaglehole 1967:1220; for Ni'ihau, Clerke in Beaglehole 1967:1322, Samwell in Beaglehole 1967:1231). Europeans also regularly saw fires (for Kaua'i, Menzies 1920:32, Vancouver 1798:170; for Kaho'olawe, Spriggs 1992:I-4; for Easter Island, Métraux 1940:162; for Tonga, Ferdon 1987:209; for New Zealand, McGlone 1983:11). Although Richard Vogl (1969) has shown that naturally occurring forest fires would have been fairly common in prehuman and, by extension, in prehistoric Hawai'i (but see Smith and Tunison 1990:395–397), anthropologists still believe that the forest removal described in Polynesia during the European contact period—and later independently verified in the archaeological, botanical, and palynological records—was mostly associated with human-induced burning linked to clearing for agriculture (figure 3.1).

Indeed removal of aboriginal forest on the smaller permanently inhabited islands and atolls appears to have occurred shortly after people arrived. The native forests on the wetter islands were replaced quickly, at least in part by Polynesian-introduced woody floras, as they were on several Polynesian Outliers, for example (for Anuta, Yen 1973:122; for Kapingamarangi, Niering 1956:4; for Tikopia, Kirch and Yen 1982:25–63), and on the atolls generally (Barrau 1957). The larger islands were less extensively deforested in relative terms apparently because of their size, although in such cases leeward dry and mesic forests were affected more than the windward more-difficult-to-burn wet forests (Cuddihy and Stone 1990:104; McGlone 1983:17).

Forest clearing most likely began in association with swidden (i.e., shifting) cultivation (see, for example, Colenso 1880:7, 11). Then, as populations grew, the Polynesians established permanent field systems. Researchers have described a range of shifting and permanent regimes, sometimes with several types occurring simultaneously, for a number of the larger high islands (for the Marquesas, Decker 1971:161; for North Island, New Zealand, McGlone 1983:19; for Hawai'i, Cuddihy and Stone 1990:28–29, citing McEldowney 1979, 1983), as well as on some smaller high islands (for Futuna and Uvea, Kirch 1982a:1; for Tikopia, Kirch and Yen 1982:38). Similarly, they used burning outside of both swidden and permanent field systems to maintain successional stages favorable to desired vegetal resources, especially edible grasses and ferns (for East Futuna, Kirch 1975, Spriggs 1985:412, citing Burrows 1936; for New Zealand, McGlone 1983:16, 18; for the Marquesas, Decker 1971:236–241; for Hawai'i, Kirch 1982b:7, McEldowney 1979:23–24).

Polynesians also used fire in war to clear lines of sight around fortifications and to attack and destroy enemy food production. War-related fire is well documented in some parts of Polynesia (for New Zealand, Vayda 1956:128, 166; for Easter Island, Métraux 1940:149, citing Roussel 1926:423). Forest trails were also kept open through burning (Best 1942:2; McGlone 1983:20; McKelvey 1958:31). During all such applications fire would have escaped into surrounding vegetation. Some researchers suggest that Polynesians used fire to drive animals while hunting (for New Zealand, Cumberland 1963:191, but see McGlone 1983:20; for the Marquesas, Decker 1971:238; see also Pawley and Green 1973:6). Fires that I and others observed in 1988 in the Marquesas (J. Florence, B. Gagné, S. Montgomery, personal communications 1988; see also Decker 1971:95) had been intentionally set. After speaking with some people responsible for setting them, we can perhaps best describe these fires as recreational in nature.

Several writers have claimed that Polynesians in upland areas set forest fires with the expectation that massive erosion would follow deforestation, resulting in the infilling of valleys and coastal plains, thereby rendering these areas more fruitful for agriculture (Drigot 1982:11; Spriggs 1985:409, 425, 429). Infilling of this nature has been documented throughout Polynesia (Kirch 1982a:4, 1982b:10, 1983:28; Kirch and Yen 1982:147–160; Kirch et al. 1992:170, 175; McGlone 1983:16, 22, but see Grant 1985:109).

The removal of prehistoric Polynesian forests had major consequences for natural floral and faunal diversity within the culture area. Although difficult to quantify for several important taxonomic groups, especially for plants and insects, substantial numbers of plant and animal species went extinct as a result of Polynesian-induced forest habitat modification. These include well-documented extinctions of smaller land birds (James and Olson 1991:86; Steadman 1989:181), nonmarine mollusks (Christensen 1983; Christensen and Kirch 1981; Kirch 1982b:11), and some reptilian and amphibian species (Cassels 1984:745–46). On the larger islands the drier leeward forests bore the brunt of the losses because they were so much easier to burn and because they regenerate much more slowly than the wetter forests (Olson and James 1984:777, but see Janzen 1988:131).

Effects of Polynesian-Introduced Organisms on Native Forest Species

Polynesians both purposefully and inadvertently introduced animals and plants wherever they settled (for Hawai'i, Cuddihy and Stone 1990:31–34; McKeown 1978; Nagata 1985). In some cases these introduced organisms had deleterious effects on native forest fauna and flora, on occasion provoking extinctions (Kirch 1983:27). In most of the Polynesian island groups native species are poorly prepared by evolution to compete with introduced species of herbivores and carnivores (Vitousek 1988:184).

Of the animals, the Polynesian rat (*Rattus exulans*) and dog (*Canus familiarus*, see especially Anderson 1981) have been shown to have been destructive, although dogs were not introduced to the Chatham Islands (Cassels 1984:753) and had been eliminated from the Marquesas before European contact (Decker 1971:88; Kirch 1973:38). If analogy to historic roof rat (*Rattus rattus*) predation can be accepted (Atkinson 1977), the Polynesian rat would have affected ground-nesting birds (Kirch 1973:37; Steadman, Pahlavan, and Kirch 1990:147), seeds, and sprouts of many plants (Cuddihy and Stone 1990:34), as well as insects, reptiles, and amphibians (Cassels 1984:746), and snails (Gagné and Christensen 1985:111). Avian diseases carried by the Polynesian-introduced chicken (*Gallus gallus*) may have spread into native bird populations with unknown, although generally believed to be negative, consequences (Atkinson 1977:120; Steadman 1989:201). The Polynesians also introduced many species of gecko, skink, and land snail (Cuddihy and Stone 1990:32–33; Ramsay 1978). Although it is often assumed that these had deleterious effects on forest plants and animals, in Hawai'i, at least, researchers have produced little quantitative data to demonstrate this (Gagné and Christensen 1985; Howarth 1985).

Feral invasions by Polynesian-introduced plants have occurred, such as the spread of *kukui* (*Aleurites molucanna* [L.] Willd.) on the windward sides of the larger Hawaiian Islands (Cuddihy and Stone 1990:32). Although such invasions may have been detrimental to native species, they probably were limited to sites already modified by human activities (Merrill

1954:212). On the other hand, the long history of agrosilviculture in Polynesia may also be viewed, in part at least, in positive terms biologically and especially so on atolls. Although the species involved were almost always Polynesian introductions and thus locally exotic, they must have made significant contributions to the ecological stability of naturally depauperate or degraded areas, perhaps even enhancing biological diversity (Clarke 1990:239–241; Klee 1980:260–262; Thaman and Clarke 1987). Elsewhere, native forest birds have been observed feeding on Polynesian-introduced tree species, an unanticipated positive result on native avifauna of human-induced habitat change (Warner 1968:102–103).

Cosmological Significance of the Forest and Its Products

Despite their pre-European forest clearing for agriculture, Polynesians gathered from forests many resources that have also been shown to be fundamentally important in a wide range of prehistoric life-sustaining and life-enriching activities. Not surprisingly, forests and their products are common elements in creation myths and other forms of Polynesian oral tradition (Alpers 1987; Beckwith 1972, 1976; Handy 1923; Henry 1928; Langridge and Terrell 1988; Métraux 1940; Valeri 1985). These oral traditions, in conjunction with ethnographic and ethnohistoric accounts, most often form the basis for contemporary analyses of Polynesian cosmology.

Polynesians viewed natural phenomena to a greater or lesser degree as manifestations of the gods and thus imbued with divinity (Handy and Pukui 1972:122–126; Henry 1928:382–92; Orliac 1990; Valeri 1985:10, 28). In Hawai'i, for example, "each deity is characterized by a number of natural manifestations, mostly species, which are said to be his kinolau, [or] 'myriad bodies'" (Valeri 1985:10). Although Polynesians saw the gods and their earthly manifestations as dangerous and requiring circumspect treatment, they also viewed plants and animals as having been "produced in order to produce the life of man . . . [and] constitute it because they are practically and intellectually appropriated by him" (Valeri:5–6; see also Burrows 1989:205; Ii 1959:85; Kepelino 1932:176). Polynesians believed that they could safely use some natural species, especially those considered economically important, if they first removed their divinity and danger by making first fruits sacrifices or engaging in other propitiatory rituals that "opened" their use to people (Best 1942:7–8; Firth 1929:135; Handy 1923:259; Valeri 1985:7, 156, but see Mead 1930:85, 123). Two excellent accounts of such rituals can be found in John Papa Ii's book on Hawaiian history (1959:39, 42). In both cases the Polynesians sacrificed pigs (and possibly humans) before they could begin harvesting the Hawaiian forest trees known as *lama* (*Diospyros sandwicensis* [A. DC] Fosb.) and *'ohi'a* (*Metrosideros polymorpha* Gaud.); this is an indication of not only the cultural importance of these activities and of the two species but also of the necessity of the ritual sacrifice in maintaining cosmological balance between humans and nature. Unfortunately, the literature is rarely explicit about the degree to which such religious behavior and beliefs affected or controlled levels of appropriation of forest products or about the range of species that required such ritual before harvesting could start (Burrows 1989:204, 208). For example, little is known about the cosmological significance or the ritual requirements associated with the many secondary species, or even the plants lama and 'ohi'a, which were used daily by Polynesian commoners.

An Assessment of the Relationship of Polynesians and the Forest

Despite a great natural range of biotic communities and substantial species diversity, major human-induced ecosystemic changes and species depletions occurred in Polynesia before European contact. These were variously the result of aboriginal actions related to gathering and hunting, the removal of vegetation for agriculture, and the destabilizing effects of invasive introduced plants and animals. The processes leading to these changes were fast acting and will be long lasting in Polynesia. However, this is not because Polynesians were somehow inherently more effective habitat modifiers than other native colonizers (Burrows 1989:207; Clarke 1990; Diamond 1990:32; Meilleur 1994). Rather, it is because of the well-known fragility, vulnerability, and—especially on the isolated high islands—high rate of endemism in the native biota that evolved for millions of years in the absence of humans, grazing mammals, and many other common continental forms of competition (Carlquist 1974:xiii; Cassels 1984:760+; Fosberg 1965:5; Kirch 1983:30; Thorne 1963:319–320; Zimmerman 1965:60).

Range extensions of some native species undoubtedly occurred after forest removal and replacement by humanized biotic communities (Cassels 1984:760; Fosberg 1953:11; McGlone 1983:22), and Polynesian agrosilviculture may have enhanced atoll plant biodiversity (e.g., some Polynesian-introduced trees do indeed provide food for native birds). However, most biological scientists believe that the prehistoric human-induced changes to Polynesian forests were exceedingly destructive, especially with regard to the fauna (Cassels 1984:741; James et al. 1987:2350; Merrill 1940:629; Olson and James 1984:777–778; Steadman 1989:201; Zimmerman 1965:59–60). Polynesian effects on floral biodiversity are much less well understood, but they are also thought to be significant (Wagner, Herbst, and Yee 1985:26), especially with regard to the leeward tropical dry and mesic forest plants.

This leads us to the question of environmental degradation and how such a concept should be understood in Polynesia. Should we look at it in biological terms alone, as we have summarized them here, or also in human terms by emphasizing, for example, some sort of concept of expanded or reduced human carrying capacity? Although Polynesian occupation led to substantial biotic changes throughout a range of island forest environments, the narrow biological position ignores almost entirely the human benefits and accomplishments involved in these processes and events. There is certainly some evidence for environmental degradation to a degree that would have provoked group dislocation and/or island abandonment, sometimes rather vaguely linked to overpopulation (for Easter Island, McCoy 1979:159–161; for the so-called Mystery Islands, Bellwood 1987:109, Kirch 1984:89; for Kaho'olawe, Hommon 1980a, 1980b, but see Spriggs 1992). However, setting these few examples aside, the Europeans found throughout Polynesia prosperous and often dense human populations, apparently enjoying fairly good nutrition and health (Bushnell 1966; Lai 1974:163; Macpherson and Macpherson 1990:28–37, 49; Miller 1974:174; Snow 1974:10; Stannard 1989:60–61, but see Norton 1992). Highly stratified and sophisticated societies had evolved on the larger islands in the absence of metal. Polynesian excellence in open sea voyaging, in economic production, in song, dance, and crafts, and in stone and wood carving are today widely recognized and celebrated. Given the technological level of Polynesians, it is unlikely that they could have realized such achievements without significant consumption of natural resources and related habitat modification.

To appreciate the ambivalence inherent in the biological and the cultural assessments, which is certainly not unique to Polynesia, the fundamental questions are not who, what, or why but rather where and (perhaps to an equally important degree) when these events occurred. Thus we should emphasize the unique vulnerability of islands to introduced organisms, *Homo sapiens* above all, and on the recency of these events in human prehistory. If Polynesians had colonized some continental landmass many millennia ago, the biological consequences of their occupation probably would not be so substantial and obvious today. Should Polynesians be blamed for what they did to their Pacific Island forests? Certainly not. Such concepts as biodiversity, endemism, rarity and endangerment, and extinction have only recently been developed and understood on a world scale. To use them retrospectively and judgmentally is contextually inappropriate and unfair. Because forest prehistory in Polynesia is now fairly well understood, we should ask instead whether we have the collective will to apply what we know about these processes and outcomes to our world.

Acknowledgments   Thanks are owed to E. Anderson, T. Headland, and L. Sponsel for ideas and suggestions, as they are to T. Tunison for kindly sending me several manuscripts and articles on fire use in Hawai'i.

References Cited

Abbott, Isabella. 1992. *La'au Hawai'i.* Honolulu: Bishop Museum Press.
Alpers, Anthony. 1987. *The World of the Polynesians, Seen Through Their Myths and Legends.* London: Oxford University Press.
Amadon, Dean. 1950. The Hawaiian honeycreepers (*Aves, Drepanidae*). *Bulletin of the American Museum of Natural History* 95: 151–262.
Anderson, Atholl. 1981. Pre-European hunting dogs in the South Island, New Zealand. *New Zealand Journal of Archaeology* 3: 15–20.
——. 1983. Faunal depletion and subsistence change in the early prehistory of southern New Zealand. *Archaeology in Oceania* 18: 1–10.
——. 1984. The extinction of moa in southern New Zealand. In *Quaternary Extinctions*, P. Martin and R. Klein, eds., pp. 728–740. Tucson: University of Arizona Press.
Atkinson, I. A. E. 1977. A reassessment of factors, particularly *Rattus rattus L.*, that influenced the decline of endemic forest birds in the Hawaiian Islands. *Pacific Science* 31 (2): 109–133.
Barrau, Jacques. 1957. Les atolls océaniens: Essai d'agronomie (The ocean atolls: An essay in agronomy). *Études d'Outre-mer* 40 (7): 253–266.
Beaglehole, J. C., ed. 1967. *The Journals of Captain James Cook on His Voyages of Discovery. Vol. 3: The Voyage of the Resolution and Discovery.* Cambridge, England: Cambridge University Press for the Hakluyt Society
Beckwith, Martha. 1972. *The Kumulipo: A Hawaiian Creation Chant.* Honolulu: University of Hawaii Press.
——. 1976. *Hawaiian Mythology.* Honolulu: University of Hawaii Press.
Bellwood, Peter. 1987. *The Polynesians: Prehistory of an Island People.* London: Thames and Hudson.
Berger, Andrew. 1972. *Hawaiian Birdlife.* Honolulu: University of Hawaii Press.
Best, Elsdon. 1942. *Forest Lore of the Maori.* Wellington, New Zealand: The Polynesian Society.
Brooker, Stanley, Richard Cambie, and Robert Cooper. 1987. *New Zealand Medicinal Plants.* Auckland: Heinemann.
——. 1989. Economic native plants of New Zealand. *Economic Botany* 43 (1): 79–106.
Burrows, Charles K. P. M. 1989. Hawaiian conservation values and practices. In *Conservation Biology in Hawaii*, C. Stone and D. Stone, eds., pp. 203–213. Honolulu: University of Hawaii, Cooperative National Park Resources Studies Unit.
Burrows, Edwin. 1936. *Ethnology of Futuna.* Honolulu: Bernice P. Bishop Museum.
Bushnell, O. A. 1966. Hygiene and sanitation among the ancient Hawaiians. *Hawaii Historical Review* 2 (5): 316–336.
Carlquist, Sherwin. 1974. Introduction to the new edition. In *The Indigenous Trees of the Hawaiian Islands*, J. Rock, pp. xi–xvi. Rutland, Vt.: Charles E. Tuttle.

Cassels, Richard. 1984. The role of prehistoric man in the faunal extinctions of New Zealand and other Pacific islands. In *Quaternary Extinctions*, P. Martin and R. Klein, eds., pp. 741–767. Tucson: University of Arizona Press.

Chock, Alvin. 1968. Hawaiian ethnobotanical studies 1: Native food and beverage plants. *Economic Botany* 22: 221–238.

Christensen, Carl. 1983. Analysis of land snails. In *Archaeological Investigations of the Mudlane–Waimea–Kawaihae Road Corridor, Island of Hawaii*, J. Clark and P. Kirch, eds., pp. 449–471. Dept. Anthropology, Report Series No. 831. Honolulu: Bishop Museum Press.

Christensen, Carl and Patrick Kirch. 1981. Non-marine mollusks from archaeological sites in Tikopia, Solomon Islands. *Pacific Science* 35: 75–88.

Clarke, William. 1990. Learning from the past: Traditional knowledge and sustainable development. *Contemporary Pacific* 2 (2): 233–253.

Colenso, W. 1880. On the vegetable food of the ancient New Zealanders before Cook's visit. *Transactions of the New Zealand Institute* 13: 3—38.

Cox, Paul and Sandra Banack, eds. 1991. *Islands, Plants, and Polynesians.* Portland, Oreg.: Dioscorides Press.

Croft, Kevin and Peaua Tu'ipulotu. 1980. A survey of Tongan medicinal plants. *South Pacific Journal of Natural Science* 1: 45–57.

Cuddihy, Linda and Charles Stone. 1990. *Alteration of Native Hawaiian Vegetation.* Honolulu: University of Hawaii, Cooperative National Park Resources Studies Unit.

Cumberland, Kenneth. 1961. Man in nature in New Zealand. *New Zealand Geographer* 17: 137–154.

———. 1963. Man's role in modifying island environments in the southwest Pacific: With special reference to New Zealand. In *Man's Place in the Island Ecosystem*, R. Fosberg, ed., pp. 187–205. Proceedings of the Tenth Pacific Science Congress. Honolulu: Bishop Museum Press.

Decker, Bryce. 1971. Plants, Man, and Landscape in the Marquesan Valleys, French Polynesia. Unpublished Ph.D. diss., University of California, Berkeley.

Diamond, Jared. 1990. Bob Dylan and moas' ghosts. *Natural History* (October 1990): 26–32.

Douglas, G. 1969. Check list of Pacific oceanic islands. *Micronesia* 5: 327–464.

Downes, T. W. 1928. Bird-snaring, etc., in the Whanganui River district. *Journal of the Polynesian Society* 37: 1–29.

Drigot, D. 1982. *Ho'ona'auao No Kawai Nui (Educating about Kawai Nui): A Multimedia Educational Guide.* Honolulu: University of Hawaii, Environmental Center.

Emerson, N. B. 1894. The bird-hunters of ancient Hawaii. *Thrum's Hawaiian Annual for 1895*. Honolulu: Thomas G. Thrum.

Emory, Kenneth. 1947. Tuomotuan plant names. *Journal of the Polynesian Society* 56: 266–277.

———. 1975. Material culture of the Tuamotu archipelago. *Pacific Anthropological Records* 22: 1–253.

Emory, Kenneth and Yoshiko Sinoto. 1961. *Hawaiian Archaeology: Oahu Excavations.* Honolulu: Bishop Museum Press.

Ferdon, Edwin. 1981. *Early Tahiti as the Explorers Saw It, 1767–1797*. Tucson: University of Arizona Press.

——. 1987. *Early Tonga as the Explorers Saw It, 1616–1810*. Tucson: University of Arizona Press.

Firth, Raymond. 1929. *Primitive Economics of the New Zealand Maori*. London: G. Routledge.

——. 1975. *Primitive Polynesian Economy*. New York: W. W. Norton.

Flenley, J. R. 1979. Stratigraphic evidence of environmental change on Easter Island. *Asian Perspectives* 22 (1): 33–40.

Fosberg, Raymond. 1953. Vegetation of central Pacific atolls. *Atoll Research Bulletin* 23: 1–26.

——. 1965. The island ecosystem. In *Man's Place in the Island Ecosystem*, R. Fosberg, ed., pp. 1–6. Proceedings of the Tenth Pacific Science Congress. Honolulu: Bishop Museum Press.

Gagné, Wayne and Carl Christensen. 1985. Conservation status of terrestrial invertebrates in Hawaii. In *Hawaii's Terrestrial Ecosystems: Preservation and Management*, C. Stone and J. Scott, eds., pp. 105–126. Honolulu: University of Hawaii, Cooperative National Park Resources Studies Unit.

Grant, Patrick. 1985. Major periods of erosion and alluvial sedimentation in New Zealand during the late Holocene. *Journal of the Royal Society of New Zealand* 15 (1): 67–121.

Handy, E. S. Craighill. 1923. *The Native Culture in the Marquesas*. Honolulu: Bernice P. Bishop Museum.

Handy, E. S. Craighill and Mary Pukui. 1972. *The Polynesian Family System in Ka-'u, Hawai'i*. Rutland, Vt.: Charles Tuttle.

Henry, Teuira. 1928. *Ancient Tahiti*. Honolulu: Bernice P. Bishop Museum.

Hommon, R. 1980a. *National Register of Historic Places Multiple Resource Nomination Form for the Historic Resources of Kaho'olawe*. Honolulu: U.S. Navy, Naval Facilities Engineering Command.

——. 1980b. *Kaho'olawe: Final Report of the Archaeological Survey*. Honolulu: U.S. Navy, Naval Facilities Engineering Command.

Howarth, Francis. 1985. Impacts of alien land arthropods and mollusks on native plants and animals in Hawaii. In *Hawaii's Terrestrial Ecosystems: Preservation and Management*, C. Stone and J. Scott, eds., pp. 149–179. Honolulu: University of Hawaii, Cooperative National Park Resources Studies Unit.

Ii, John Papa. 1959. *Fragments of Hawaiian History*. Honolulu: Bishop Museum Press.

James, Helen and Storrs Olson. 1991. Descriptions of thirty-two new species of birds from the Hawaiian Islands: Part 2. Passeriformes. *Ornithological Monographs* 46: 1–88.

James, Helen, Thomas Stafford, David Steadman, Storrs Olson, Paul Martin, A. J. T. Jull, and Patrick McCoy. 1987. Radiocarbon dates on bones of extinct birds from Hawaii. *Proceedings of the National Academy of Sciences* 84: 2350–2354.

Janzen, Daniel. 1988. Tropical dry forests: The most endangered major tropical ecosystem. In *Biodiversity*, E. O. Wilson, ed., pp. 130–138. Washington, D.C.: National Academy Press.

Kaaiakamanu, D. M. and J. K. Akina. 1922. *Hawaiian Herbs of Medicinal Value.* Honolulu: Pacific Book House.

Kamakau, Samuel. 1976. *The Works of the People of Old.* Honolulu: Bishop Museum Press.

Kepelino, Keauokalani. 1932. *Kepelino's Traditions of Hawaii.* Honolulu: Bernice P. Bishop Museum.

Kirch, Patrick. 1973. Prehistoric subsistence patterns in the Northern Marquesas Islands, French Polynesia. *Archaeology and Physical Anthropology in Oceania* 89 (1): 24–40.

——. 1975. Cultural Adaptation and Ecology in Western Polynesia: An Ethnoarchaeological Study. Unpublished Ph.D. diss., Yale University.

——. 1982a. Ecology and the adaptation of Polynesian agricultural systems. *Archaeology in Oceania* 17: 1–6.

——. 1982b. The impact of the prehistoric Polynesians on the Hawaiian ecosystem. *Pacific Science* 36 (1): 1–14.

——. 1983. Man's role in modifying tropical and subtropical Polynesian ecosystems. *Archaeology in Oceania* 18: 26–31.

——. 1984. *The Evolution of the Polynesian Chiefdoms.* Cambridge, England: Cambridge University Press.

Kirch, Patrick and Douglas Yen. 1982. *Tikopia: The Prehistory and Ecology of a Polynesian Outlier.* Honolulu: Bishop Museum Press.

Kirch, Patrick, John Flenley, David Steadman, Frances Lamont, and Stewart Dawson. 1992. Ancient environmental degradation. *National Geographic Research and Exploration* 8 (2): 166–179.

Klee, Gary. 1980. Oceania. In *World Systems of Traditional Resource Management,* G. Klee, ed., pp. 245–281. London: Edward Arnold.

Krauss, Beatrice. 1975. Ethnobotany of the Hawaiians. *University of Hawaii, Harold Lyon Arboretum Lecture* 5: 1–32.

——. 1993. *Plants in Hawaiian Culture.* Honolulu: University of Hawaii Press.

Lai, L. 1974. An oral examination of early Hawaiians. In *Early Hawaiians,* C. Snow, ed., pp. 159–164 (Appendix C). Lexington: University Press of Kentucky.

Langridge, Martha, trans., and Jennifer Terrell, ed. 1988. *von den Steinen's Marquesan Myths.* Canberra: Australian National University Printing and Publishing Service.

Lyons, Curtis. 1903. *A History of the Hawaiian Government Survey with Notes on Land Matters in Hawaii.* Honolulu: Hawaiian Gazette.

MacCaughey, Vaughan. 1918. History of botanical exploration in Hawaii. *Hawaiian Forester and Agriculturist* 15 (9): 388–396.

Maclet, Jean-Noël and Jacques Barrau. 1959. Catalogue des plantes utiles aujourd'hui présentes en Polynésie Française (Catalogue of useful plants present today in French Polynesia). *Journal d'agriculture tropicale et botanique appliquée* 6: 1–21, 161–184.

Macpherson, Cluny and La'avasa Macpherson. 1990. *Samoan Medical Belief and Practice.* Auckland, New Zealand: Auckland University Press.

Malo, David. 1951. *Hawaiian Antiquities.* Honolulu: Bishop Museum Press.

Manning, Anita. 1981. Hawelu: Birdcatcher, innkeeper, farmer. *Hawaii Journal of History* 15: 59–68.

McCoy, Patrick. 1979. Easter Island. In *The Prehistory of Polynesia*, J. Jennings, ed., pp. 135–166. Cambridge, Mass.: Harvard University Press.

McEldowney, Holly. 1979. *Archaeological and Historical Literature Search and Research Design, Lava Flow Control Study, Hilo, Hawaii*. Honolulu: Bishop Museum Anthropology Department.

——. 1983. A description of major vegetation patterns in the Waimea–Kawaihae region during the early historic period. In *Archaeological Investigations of the Mudlane–Waimea–Kawaihae Road Corridor, Island of Hawaii*, J. Clark and P. Kirch, eds., pp. 407–448. Dept. Anthropology, Report Series No. 831. Honolulu: Bishop Museum Press.

McGlone, M. 1983. Polynesian deforestation of New Zealand: A preliminary synthesis. *Archaeology in Oceania* 18: 11–25.

McKelvey, P. J. 1958. Forest history and New Zealand prehistory. *New Zealand Science Review* 16: 28–32.

McKeown, S. 1978. *Hawaiian Reptiles and Amphibians*. Honolulu: Oriental Publishing.

Mead, Margaret. 1930. Social organization of Manu'a. *Bishop Museum Bulletin* 76: 1–237.

Meilleur, Brien. 1994. In search of 'keystone societies.' In *Eating on the Wild Side: The Pharmacologic, Ecologic, and Social Implications of Using Noncultigens*, N. Etkin, ed., pp. 259–279. Tucson: University of Arizona Press.

Menzies, Archibald. 1920. *Hawaii Nei One Hundred Twenty-Eight Years Ago*. Honolulu: The New Freedom.

Merrill, Elmer. 1940. Man's influence on the vegetation of Polynesia, with special reference to introduced species. *Proceedings of the Sixth Pacific Science Congress*, pp. 629–639. Berkeley: University of California Press.

——. 1954. The botany of Cook's voyages. *Chronica Botanica* 14 (5–6): 161–384.

——. 1981. *Plant Life of the Pacific World*. Rutland, Vt.: Charles E. Tuttle.

Miller, Carey. 1974. The influence of foods and food habits upon the stature and teeth of the ancient Hawaiians. In *Early Hawaiians*, C. Snow, ed., pp. 167–175 (Appendix E). Lexington: University Press of Kentucky.

Métraux, Alfred. 1940. Ethnology of Easter Island. *Bishop Museum Bulletin* 160: 1–432.

Mulloy, William. 1970. A speculative reconstruction of techniques of carving, transporting and erecting Easter Island statues. *Archaeology and Physical Anthropology in Oceania* 5: 1–23.

Murakami, Gail. 1983a. Analysis of charcoal from archaeological contexts. In *Archaeological Investigation of the Mudlane–Waimea–Kawaihae Road Corridor, Island of Hawaii*, J. Clark and P. Kirch, eds., pp. 514–524. Dept. Anthropology, Report Series No. 831. Honolulu: Bishop Museum Press.

——. 1983b. Identification of charcoal from Kaho'olawe archaeological sites. In *Kaho'olawe Archeologicl Excavations 1981*, R. Hommon, ed., Appendix B. Honolulu: U.S. Navy.

———. 1989. Identification of charcoal from Kuololo Rockshelter. In *Prehistoric Hawaiian Occupation in the Anahulu Valley, O'ahu Island: Excavations in Three Island Rockshelters*, P. Kirch, ed., pp. 103–110. Berkeley: University of California, Archaeological Research Facility No. 47.

Nagata, Kenneth. 1971. Hawaiian medicinal plants. *Economic Botany* 25 (3): 245–254.

———. 1985. Early plant introductions in Hawaii. *Hawaiian Journal of History* 19: 35–61.

Niering, William. 1956. *Bioecology of Kapingamarangi atoll, Caroline Islands: Terrestrial aspects*. New London, Conn.: Connecticut College.

Norton, Scott. 1992. Salt consumption in ancient Polynesia. *Perspectives in Biology and Medicine* 35 (2): 160–181.

Office of Comprehensive Health Planning. 1974. *Samoan Medicinal Plants and Their Usage*. American Samoa: Department of Medical Services.

Olson, Storrs and Helen James. 1982. Fossil birds from the Hawaiian Islands: Evidence for wholesale extinction by man before Western contact. *Science* 217: 633–635.

———. 1984. The role of Polynesians in the extinction of the avifauna of the Hawaiian Islands. In *Quaternary Extinctions*, P. Martin and R. Klein, eds., pp. 768–780. Tucson: University of Arizona Press.

———. 1991. Descriptions of thirty-two new species of birds from the Hawaiian Islands: Part 1. Nonpasseriformes. *Ornithological Monographs* 45: 1–88.

Orliac, Catherine. 1990. Des arbres et des dieux: Choix des matériaux de sculpture en Polynésie (Trees and gods: Choices of material used in sculpture in Polynesia). *Journal de la Société des Océanistes* 90 (1): 35–42.

Pawley, Andrew and Roger Green. 1973. Dating the dispersal of the oceanic languages. *Oceanic Linguistics* 12: 1–67.

Pétard, Paul. 1986. *Quelques plantes utiles de Polynésie Française* (Some useful plants of French Polynesia). Papeete, Tahiti: Editions Haere Po No Tahiti, Ra'au Tahiti.

Ramsay, G. 1978. A review of the effect of rodents on the New Zealand invertebrate fauna. In *The Ecology and Control of Rodents in New Zealand Native Reserves*, P. Dingwall, I. Atkinson, and C. Hay, eds. Wellington, New Zealand: Department of Lands and Survey.

Ranapiri, Tamati. 1895. Ancient methods of bird-snaring amongst the Maoris. *Journal of the Polynesian Society* 4: 143–152.

Rock, Joseph. 1913. *The Indigenous Trees of the Hawaiian Islands*. Rutland, Vt.: Charles E. Tuttle.

Roussel, Hippolyte. 1926. Ile de Pâques (Easter Island). *Annales des Sacrés Coeurs* 305–309: 423.

Sinoto, Yosihiko. 1979. Excavations on Huahine, French Polynesia. *Pacific Studies* 3: 1–40.

———. 1988. A waterlogged site on Huahine Island, French Polynesia. In *Wet Site Archaeology*, B. Purdy, ed., pp. 113–130. Caldwell, N.J.: Telford Press.

Smith, Clifford and Timothy Tunison. 1990. Fire and alien species in Hawaii: Research and management implications for native ecosystems. In *Alien*

Plant Invasions in Native Ecosystems of Hawaii: Management and Research, C. Stone, C. Smith, and J. T. Tunison, eds., pp. 394–408. Honolulu: University of Hawaii, Cooperative National Park Resources Studies Unit.

Snow, Charles. 1974. *Early Hawaiians: An Initial Study of Skeletal Remains from Mokapu, Oahu*. Lexington: University Press of Kentucky.

Spriggs, Matthew. 1985. Prehistoric man-induced landscape enhancement in the Pacific: Examples and implications. Prehistoric Intensive Agriculture in the Tropics (Part 1). *British Anthropological Records* 232: 409–434.

——. 1992. Preceded by forest: Changing interpretations of landscape change on Kaho'olawe. In *Kaho'olawe Excavations, 1982–1983 Data Recovery Project*, P. H. Rosendahl, A.E. Haun, J.B. Halbig, M. Kaschko, and M. S. Allen, eds. Appendix I. Pearl Harbor: U.S. Navy, Naval Facilities Engineering Command.

Stannard, David. 1989. *Before the Horror: The Population of Hawaii on the Eve of Western Contact*. Honolulu: University of Hawaii Press.

Steadman, David. 1989. Extinction of birds in Eastern Polynesia: A review of the record, and comparisons with other island groups. *Journal of Archaeological Science* 16: 175–205.

Steadman, David, Dominique Pahlavan, and Patrick Kirch. 1990. Extinction, biogeography, and human exploitation of birds on Tikopia and Anuta, Polynesian outliers in the Solomon Islands. *Bishop Museum Occasional Papers* 30: 118–153.

Sutton, Douglas. 1980. A culture history of the Chatham Islands. *Journal of the Polynesian Society* 89: 67–94.

Sutton, Douglas and Y. M. Marshall. 1980. Coastal hunting in the subantarctic zone. *New Zealand Journal of Archaeology* 2: 25–49.

Thaman, R. R. and W. C. Clarke. 1987. Pacific Island agrosilviculture: Systems for cultural and ecological stability. *Canopy International* 13 (1): 6–7; 13 (2): 8–10; 13 (3): 6–9.

Thorne, Robert. 1963. Biotic distribution patterns in the tropical Pacific. In *Pacific Basin Biogeography*, J. Gressitt, ed., pp. 311–350. Honolulu: Bishop Museum Press.

Trotter, Michael and Beverley McCulloch. 1984. Moas, men, and middens. In *Quaternary Extinctions*, P. Martin and R. Klein, eds., pp. 708–727. Tucson: University of Arizona Press.

Uhe, George. 1974. Medicinal plants of Samoa: A preliminary survey of the use of plants for medicinal purposes in the Samoan Islands. *Economic Botany* 28: 1–30.

Utanga, A. 1989. Customary tenure and traditional resource management in the Cook Islands. In *South Pacific Regional Environment Program Report on the Workshop on Customary Tenure, Traditional Resource Management and Nature Conservation*, pp. 101–105. Noumea, New Caledonia: South Pacific Commission.

Valeri, Valerio. 1985. *Kingship and Sacrifice: Ritual and Society in Ancient Hawaii*. Chicago: University of Chicago Press.

Vancouver, George. 1798. *A Voyage of Discovery to the North Pacific Ocean and Around the World in the Years 1790–1795*. London: G. & J. Robinson.

Vayda, Andrew. 1956. Maori Warfare. Unpublished Ph.D. diss., Columbia University.

Vitousek, Peter. 1988. Diversity and biological invasions of oceanic islands. In *Biodiversity*, E. O. Wilson, ed., pp. 181–189. Washington D.C.: National Academy Press.

Vogl, Richard. 1969. The role of fire in the evolution of the Hawaiian flora and vegetation. *Proceedings of the Annual Tall Timbers Fire Ecology Conference* 9: 5–60.

Wagner, Warren, Derral Herbst, and Rylan Yee. 1985. Status of the native flowering plants of the Hawaiian Islands. In *Hawaii's Terrestrial Ecosystems: Preservation and Management*, C. Stone and J. Scott, eds., pp. 23–74. Honolulu: University of Hawaii, Cooperative National Park Resources Studies Unit.

Walls, Geoff. 1988. *Traditional Uses of Plants in New Zealand and the Pacific*. Commonwealth Science Council Biological Diversity and Genetic Resources Project 1986–1988 Summary Report. Havelock North, New Zealand: Department of Scientific and Industrial Research, Botany Division.

Warner, Richard. 1968. The role of introduced diseases in the extinction of the endemic Hawaiian avifauna. *Condor* 70: 101–120.

Whistler, W. Arthur. 1985. Traditional and herbal medicine in the Cook Islands. *Journal of Ethnopharmacology* 13: 239–280.

——. 1988. Ethnobotany of Tokelau: The plants, their Tokelau names, and their uses. *Economic Botany* 42 (2): 155–176.

Williams, Michael. 1989. *Americans and Their Forests: A Historical Geography*. Cambridge, England: Cambridge University Press.

Yen, Douglas. 1973. Agriculture in Anutan subsistence. In *Anuta: A Polynesian Outlier in the Solomon Islands*, D. Yen and J. Gordon, eds., pp. 113–155. Department of Anthropology, Pacific Anthropological Records 21. Honolulu: , Bishop Museum Press.

Zimmerman, Elwood. 1965. Nature of the land biota. In *Man's Place in the Island Ecosystem*, R. Fosberg, ed., pp. 57—63. Proceedings of the Tenth Pacific Science Congress. Honolulu: Bishop Museum Press.

III

Colonial and Neocolonial History

Part 3, Colonial and Neocolonial History, samples two countries in which European colonization greatly influenced the forest history and related human ecology—India and Kenya, which by coincidence both involved the British as colonizers. Analysis of the political economy of forests in the context of environmental history suggests that national forest management has often led to deforestation, whereas local forest management has often conserved forests, even though some profound changes may be involved as well.

In chapter 4 Janis Alcorn and Augusta Molnar argue that the history of deforestation in India offers lessons for environmentalists and for other countries considering options for managing forest resources. Alcorn and Molnar conclude that the effects of deforestation on subsistence economies, as well as the role of subsistence economies in promoting sustainable development, have been ignored. Although India had experienced millennia of heavy use of forests, as well as famine and warfare, the country was densely forested when the British began to colonize South Asia. During the colonial period deforestation accelerated because of British policies that governed forest management, the expansion of commercial interests, and the consequent decline of local community regulation of forests. At that time, as today, many rural people of India depended for survival on the exploitation of the natural resources in their environment. Powerful outsiders brutally suppressed local communities' resistance to outsiders' extraction of forest resources. Under pressure from outside forces local people had no choice but to degrade the remaining forest, with resultant deterioration of their own economy, nutrition, and health.

After independence the neocolonial government continued the trends of the colonial period. As population continued to grow and available forest resources continued to shrink, pressures on the resource base grew, and the increase in fuelwood demands alone became devastating. Many people have been forced to migrate to city slums as environmental refugees from deforestation. However, since the mid-1980s local communities and nongovernmental organizations (NGOs) have been able to reforest where national forestry department programs have failed. In recent years the government has begun to share with local communities the management and ecological restoration of deforested and degraded lands with severely depleted biodiversity. A century after the government initiated formal policies usurping community rights over forests, community rights are being reinstated and forests are growing.

The India case demonstrates a trajectory of government assumption of forest ownership and related human rights abuses that is linked to accelerated deforestation globally. Other countries are following this trajectory, although deforestation is not as far along as it was in India—the countries' forests are still healthy and intact but shrinking, and community rights are still intact but shrinking. Alcorn and Molnar conclude that the lessons from the India case are twofold. First, if nations do not respect local communities' rights to forests and instead build paramilitary forest departments to protect government-owned forests, deforestation will accelerate. Second, deforestation can be arrested if governments choose to experiment with innovative options to assist communities in managing their forests.

In chapter 5 Alfonso Peter Castro examines both the successes and failures of the colonial government's farm forestry strategies in Kenya, again from the perspective of the political economy of forests in the context of environmental history. He suggests that farm forestry, promoting tree growing by rural families on their own land, has implications for using farm forestry to combat deforestation.

Early in this century the British colonials restricted local communities from using the forests of Mount Kenya in order to establish the forests as a reserve. However, the problem of wood scarcity was widespread, and colonial officials realized that indigenous agroforestry might be a solution. Unfortunately, they introduced more exotic than native tree species, and some, like eucalyptus, competed with crops for water and nutrients. Because local residents recognized the problem and feared the loss of land, they resisted the colonial programs. By the 1930s the production of wattle trees (*Acacia mearnsii*) had emerged as a major cash crop with multiple uses, subsistence and commercial value, and low labor requirements. However, the system deteriorated and collapsed, because the government cleared vegetation to reduce cover for Mau Mau forces.

4

Deforestation and Human-Forest Relationships: What Can We Learn from India?

Janis B. Alcorn and Augusta Molnar

The history of deforestation in India offers some disturbing lessons about the consequences of state control of forests. On the government's watch the state forests have been destroyed, and its reforestation programs have been less than successful. The Indian case serves as a warning to environmentalists who seek to save forests elsewhere by supporting government policies to usurp forests from local communities.

In the late 1940s a third of India was still forested (Morris 1986; Arnold and Stewart 1989). India's forests include the full range of forest types, from coastal mangroves to semiarid open woodlands, from temperate evergreen forests to lowland tropical rain forests. Indian forests harbor thousands of unique endemic species (Gadgil and Meher-Homji 1986). However, in many areas natural vegetation is stabilized at the lowest level of succession (figure 4.1) because of incessant disturbance by humans and livestock (Kendrick 1989). Forests of all types are steadily being replaced by wastelands.[1] The deforestation rate is estimated at 1.5 million hectares per year (Nadkarni 1987; Myers 1988).

Twenty-three percent of India's total land area is officially designated as state forest land (land under the control of state-level forest departments and therefore not under local communities' control and management),[2] but only 11% of the country has forest cover today (Karaosmanoglu 1989), and

Figure 4.1   Severely degraded forest, shown here in the buffer
zone of Ranthambhore National Park, India, is typical of Indian
state forest lands today.

probably less than 2% is covered by natural forest (Guha, Prasad, and
Gadgil 1984). Deforestation and other unsustainable use of marginal lands
has left more than 40% of the country in wasteland, silted up major dams,
increased losses from flooding, and reduced dry season river flows, thus
causing salinization of coastal agricultural lands.

Rural people are particularly dependent on forests for meeting part of
their basic needs. India has a population of 850 million people, 75% of
whom live in rural areas and 50% of whom live below the poverty line

(Gadgil 1989). Millions of rural people live on incomes below the minimum on which they could be expected to survive, because their direct access to subsistence goods harvested from nature allows them to survive on their current incomes (Jodha 1990). Deforestation is pushing them to the edge of survival and driving environmental refugees into city slums. Loss of forest (and forest products such as bamboo) is also leading to the disruption of traditional occupations (Gadgil 1989) and greater impoverishment.

Ninety percent of Indian households are dependent on firewood, dung, and agricultural residue (e.g., straw, husks, sugarcane bagasse) as fuel to cook their food (People's Union for Democratic Rights [PUDR] 1982). The price of fuelwood has skyrocketed (Bowonder, Prasad, and Unni 1988), and agricultural residues have entered the marketplace as fuel. Per capita fuelwood consumption is on the rise in urban areas because of rising standards of living. Although many urbanites would use greater quantities of substitutes, such as kerosene, they continue to use firewood, because substitutes are in scarce supply. Projections of wood demand show India will be unable to meet fuelwood and other wood needs by the end of this century (Forest Research Institute 1988). Although annual fuelwood and industrial consumption are 240 million cubic meters and 25 million cubic meters, respectively, official production is only one-tenth of those figures (Contreras 1991). The projected increase in demand for roundwood by the year 2000 is 51 million cubic meters (Food and Agriculture Organization [FAO] 1990). The value of the incremental increase in foreign exchange is on the order of U.S.$ 2.5 billion, clearly not easily provided through imports.

How did this come to pass? India recognizes that forests are economic assets, government policy has favored natural regeneration and native species over exotics, apparently enlightened policies give usufruct rights to people, and the country has tried major social forestry projects.[3] The forest department has paramilitary control over 98% of the nation's forests (Bowonder, Prasad, and Unni 1987). Yet deforestation continues.

Is it the fault of the poor who sneak past guard patrols or pay forest guards the going rate to carry the forest away in pieces by night? Or have commercial interests, in collusion with the forest department, been to blame for removing more trees than the management plans allow? Is it the fault of the forest guard and his bosses that they participate in this shadow economy? Will the rural poor, the commercial interests, or the forest department prove most able to turn around the deforestation? What kind of collaboration between these interests is possible? India is debating these questions today.

Deforestation in India, as in most developing countries, is the manifestation of a long political struggle between two interest groups: commercial interests and subsistence interests. The commercial interest group has used forests to generate capital, as if nature were just another asset to be converted to some other capital asset without penalty (Daly and Cobb 1989).

Members of the subsistence interest group, on the other hand, view nature as an irreplaceable asset, because they depend on nature to replenish their limited resource base. For the subsistence interest group destruction of nature means the end of benefit-cash flows. Through its rules, policies, and price supports designed to promote industrialization, as well as through budget allocation and economic analyses, the state has generally supported the commercial interests allied with the political elites. Those dependent on nature for subsistence have exercised little political power. Although Indian communities have long fought to retain or regain rights to make decisions about forest management, the state has usurped their rights in the name of modern management and conservation. During the 1970s and 1980s people whose livelihoods depend on subsistence subsidies from nature increasingly turned to civil disobedience to fight for their right to once again make forest management decisions.

The relative scarcity of forest resources will probably continue to make political struggle the most significant determinant of forest-human relationships on the subcontinent.

Regulation of Forest Use:
Shift from Community to State

India's forests, like those of many other tropical countries, were used by large populations of people and livestock for thousands of years. Historical records demonstrate that forests were under so much pressure that they were protected by local rulers more than two thousand years ago. Armed guards maintained breeding elephant populations in forest patches (Basham 1967). India's forests have been converted into crop lands and used as cattle-grazing lands for thousands of years. The Indian subcontinent has a long history of anarchy and wars; rulers and their laws have come and gone, agricultural lands have been reclaimed by forests, and the forests subsequently were converted to agricultural use. Ozymandian ruins literally dot the landscape; tigers in modern forest reserves sleep in the overgrown ruined forts of rulers past.

Rajas, who controlled most forests, had limited capacity and little desire to extract forest resources for trade. They viewed forests as royal hunting grounds, and royal gamekeepers prevented encroachment (Tucker 1986). Rulers taxed forest products exported from villages but not those used for local subsistence (Guha 1983a, 1983b; Chaturvedi and Sahai 1988). Rulers also demanded some forest products in tribute, and substantial amounts of Indian forest products have been transported over East-West trade routes for more than three thousand years (Westland 1979).

Yet the British waxed eloquent over the immense wealth of India's forests three hundred years ago. Exactly why forests remained in India through centuries of war, famine, and heavy use is not clear. Indications are that authoritarian rules regarding forest use and maintenance were established and enforced through tribal authorities, local elites, or princes. These included taxes, land-use zoning, fines, labor used to maintain the resource base, village forest protection committees, and forest guards (Jodha 1990). Traditional subsistence "scripts" allowed natural regeneration of resources (e.g., seasonal grazing routes [Tucker 1986]; swidden [or shifting] cultivation methods [Ramakrishnan 1984]).[4] The forests were used heavily, but they were not eliminated. Human and livestock populations have grown markedly during the twentieth century, but deforestation cannot be blamed solely on population increase (Jodha 1986, 1990; Gadgil 1989), although some continue to blame overpopulation of livestock and people for the loss of forests (Shyamsunder and Parameswarappa 1987). Two other factors have been critically important: increased pressure to supply commercial products and decline in local regulation of forest use.

The precipitous decline of India's forests began in the seventeenth and eighteenth centuries when the British East India Company plundered India's wealth and dismantled the domestic economy (Brockway 1979). The East India Company ignored local rules for forest conservation and clearcut sacred groves and princely teak plantations (Gadgil 1989). In the mid-nineteenth century when the Crown took over administration of India, the British government dissolved the authority of the local systems that had traditionally regulated forest use. The British then established the Indian Forest Service with German assistance. The regulation of India's forests and the Indian Forest Service were built along the lines of the German version of the enclosure movement (Chowdhry 1989). The British then proceeded to establish a set of policies and legislation that established the state's ownership of forest assets. The British progressively reduced community rights to forests and promoted species mixes useful for infrastructure and commercial use rather than the diverse species important in the local economy.

Beginning in 1865 the state passed the series of Forest Acts and Policies that slowly shifted from granting communities "rights" to forest products (products that had up to then been managed under local authority) to complete state control of forest products and forest management. States can either grant or recognize rights to forests (Lynch 1991). In this case, rather than recognize Indian communities' preexisting rights to forests, the state retained the power to grant such rights. The Forest Act of 1865 demarcated forests reserved for the state and forests to meet local needs (Guha 1983a, 1983b), but it also stated that the state's right to declare any forest a government forest should not abridge the existing rights of communities (PUDR 1982). In contrast, the Forest Act of 1878 was designed, in the words of its chief architect, to "secure the 'best possible legal title' to the forest [the state] sought to control" (Chaturvedi and Sahai 1988:25). Rights and privileges were codified by a political process that used printed announcements (that could not be read by illiterate rural people) and relied on district officials to act as arbiter between the forest department and the people (Guha 1983a, 1983b). In areas in which resistance to state usurpation of rights was strong and organized, more rights and privileges were granted than in areas such as tribal lands in which the politically active segments of society (zamindars) did not represent the interests of those dependent on forests.

The Forest Policy of 1894 further weakened local rights as the language was changed from traditional rights to "rights and privileges." Under this policy the British also claimed the right to take over tribal forests and declare them reserved forests. This slow decrease in people's rights to forests was accompanied by a decrease in forests set aside for villagers' use. Between 1860 and 1914 the area of reserved forests continued to expand at the cost of reduction in area of forests set aside to meet village needs (Gadgil 1989). The Forest Act of 1927 no longer recognized community rights, only the rights and privileges of individuals, it gave the government the right to tax forest produce harvested from all forests, and it increased the number of offenses and fines (PUDR 1982).

After independence the new Indian government followed the same trend in extinguishing local rights (Singh 1986). The 1952 Forest Policy was changed so that "rights and privileges" to forest products became "rights and concessions." The parts of India previously ruled for the British by princely states and zamindars (approximately half of India) were formally brought under Indian forest law as states enacted new Forest Acts in line with the 1952 Forest Policy (PUDR 1982). The trend in more recent policy and legislation has been toward "concessions" and no local rights. Between 1952 and 1982 the forest department's staff increased from 10,000 to 93,000 (PUDR 1982) in order to protect state forests from unsanctioned use. Forest departments have continued to take control of forests where tribal peoples farm (von Furer-Haimendorf 1982).

The Forest Conservation Act of 1980 allows the state government to declare any area as "reserved forest," and vests forest department officials with magisterial powers. State governments can regulate collection of grass and prescribe management practices in village forests and , can assume management of private forests (PUDR 1982). The 1988 amendment to the Forest Act of 1980 clearly denies people a role in forest management and explicitly discourages the planting of trees or other crops to which people have usufruct rights on forest department lands (Saxena 1991).

The Subsistence Economy Losers: Alienation from Forest Niches

The subsistence base of three major sociopolitical groups in India has been particularly affected by this progressive loss of rights and alienation from

Figure 4.2   Graziers, shown here in a farmer's fallow field, are among the sociopolitical groups suffering the effects of forest degradation.

forest management: pastoralists, tribals, and sedentary farmers. The three groups find themselves in increasing competition for natural resources that could serve for commercial or subsistence purposes.

Pastoralists of two hundred different castes make up 6% of the Indian population (Agarwal 1985), and they are politically powerful in several states. They largely depend on open woodlands for fodder to supplement fodder consumed when herders are paid to graze or fertilize farmers' fallow fields (figure 4.2). Their rights to seasonally graze their animals in forests were documented by the British, whose attempts to limit graziers' migrations were opposed by farming interests that wanted the annual accumulations of manure (Tucker 1986). Farmers are now stall feeding more livestock and have less need for the graziers' manure services. As more and more sedentary farmers have begun to raise dairy cattle, graziers have come into increasing competition with sedentary cattle owners over traditional pastures. The political will has arisen in some states to prevent graziers from migrating across state lines (Agarwal 1985; Tucker 1986). Graziers have also lost common property grazing lands to privatization (Brara 1989; Jodha 1986). Faced with barriers to traditional routes to seasonal pastures, diminished fodder availability, and expulsion from nature reserves (Patel 1991), members of grazier groups are changing professions.

The second group, tribals, make up more than 7% of the population: fifty-two million people from more than four hundred different groups, ranging from hunter-gatherers to swidden farmers to sedentary agricultural villagers (Basu 1987). Tribals in India have been concentrated in the hill forests stretching across the country into the northeastern border with Myanmar (Burma) and China; today 94% of tribals live in and around forests (Basu 1987). Tribal cultures had the most rules regarding forest management. These ranged from proscriptions against cutting down a tree without replanting a seed of the same species (Bose and Pradhan 1989), to taboos against cutting certain species, to the maintenance of sacred groves that continue to serve as habitat for rare species today (Chaudhuri 1989; Ramakrishnan 1989; Hajra 1981).

Of the three groups tribals are most dependent on forests for income and food. For example, the five million tribals living on Chotanagpur Plateau in Bihar, depend on forests for 50% to 80% of their subsistence. The sale of minor forest products provides 10% to 50% of family income for tribals in Orissa, Madya Pradesh, Bihar, and Andhra Pradesh (Pathy 1989). Yet the forest department has moved to replace natural mahua (*Bassia* spp.) and sal (*Shorea robusta*) forests important for tribal subsistence with teak plantations. Tribals have resisted government efforts by cutting down teak plantations, destroying teak nurseries, and demanding that useful trees be planted (PUDR 1982). But tribal resistance has been brutally suppressed.

The British instigated a system whereby tribals could no longer trade or sell on the open market the "minor forest products" (also known as nontimber forest products, or NTFPs) from their ancestral forests; a series of rules made them wards of the forest department to be employed in collecting forest products under politically powerful contractors. New legislation was introduced after independence to protect tribal rights over land and forest products, but these measures, although they protected tribals from certain elite forces, have nonetheless increased the powerlessness of these groups (PUDR 1982; Morris 1986; Chambers, Saxena, and Shah 1989; Saxena 1989).

Tribals, particularly those of central India, are politically the weakest of the three groups affected by deforestation, and they have resorted to the most extreme measures of violence to regain rights to forests. Their resistance was recorded by the British beginning in 1772 (Basu 1987). Demands for *Jharkand* (literally, 'land of forests'), a state to include the tribal lands from Orissa, Bihar, Madya Pradesh, and West Bengal, have been made for generations (Gupta, Banerji, and Guleria 1981).

Under these pressures tribals have shifted from a positive interdependent relationship with forests to a negative dependence on a diminishing resource. Many tribals are now dependent on selling "headloads" of wood into the market economy. Studies of tribal groups indicate that with loss of access to forest products, average caloric intake is falling, nutritional status is deteriorating, income from cottage industries based on bamboo and other forest materials is decreasing, and women are suffering increased work loads and worsening health (Fernandes and Menon 1987). Failure to recognize tribal authorities and customary rights when tribal lands are brought into reforestation schemes has further undermined traditional resource management systems (Burman 1989). At the same time displacement from traditional farming lands by more powerful nontribal farming interests has forced tribals to be more dependent on forests. Spokesmen for industrial interests have nonetheless claimed that full blame for deforestation should be placed on tribals who live on the edges of forests.

The third group is the largest and most politically powerful of the three: sedentary nontribal villagers who depend on forests for fodder for their livestock, for cooking fuel, for house timbers, for green manure, and for nontimber forest products. This group has had the most success in resisting the usurpation of their forests, in part because of the Gandhian tradition of peaceful resistance and a religious tradition that values *Vasudhaiva Kutumbakam*—peace for all living organisms (Shiva 1986). The world-renowned Chipko movement, in which women hugged trees to keep them from being felled for commercial use, began in 1972 and in 1981 achieved a ban on felling trees for commercial purposes in an area of 40,000 square kilometers (Bahuguna 1988). In Satyagraha protests in Karnataka, villagers removed the eucalyptus and casuarina seedlings and replanted mango, jackfruit, tamarind, neem (*Azadirachta indica*), and other useful trees in their place. Their protest stopped the reforestation scheme and raised people's consciousness about taking the initiative to experiment with ways to make village lands more productive (Hiremath 1989).

The Indian government and the forest departments are beginning to recognize the political demands of the large numbers of people in the subsistence economy. Yet government responses have been mixed. The chief inspector has sought more police power to keep people and livestock off the state's land (Shyamsunder and Parameswarappa 1987).

The Commercial Economy Losers:
The Failure of Modern Forest Niches

In the industrial sector shortages of raw materials and obsolete equipment are causing forest-based processing enterprises to operate at a fraction of their capacity (Contreras 1990). Shortages are a problem for both large- and small-scale enterprises. Government plantations cover only 6 million hectares of forest land, so these will not meet a significant part of the industrial demand when they mature. The government has established preferential policies for small-scale enterprises. Many enterprises remain small to take advantage of these policies, although an economic and environmental (pollution-related) advantage accrues to larger economies of scale.

In India the demand for pulp, paper, and manufactured wood products is spiraling as urban and middle-class incomes rise and consumer demands for wood and paper products increase. This trend is unlikely to abate in the near future. To control overcutting and encourage the development of nonforest wastelands the 1988 Forest Policy banned the leasing of degraded forest land to industries for establishing high-value plantations. Whether land is available outside the state forest estate to meet industrial demands is a real question.

Environmentalists take the position that most of India's forests should be managed as a multiple-use forest to sustain local as well as industrial needs. V. Shiva has characterized India's ecology movements as movements for peace and justice "through a restructuring of man's relationship with nature"(Shiva 1986:255). Environmentalists favor naturally regenerated forests. Such forests are more suitable to local needs but may be less productive than plantations. Whether such a strategy can meet the rapidly increasing commercial demand within the country is not known.[5]

Development of new infrastructure and new nonforest-based industries (i.e., large-scale irrigation projects, chemical and other manufacturing plants, roads, and large-scale power generation projects) also threatens to put continued pressure on forest lands. There is still little political support for alternative investments.

There are no immediate indications that forest-based industries are failing. Although some factions would force industries to pay market prices for raw materials, they are countered by strong political lobbies to protect industry interests. Social forestry is viewed by some as a panacea for meeting the wood deficit, but the incentives are not clearly in place for this to happen. It is not clear which industries would survive if subsidies were cut. The high costs of import substitution and the power of commercial wood-product groups suggest that government will continue to favor this sector, although probably in a modified way.

The supply of nontimber forest products (NTFPs) is also declining with deforestation, affecting both revenues and employment opportunities for a large segment of the population. This is an area that is receiving increasing attention from environmentalists. They are pushing for management of natural and secondary forests for NTFPs and for increasing local people's employment in the higher-value end of NTFP marketing and processing. Meeting the revenue- and income-generating potential of the sector would require considerable investment and planning. As in many countries the trend in India has been for formal businesses to take over processing, leaving local people as collectors, cut out of chances to engage in marketing or processing.

Finally, urban workers are hurt by deforestation in so far as they have less income to spend on consumer goods or health care, because the price of essential fuelwood continues to rise.

Efforts to Combat Deforestation in the '70s and '80s

Assisted by a variety of multilateral and bilateral donors, including the World Bank, almost all the Indian states have had extensive afforesting programs outside state-owned lands to meet the rising demand for forest products and to help check deforestation. About 6 million hectares have been planted, and about 1 billion seedlings are produced annually. In addition, forest departments have established industrial and fuelwood plantations on degraded forest areas. These range from hardwood to fast-growing pulp and urban fuel plantations of eucalyptus, acacia, and casuarina. The forest departments' program has been somewhat checked by a ban on clear-felling in all areas with expired working plans. This prevents forest departments from clear-felling degraded (but not denuded) areas for replanting until they have up-to-date working plans and the government of India approves the clear-felling. This has affected the size of plantation programs in state forest lands.

Social forestry programs have been implemented in India since the Gujarat Forest Department's pioneering experiments with community woodlots began in 1974. In 1976 the National Commission on Agriculture recommended a national social forestry effort, and government, bilateral donors, and multilateral development banks funded it. Originally directed at addressing the fuelwood crisis and helping to relieve pressure on rapidly degraded forests, social forestry programs have evolved over time. Initially, foresters (and the World Bank as donor) promoted fast-growing species that generated large quantities of biomass quickly, regardless of fuelwood quality. Eucalyptus quickly proved popular for planting on all kinds of land and with both farmers and foresters, because it is easy to raise in nurseries, easy to transport to planting sites, not browsable and therefore easy to protect from cattle, and quite valuable in markets in which pole material is scarce.

The early eighties saw the emergence of a strong backlash to the wide-spread planting of eucalyptus in both forest and nonforest lands. Environmentalists said the planting of eucalyptus supplied rayon factories' needs for pulp at the cost of local needs, caused environmental damage by reducing biodiversity on forest lands, lowered the water table, and drained the soil of nutrients (Shiva 1987, 1989). The argument was one of alternative development as much as of environmental concerns. At the same time the pole market became saturated in many places. Hence the mid eighties saw a dramatic shift to plantation of greater numbers of leguminous and multipurpose trees (fruit, fodder, fuel) and grasses in *panchayat* (the term for a type of local government system), revenue, and forest lands. Many farmers continue to plant eucalyptus in private wastelands, and wealthier farmers are switching to trees from other cash crops in areas where labor costs have shot up. A much wider variety of trees is being raised in farm forestry in eastern Uttar Pradesh. Farmers who once planted eucalyptus are now raising fruit trees for cash income. In Gujarat some farmers are raising acacia as a fuel for sale and local use, and in Maharastra farmers are planting Jatropha oilseed.

Most state forest departments now promote a much wider variety of species for grass, fodder, fuel, and fruit production for roadside, community, and government plantations. This potentially increases the range of benefits from plantations for different segments of the community. But panchayat and many village-level bodies often find it difficult to devise enforceable systems to distribute benefits in kind. Although some villages allow cutting of grass from the plantations or distribute fuelwood thinnings free, many auction even these intermediate products to generate cash to pay for forest watch and ward, or a development fund. Such a system of selling forest products at a subsidized price locally can benefit the poor and marginal if they are already purchasing grass and fuel in the market (which is not true of most marginal groups). In other cases forest departments raised trees on village common lands as though they were forest department lands, and villagers treat them as another category of reserved forest (Saxena 1989).

In India there is scope to experiment with new distribution systems and plantation models to increase the benefits to a wider group of villagers. One recommendation has been to establish alternate rows of salable trees and shrubby fuel trees, so that the shrubby trees can be lopped by the poorer villagers as a regular source of fuel. But few foresters or village forest committee leaders are yet convinced that the subsistence biomass thereby produced is worth the investment required.

Most state social forestry programs have not included the level of extension needed to encourage participatory forestry committees at the panchayat or subpanchayat level. Although the panchayat may retain control of the woodlot in some of the plains regions, village-level forest management with broader participation could be successful in some areas. Questions remain about the benefits of keeping village lands as a villagewide resource versus giving village lands to smaller interested groups to reforest or repasture in a system of group tree tenure.

Evaluations of the social forestry program have varied according to the criteria for success. Industries measure success by the generation of a supply of wood or pulp that is an alternative to the natural forest areas that government policies have closed to extraction. In some areas industry has found that plantations outside state forest lands are meeting its needs; in other areas production from plantations has not met industrial needs, so industry is eager for the state to reopen state lands to industrial use. Using as criteria the size of areas successfully covered with plantation or regeneration and the numbers of seedlings successfully distributed, forest department administrators deem the program a success. But forest departments recognize that actual production rates have been disappointing. Survival rates of seedlings have been low (20% to 35%) in many of these social forestry projects. A key problem in all social forestry programs, including many of those sponsored by nongovernmental organizations (NGOs), is the generally poor quality of seedlings being planted.

However, those concerned with social equity issues measure success by the contribution of the program to balanced social and economic development, which means that the poor and marginal must be major beneficiaries. The evaluation is mixed. In many cases poor people were hurt when common property resources were closed to them in order to create plantations whose products have been auctioned to the resource rich. The social forestry program provided free eucalyptus seedlings, which large farmers trucked away to raise for sale as poles or to the paper industry (Chowdhry 1989), whereas poor farmers received limited benefits. Independent evaluations of the social forestry program before 1988 demonstrated that the major beneficiaries were large farmers and the paper, pulp, and building industry (Chowdhry 1989). On the other hand, social forestry programs have frozen the common property status of land and thereby prevented further privatization. A poorly documented benefit to the poor has also been the increased availability of reasonably priced eucalyptus poles in local markets. Village elites view production of salable forest products as success. But marginal graziers see less success in the social forestry program, because they seek increased grass production and access to that grass.

Similarly, ecologists see mixed success. Social forestry plantations have in some cases improved soil fertility and water conservation and checked degradation caused by overgrazing, but in other cases they have had no effect on the trend toward land degradation.[6] Plantations have had the negative ecological effect of reducing the native biodiversity of these sites.

After satellite imagery showed that deforestation, despite social forestry efforts, had proceeded far beyond the situation described by forest departments' official statistics, India established the National Wastelands Development Board (NWDB) in 1985. But the NWDB was unable to meet its targets for reforestation, because many areas identified as wastelands had been privatized or diverted to other uses, the board had no decentralized network for identifying a wide range of nongovernmental groups and individuals to receive grants, and grant procedures were too complicated and poorly understood by local officials participating in funding and land allocation. The board was officially transformed into a technical mission in 1989 in order to cut red tape; and it has designated local institutions for monitoring its program and has begun to establish state-level implementation cells.

The bright prospects for a massive assault on deforestation from the central government authority, as initially promoted, have faded. Hopes for reforestation are now pinned on efforts to return forest management to communities.

Visions for the '90s: "Rebuilding Nature"

Hundreds of grassroots groups are concerned with conserving the environment for the benefit of local communities. These grassroots groups are not concerned with environmental protection per se but with use of the environment and who should benefit from it and with rebuilding nature (Agarwal 1985). For example, the vision of the strategy that the Centre for Science and Environment (CSE, a major NGO based in New Delhi) calls "Towards Green Villages" is for each rural settlement in India to have a clearly and legally defined environment to protect, improve, care for, and use (CSE 1990, 1991).

Others envision forestry for the poor as a joint effort of the state and the community to meet subsistence needs and increase incomes (Chambers, Saxena, and Shah 1989). Under this vision forest departments and communities would cooperate to create more diverse forests instead of fast-growing plantations of a single species. These diverse forests would develop more slowly and produce multiple products for local use and commercial sale. Communities would be responsible for making management decisions within the parameters established by forest department rules and regulations. Communities would gain increased benefits, but the state would continue to own the forests and claim revenue from the sale of forest products.

Although the government has not moved to return forest lands to communities, by the end of the 1980s experiments designed for greater participation of local people in afforestation and forest management had proliferated. These experiments stressed not only protection of forests as a future source of timber and fuel and as environmental resource but also the value of NTFPS to local people and the national economy.

Five states have set up joint management systems (West Bengal, Gujarat, Haryana, Orissa, and Himachal Pradesh) whereby local forest committees protect forest areas in return for usufruct rights to intermediate products (NTFP) and in some states a share as great as 25% of the final yield (e.g., Malhotra and Poffenberger 1989). In September 1990 representatives from the five states met in New Delhi to exchange information about their experiences in implementing programs in which local communities and the forest department jointly manage natural regenerated forest. The Ministry of Environment and Forests endorsed these joint management experiments in a June 1990 resolution that encourages other states to experiment with joint forest management approaches to natural regeneration to reforest degraded forest lands. Questions remain, however, as to the types of forests that can be managed in this fashion and whether resulting high-value forests can withstand pressures for illegal logging.

Another set of experiments involves the development of pasture or forest areas outside state forest lands by small groups or by local communities. Some groups of landless, women, or dairy farmers have developed revenue lands under different species of trees and/or pasture with the support of local authorities (e.g., the Behavioral Science Center charcoal project in Gujarat, tassar silk schemes of women NGOs in Bihar, dairy farmer cooperative groups in Maharastra, and a plantation scheme for producing mahiti oilseed). Some local communities have protected corporate (e.g., *shamlat*), panchayat, or revenue lands that have enough rootstock to produce good forest or grass stock once they are protected. Other communities have used money allocated for various social forestry or wasteland projects to develop plantations. In most cases formal plantations have tended to stress products that can be communally sold for their commercial value, such as timber or fruit, rather than subsistence products for local distribution. The village uses the profits to generate community assets. The panchayat sets the profits aside as community assets or uses them to develop panchayat infrastructure, a more distant benefit from the perspective of the average villager. In contrast, local village councils that manage lands for grass production have distributed to all members profits from the sale of rights to hand-harvest fodder (for example, an NGO called AKRSP in Gujarat, dairy cooperatives in Maharastra), with the formula skewed somewhat to give more to those with less of their own land.So far forest development on nonforest lands at the subpanchayat level represents only a small portion of total forest development on nonforest lands.

Another set of experiments is attempting to provide alternative sources of income to people living in or adjacent to conservation or protected areas by creating a buffer zone that they can exploit or by establishing an extractive reserve (e.g., VIKSAT 1990).

NGOs and local communities are now showing success in reforestation where central government efforts failed. Yet even with massive implementation of village forestry, there may not be enough forest to meet the subsistence demands from projected population growth. Even now West Bengal foresters are grappling with problems that have arisen because some villages have been allocated 200 hectares of forest whereas other villages must share 20 meager hectares with another village.

A Marginal Third Interest Group—Biodiversity

The return of forest management to community control fails to reproduce the traditional relationship between India's people and nature. Before colonial and neocolonial usurpation of communities' forests, an original goal of traditional community-based management of forests was maintenance of species-diverse lands. Today this goal is more difficult to achieve given that, under new community forestry initiates, communities are assuming authority over wastelands. The new community forests are often manmade assemblages of relatively few species, a faint shadow of what inhabited community forests before their usurpation by the state. However, Mark Poffenberger (personal communication 1992) reports that community forests created by natural regeneration in West Bengal include more than two hundred plant species, and Jeffrey Campbell (personal communication 1993) reports that elephants are expanding their ranges into the older regenerating community forests in West Bengal.

Chatrapati Singh notes, "The issues actually at stake in the forest question . . . are three: (a) justice to the people, forest dwellers and non-dwellers; (b) justice to nature (trees, wildlife, etc.); and (c) justice to coming generations" (1986:7). Singh further relates the loss of biodiversity to injustice as seen through traditional Hindu doctrine: "*Prakriti* or nature is a living organic force, like man. It is a self-regulating, self-healing organism—so long as externalities do not disturb or destroy it. Amongst these externalities the most destructive is injustice or *adharma*. For the past two centuries or so the consequences of such adharma have been borne by the rural poor, the tribals, and the flora and fauna of India" (1986:1).

India's biodiversity and subsistence security have been severely eroded by the process of state usurpation of community rights. The tribals and other poor rural people dependent on the maintenance of biodiversity for medicines, food, and raw materials have been marginal political actors, too weak to halt the political processes that led to the erosion of biodiversity.

Although India has a rich biodiversity heritage (Gadgil and Meher-Homji 1986 and others) and an ancient religious tradition of protecting nature (e.g., Gadgil and Vartak 1981), the politically powerful have not engaged in strong activism for nature preservation. The Western-style conservation movement was founded in India by gentlemen hunters; it has focused on saving tigers (Gadgil 1984). The Silent Valley Dam debate brought the issues of rain forest biodiversity to public attention. In northeastern India tribals continue to fight to defend their natural forests and the biodiversity upon which they depend. But India has no strong, nationwide, broad-based political movement to conserve biodiversity per se.

The future political viability of biodiversity's interests and the future of the few remaining wild areas are uncertain. Given the power base of entrenched interests, the government is unlikely to move quickly to empower local communities to maintain the remaining natural forests.

How will biodiversity conservation fare as modern villagers take on a new responsibility for rebuilding nature on wasteland and maintaining biodiversity in neighboring reserves that are not yet reduced to wasteland? Will modern villagers accept the cost of maintaining wildlife that can damage their crops and/or livestock and threaten their children? For example, villagers living on the boundary of world-famous Project Tiger reserve of Ranthambhore National Park agree that tigers need a home. But these same park neighbors are not eager to have tigers in their neighborhood if they must suffer economic losses from crop and livestock depredation while tourist hotels rake in money from the tigers (Groenfeldt et al. 1990). On the other hand, graziers of the Gir Forest, home of the last population of Asiatic lions, accept livestock losses to lion predation as part of the system that provides their livelihood (Patel 1991).

Does India represent the full cycle other countries will follow— from close relationships between communities and natural forests to usurpation of forest by the state, followed by the overexploitation and mismanagement of state-owned forests for short-term profits, and culminating in the return of degraded state land to the rural poor to rebuild nature without most species that once made up the forest and supported their subsistence needs?

The Role of Forests in Sustainable Economic Development

The political battle between the commercial and subsistence interests over the forest niche in India raises a basic question about the role of the subsistence economy in supporting sustainable economic growth (c.f. Alcorn 1991). There have been numerous calls to reexamine the pre-analytic bias of cost-benefit analyses used by economists to program development (e.g., Daly and Cobb 1989; Dahlberg 1987). Analysts are focusing on the costs of environmental degradation that could hurt industrial development and economic growth. R. Repetto and colleagues (1989) and others have focused attention on reforming national income accounts. But the effect of environmental degradation on the subsistence economy and the subsequent effect on economic growth have been ignored. Anil Agarwal's 1985 proposal that India develop a gross nature product indicator is a major step toward integrating a concern for subsistence at the national level. We encourage development thinkers to turn their attentions to this question and give further focus to stabilizing subsistence production as part of the development process.

The goal of the Green Villages strategy is not only good for villagers and a commendable, socially conscious nod toward fairness. It may make good economic sense at the national level as well. A stable rural subsistence support system, a "safety net" for the poor (World Bank 1990), would support continued economic growth for the hundreds of millions of people who live in India's cities today and in the future (predicted to have the largest urban population in the world in the year 2000). If the subsistence economy is no longer considered archaic and something to be despised and instead becomes valued for its support of economic development, the political clout of subsistence interests may ally with commercial interests and the state to halt deforestation and conserve the biodiversity remaining in natural forests.

Lessons from the India Case

The slow erosion of communities' rights to manage their forests has been part of the colonial process in many countries and continues today under independent states. During the 1600s in Guatemala, for example, indigenous people unsuccessfully argued in court that nearby forests were integral to their production systems and critical for their survival and hence should not be turned over to others for pasture development (Murdo J. MacLeod, personal communication 1992).

The loss of community rights to forests in Sri Lanka also proceeded rapidly. When the British Empire assumed control over Sri Lanka in 1815, the Crown agreed to continue administration under existing laws and traditions, including community rights to forests. But by 1840 colonial administrators had expropriated the forests and changed tenure legitimacy rules to favor the rights of foreign coffee planters (Karunaratna 1987) despite vigorous protest by local communities. As demand for tropical timber increases, similar processes of forest usurpation are underway in Thailand (Lynch and Alcorn 1994), Papua New Guinea, Amazonia, and other areas where forest-dwelling communities are fighting to retain their rights, just as India's communities fought for their rights to their forests a century ago or more. Globally, an estimated 200 to 500 million people are forest dwellers (Lynch 1990), many of whom belong to communities that maintain traditional forest management under common property systems.

Other countries and interested environmentalists might learn a lesson from the Indian experience and avert a repetition of Indian forest history in earth's remaining forests. If the countries still rich in biodiversity choose to use the power of the state to support local peoples' rights to manage natural forests and defend remaining enclaves of biodiversity, natural forests and mutually beneficial forest-human relationships will be more likely to endure.

Acknowledgments   The authors alone accept full responsibility for the contents of this paper and thank all those who have commented on the manuscript. For teaching her lessons about deforestation, Janis Alcorn also thanks the people of Matala and Talmadgi, Karnataka, families living around Hingolgad Sanctuary in Gujarat and Ranthambhore National Park in Rajasthan, and Indian NGOs, including the Centre for Science and Environment and the Nehru Foundation. The views and conclusions expressed herein should not be attributed to the World Bank, U.S. Agency for International Development (USAID), or the World Wildlife Fund (WWF).

Notes

1. Wastelands have been defined in various ways during India's history. *Wasteland* is a term initially used by the British during the colonial period to classify lands of low productivity or low tax generation. The National Wastelands Development Board, as it was called in 1985, defined wasteland as land that has deteriorated from lack of appropriate water and soil management but that can be brought under vegetative cover with reasonable effort (Saxena 1989). Of India's approximately 329 million hectares, 175 million hectares are classed as wasteland. Culturable wasteland (land that could be cultivated) is estimated to cover 84 to 94 million hectares, including 12 million hectares of revenue lands. (Revenue lands are public lands that are supposed to be used to generate tax revenue for various government agencies but are usually wastelands used as commons for open access grazing and do not generate revenue.) State forest lands officially cover 75 million hectares (Chowdhry 1989), and 36 to 40 million hectares of official forest lands have been classed as wastelands (Dey 1989).

2. Under India's political system the individual state governments have strong powers in relation to the central government. Forests belonging to India are not managed by any national forest agency but rather by the state forest departments of the individual states, each of which interprets national law and follows state-specific laws and regulations. Senior forestry officials are drawn from state cadres, and foresters are drawn from a national administrative forestry cadre. At the national level is the Ministry of Forests and Environment, but it has no field jurisdiction. Within the ministry is the National Wasteland Development Mission (which succeeded the NWDB in 1989),which allocates substantial grant money to state and local governments and nongovernmental institutions for afforestation programs.

3. *Social forestry* generally refers to forestry programs that involve local peo-

ple. Social forestry in India specifically refers to afforestation in areas outside state forest lands in order to benefit local people. Although local people were supposed to be key actors in establishing forests and trees, social forestry was implemented through the forest department, a government agency with a long history of protecting forests from people (Chowdhry 1989).

4. Scripts are internalized plans. Farmers carry out routine activities by using scripts that contain ecological information of which the farmer often is not consciously aware (Alcorn 1989). Such customary ways of farming are passed from generation to generation through enculturation.

5. Questions that arise for development planners include the following: Should some state forest land be leased to industry again for production purposes? Should forest departments raise captive plantations for industries? Or should forest departments remain the exclusive afforestation agent on public land? How much area should be retained under a multiple-use forest? Should management systems involving local people be limited to areas of natural regeneration, or should local communities also be given a stake in commercial plantations and high forest? The government of India has taken a strong position against any divestment of the state's tenure over forest lands, thereby removing that option from discussion.

6. Summary of results of World Bank evaluations.

References Cited

Agarwal, A. 1985. *Human-Nature Interactions in a Third World Country.* Fifth World Conservation Lecture, London. Surrey, England: World Wildlife Fund–U.K.

Alcorn, J. B. 1989. Process as resource: The traditional agricultural ideology of Bora and Huastec resource management and its implications for research. In *Resource Management in Amazonia: Indigenous and Folk Strategies,* D. A. Posey and W. Balée, eds., pp. 63–77. New York: New York Botanical Garden.

——. 1991. Ethics, economies, and conservation. In *Biodiversity: Culture, Conservation, and Ecodevelopment,* M. L. Oldfield and J. B. Alcorn, eds., pp. 317–349. Boulder, Colo.: Westview.

Arnold, J. E. M. and W. C. Stewart. 1989. Common property resource management in India. *Report to World Bank.* Mimeo.

Bahuguna, S. 1988. Chipko: The people's movement with a hope for the survival of humankind. *ifda dossier* 63: 3–14.

Basham, A. L. 1967. *The Wonder That Was India: A Survey of the History and Culture of the Indian Subcontinent Before the Coming of the Muslims.* Bombay: Fontana Books.

Basu, N. G. 1987. *Forests and Tribals.* Calcutta: Manisha.

Bose, A. and A. Pradhan. 1989. Social forestry. In *Forests and Forest Development in India,* B. Chaudhuri and A. K. Maiti, eds., pp. 305–312. New Delhi: Inter-India Publications.

Bowonder, B., S. S. R. Prasad, and N. V. M. Unni. 1987. Afforestation in India. *Land Use Policy* (April): 133–146.

——. 1988. Dynamics of fuelwood prices in India: Policy implications. *World Development* 16: 1213–1229.

Brara, Rita. 1989. Commons' policy as process: The case of Rajasthan, 1955–1985. *Economic and Political Weekly*, October 7, pp. 2247–2254.

Brockway, L. H. 1979. *Science and Colonial Expansion*. New York: Academic Press.

Burman, B. K. Roy. 1989. Development of forestry in harmony with the interest of the tribals. In *Forests and Forest Development in India*, B. Chaudhuri and A. K. Maiti, eds., pp. 361-379. New Delhi: Inter-India Publications.

Centre for Science and Environment (cse). 1990. *Towards Green Villages*. New Delhi: Centre for Science and Environment.

——. 1991. *Towards a Self-Reliant India: Statement of the National Seminar on Village Ecosystem Planning*, December 4–6, New Delhi.

Chambers, R., N. C. Saxena, and T. Shah. 1989. *To the Hands of the Poor*. New Delhi: Oxford and ibh Publications.

Chaturvedi, S. and V. Sahai. 1988. *Institutionalization of Forest Politics*. Delhi: Eastern Book Linkers.

Chaudhuri, Buddhadeh. 1989. Forest, forest-dwellers, and forest development. In *Forests and Forest Development in India*, B. Chaudhuri and A. K. Maiti, eds., pp. 19–33. New Delhi: Inter-India Publications.

Chowdhry, Kamla. 1989. Social forestry: Roots of failure. *Indian Journal of Public Administration* 35: 437–443.

Contreras H., Arnoldo. 1990. *India Forest Sector Overview*. Draft report to India Country Department, Asia Region. Washington, D.C.: World Bank.

Dahlberg, K. A. 1987. Redefining development priorities: Genetic diversity and agroecodevelopment. *Conservation Biology* 1: 311–322.

Daly, H. E. and J. B. Cobb. 1989. *For the Common Good*. Boston: Beacon Press.

Dey, Gautam. 1989. Approaches of wasteland development. *Indian Journal of Public Administration* 35: 497–504.

Fernandes, W. and G. Menon. 1987. *Tribal Women and Forest Economy*. New Delhi: Indian Social Institute.

Food and Agriculture Organization (fao). 1990. Unpublished data.

Forest Research Institute. 1988. *The State of India's Forests*. New Delhi: Government of India.

Gadgil, M. 1984. Conserving biological diversity: The Indian experience. In *UNESCO, Ecology in Practice*, pp. 485–491. Paris: Tycooly International.

——. 1989. Deforestation: Problems and prospects. Foundation Day Lecture, Society for Promotion of Wastelands Development, New Delhi. Supplement to *Wastelands News* 4 (May–June): 1–19.

Gadgil, M. and V. M. Meher-Homji. 1986. Role of protected areas in conservation. In *Conservation for Productive Agriculture*, V. L. Chopra and T. N. Khoshoo, eds., pp. 143–159. New Delhi: Indian Council of Agricultural Research.

Gadgil, M. and V. D. Vartak. 1981. Sacred groves of Maharashtra: An inventory. In *Glimpses of Indian Ethnobotany*, S. K. Jain, ed., pp. 279–294. New Delhi: Oxford and ibh Publications.

Groenfeldt, D., J. Alcorn, S. Berwick, D. Flickinger, and M. Hatziolos. 1990. *Opportunities for Eco-Development in Buffer Zones: An Assessment of Two Cases in Western India.* Report to the U.S. Agency for International Development–India. Washington, D.C.: World Wildlife Fund.

Guha, Ramachandra. 1983a. Forestry in British and post-British India. *Economic and Political Weekly,* October 29, pp. 1882–1897.

———. 1983b. Forestry in British and post-British India. *Economic and Political Weekly,* November 5–12, pp. 1940–1946.

Guha, Ramachandra, S. N. Prasad, and Madhav Gadgil. 1984. Deforestation and degradation of natural plant resources in India. *Journal of the Indian Anthropological Society* 19: 246–253.

Gupta, R., P. Banerji, and A. Guleria. 1981. *Tribal Unrest and Forestry Management in Bihar.* Ahmedabad, India: Indian Institute for Management.

Hajra, P. K. 1981. Nature conservation in Khasi folk beliefs and taboos. In *Glimpses of Indian Ethnobotany,* S. K. Jain, ed., pp. 149–152. New Delhi: Oxford and IBH Publications.

Hiremath, S. R. 1989. People's participation in protection and sustainable use of environment. *Indian Journal of Public Administration* 35: 388–396.

Jodha, N. S. 1986. Common property resources and rural poor in dry regions of India. *Economic and Political Weekly,* July 5, pp. 1169–1181.

———. 1990. *Rural Common Property Resources: Contributions And Crisis.* Foundation Day lecture, May 16. New Delhi: Society for Promotion of Wastelands Development.

Karaosmanoglu, A. 1989. Environment, poverty, and growth: The challenge of sustainable development in Asia. In *World Bank, Poverty, and Prosperity: The Two Realities of Asian Development,* pp. 34–47. Washington, D.C.: World Bank.

Karunaratna, Nihal. 1987. *Forest Conservation in Sri Lanka from British Colonial Times,* pp. 1818–1982. Talangama, Sri Lanka: Trumpet Publishers.

Kendrick, Karolyn. 1989. India. In *Floristic Inventory of Tropical Countries,* David G. Campbell and H. David Hammond, eds., pp. 133–140. New York: New York Botanical Garden.

Lynch, O. J. 1990. *Whither the People? Demographic, Tenurial, and Agricultural Aspects of the Tropical Forest Action Plan.* Washington, D.C.: World Resources Institute.

———. 1991. Community-Based Tenurial Strategies for Promoting Forest Conservation and Development in South and Southeast Asia. Paper presented at U.S. Agency for International Development Environment and Agriculture Conference, September 10–13, Colombo, Sri Lanka.

Lynch, O. J. and J. B. Alcorn. 1994. *Empowering Local Forest Managers.* Washington, D.C.: World Resources Institute.

Malhotra, K. C. and M. Poffenberger, eds. 1989. *Forest Regeneration through Community Protection.* Calcutta: West Bengal Forest Department.

Morris, B. 1986. Deforestation in India and the fate of forest tribes. *Ecologist* 16 (6): 253–257.

Myers, W. L. 1988. Tropical Forestry and Biological Diversity in India. Unpublished report to U.S. Agency for International Development–India. New Delhi: USAID.

Nadkarni. M. V. 1987. *Political Economy of Forest Use and Management in the Context of Integration of a Forest Region into the Larger Economy.* Bangalore, India: Institute for Social and Economic Change. Mimeo.

Patel, H. G. 1991. Forest Ecology and Cattle-Graziers: Where Do They Stand? Paper presented at the Conference of the International Association for the Study of Common Property, September 26–29, University of Winnipeg.

Pathy, J. 1989. Shifting cultivators: Bearing the brunt of 'development.' In *Forests and Forest Development in India*, B. Chaudhuri and A. K. Maiti, eds., pp. 93–109. New Delhi: Inter-India Publications.

People's Union for Democratic Rights (PUDR). 1982. *Tribal Rights.* New Delhi: People's Union for Democratic Rights.

Ramakrishnan, P. S. 1984. The science behind rotational bush fallow agriculture system (*jhum*). *Proceedings of the Indian Academy of Sciences (Plant Science)* 93: 379–400.

——. 1989. Tribal man in the humid tropics of the north-east. In *Forests and Forest Development in India*, B. Chaudhuri and A. K. Maiti, eds., pp., 57–92. New Delhi: Inter-India Publications.

Repetto, R., W. Magrath, M. Wells, C. Beer, and F. Rossini. 1989. *Wasting Assets: Natural Resources in the National Income Accounts.* Washington, D.C.: World Resources Institute.

Saxena, N. C. 1989. Wasteland development, environmental protection, and the poor. *Indian Journal of Public Administration* 35: 487–496.

——. 1991. *Forestry Incentives in India.* Paper submitted to the World Bank, Asia Region, India Country Department, Washington, D.C. Mimeo.

Shiva, V. 1986. Ecology movements in India. *Alternatives* 11: 255–273.

——. 1987. Forestry myths and the World Bank. *Ecologist* 17: 142–145.

——. 1989. Conserving India's forests, protecting India's people. *Indian Journal of Public Administration* 35: 380–388.

Shyamsunder, S. and S. Parameswarappa. 1987. Forestry in India: The forester's view. *Ambio* 16: 332–337.

Singh, C. 1986. *Common Property and Common Poverty: India's Forests, Forest Dwellers, and the Law.* Oxford, England: Oxford University Press.

Tucker, R. P. 1986. The evolution of transhumant grazing in the Punjab Himalaya. *Mountain Research and Development* 6: 17–28.

VIKSAT. 1990. *Buffer Zone Restoration in India.* Ahmedabad, India: VIKSAT/Center for Environment Education.

von Furer-Haimendorf, Christoph. 1982. *Tribes of India.* Berkeley: University of California Press.

Westland, P. 1979. *The Encyclopedia of Spices.* Secaucus, N.J.: Chartwell Books.

World Bank. 1990. *Poverty: World Development Report 1990.* Washington, D.C.: Oxford University Press.

5

The Political Economy of Colonial Farm Forestry in Kenya: The View from Kirinyaga

Alfonso Peter Castro

This article presents a historical analysis, but it stems in many ways from my experience as an applied anthropologist concerned with contemporary tropical deforestation. By the late 1970s it had become apparent that conventional forestry based on centralized government management and commercial fiber production could not deal with the rapid destruction of the world's forests (Fortmann and Fairfax 1989). Indeed in many places it was evident that state forestry had accelerated land degradation, often with disastrous consequences for rural communities (for example, see Anderson and Huber 1988; Hecht and Cockburn 1989; Guha 1990; Martin 1991; Peluso 1992; Colchester and Lohmann 1993). Thus foresters and planners pursued new social, or community-oriented, strategies in an attempt to address deforestation (see Arnold 1992; Sharma 1992).

Over the years I have served as a consultant to a number of international agencies engaged in social forestry. Many projects involved what is called farm forestry: promoting tree growing by rural families on their own land (see Food and Agriculture Organization [FAO] 1985). This approach arose in recognition that trees on farms, in pastures, or near homesteads, although important in meeting local needs, often receive little or no official attention (Castro 1992). Social forestry efforts during the late 1970s and early 1980s in such places as India and Tanzania sometimes involved organizing families to plant trees communally. These attempts frequently failed for a variety of reasons (FAO 1985). The focus increasingly shifted to the individual family farm as the unit for social forestry (Cernea 1992). In many cultures people had long practiced agroforestry (combining trees and food production), managed woodlots, or otherwise incorporated trees in their local production systems.

In the course of carrying out social research on forestry, especially in Kenya, I became aware that many problems occurring in contemporary projects had been encountered decades ago in similar efforts by colonial governments. Unfortunately, the lessons of their experience had been overlooked. This essay is based on archival material and ethnographic interviews and seeks to uncover some hidden history of colonial forestry in Kirinyaga, Kenya. Among the colonial documents that I examined at the Kenya National Archives in Nairobi were annual reports and political record books from Central Province, Embu District, and Nyeri District. At the Embu District Archives in Embu I consulted daily correspondence files on tree planting and soil conservation. Also in Embu the county council allowed me to examine the minutes of the local native council and the African district council.* The purpose is not simply to provide a historical description but to consider the implications for contemporary reafforestation efforts. I have explored related themes about forestry in Kirinyaga in other works (Castro 1983, 1988, 1990, 1991a, 1991b, 1991c, 1995).

The Setting

Kirinyaga is situated on the southern slopes of Mount Kenya, the second-highest mountain in Africa. It is Kenya's smallest rural district, covering 1,437 square kilometers in Central Province. A thick mantle of afro-alpine moorland, bamboo thicket, and moist evergreen forest covers Mount Kenya. The forest once reached farther down the slopes, but centuries of clearing have converted it into farms and settlements. Kirinyaga possesses some of the most productive farmland in the country, with fertile volcanic soils and abundant rainfall. It is a major producer of tea, coffee, rice (at the Mwea Scheme), maize, beans, and other crops. The population of Kirinyaga exceeds 400,000 today, having more than tripled in the twentieth century (Ministry of Finance and Planning 1984).

Kirinyaga constitutes both a political and sociocultural unit. It is the traditional homeland of the Ndia and Gichugu Kikuyu. They descend from Bantu speakers who settled in the region many years ago. As they fanned out across central Kenya, their language gradually differentiated into Ki-Ndia, Ki-Gichugu, Ki-Mathira, Ki-Embu, and related dialects. They came to regard themselves as distinct groups that lived in particular areas. The inhabitants of precolonial southern Mount Kenya were "peoples" rather than "tribes." Their socioeconomic and political relations transcended local identities and boundaries (Lonsdale 1992a).

In the late nineteenth century the Ndia and Gichugu were similar to

Figure 5.1   A large mururi tree (*Trichilia* sp.) dominates this
sacred grove in Ndia. Communal worship, sacrifices, and
ceremonies were held at such groves before widespread
conversion to Christianity.

other acephalous agriculturalists of central Kenya (see Middleton and Kershaw 1965; Muriuki 1974; Ambler 1988). They lived in small independent communities linked through trade and occasional alliances to neighboring groups. Social life revolved around local kinship (clan, lineage, and family), territorial (homestead and neighborhood), and generational (age- and generation-set) relationships. Families resided in scattered homesteads. Agricultural land consisted of fragmented plots, plus pasture and woodland that were common property. Primordial land rights were held by descent groups, but households possessed usufruct (legal) rights to the clan land they occupied. The Kikuyu also recognized several forms of temporary land occupancy. The household was the sole unit of production and consumption, but nearby homesteads were linked through mutual assistance such as labor exchange. Their food production techniques included having gardens in a range of microenvironments, intercropping, short-term fallowing, and even keeping some plots in continuous production.

The Kikuyu developed an extensive knowledge of the botanical properties and practical uses of local trees. In the 1930s Louis Leakey (1977) recorded more than four hundred different tree and plant species used in precolonial times. The Kikuyu's dependence on trees and forest resources for many items of material culture, including timber, fuel, medicines, and food, was closely tied to their incorporation of certain groves and trees in religious and magical activities. For example, communal worship and ceremonies took place at sacred groves (see figure 5.1) scattered throughout Kirinyaga (Castro 1990, 1995).

The Ndia and Gichugu Kikuyus' sense of themselves as a people was closely related to their forest environment and their transformation of it. Local oral traditions closely connected land clearing with the emergence of a distinct local sociocultural identity. Ndia and Gichugu were the names of brothers who first cleared land in the district (Lambert 1949) and who were remembered for their industrious land clearing. A Ndia elder from Nduine stated, "*Ndemi* [the cutters] was the first known generation in Kirinyaga. They were the first Kikuyus to clear the bush, and they gave birth to *Mathathi*, the second generation." John Lonsdale has written, "[The Kikuyu] were brought together by the demands and opportunities of forest clearance. Only then did they become *Agikuyu*, less a boast of ancestral descent than a claim to farming skill" (1992b:334).

The people of Kirinyaga attempted to reconcile competing pressures for retaining and removing trees (see Castro 1991a, 1991b, 1995). Their means for conserving trees were embedded in their local system of land tenure, religious beliefs, and agroforestry practices. Lineage elders, for example, could place restrictions on the clearing of woodland or large multipurpose trees on clan lands. They maintained certain groves as ceremonial sites and protected other trees because they held powerful spirits. Their agroforestry practices included selective clearing, intercropping food crops and trees,

Figure 5.2   A pollarded muu tree (*Markhamia hilde-brandtii*) intercropped with maize and other food crops in a Ndia Kikuyu smallholding. Pollarding reduces shade, provides firewood and poles, and stimulates regrowth from the tree's stem.

multiple harvesting of trees through heavy pruning of branches (pollard-ing) or the trunk (coppicing), and propagating trees from direct sowing, cuttings, and seedling transplantations (see figure 5.2).

The foregoing strategies were not aimed at reversing deforestation; instead the Kikuyu sought to mitigate the effects of deforestation by incor-porating their valued multipurpose trees in their local sociocultural and household production systems. Although people understood how to prop-agate trees, they did not do so for reforestation per se but to meet specific needs such as setting boundaries or supporting vines. Kikuyu leaders told a 1929 colonial committee, "Trees have only been planted since Europeans came" (Maxwell, Fazan, and Leakey 1929a:6)—in other words, planting trees in large numbers and for the purpose of reforesting farmland were colonial innovations.

Local wood shortages emerged by the late 1800s, especially in places that supplied food to caravans from the coast (Castro 1991a). Ernest Gedge (1892:526) observed in southern Mount Kenya that "extensive clearances" caused the forest to be some distance from settlements and fields. Chauncey Stigand described Kikuyuland as a "rolling, almost treeless, cultivated country" (1913:235). The decrease in forest cover was not necessarily synonymous with increased land degradation, because farming techniques such as intercropping and mulching made it possible to expand production without undermining soil management.

Colonial Conquest and State Forestry

Great Britain declared Kenya a protectorate in 1895, and imperial forces invaded Kirinyaga in 1904. The British adopted the local names of Ndia and Gichugu for the newly conquered "divisions." and split their jurisdiction between neighboring administrative districts. Ndia was part of Nyeri (later South Nyeri) until 1933, when it joined Embu. Gichugu was administered by Embu until 1922, when it was transferred to South Nyeri. In 1933 it reverted to Embu's jurisdiction. Ndia and Gichugu, along with Mwea, were allowed to form their own district at independence in 1963.

The Ndia and Gichugu, along with other Africans, became second-class citizens in a racially and class-stratified colony. Whites controlled key administrative posts, whereas Africans served in positions of strictly local authority, such as chief. Conquest led to other social changes, including imposition of taxes, labor migration, religious conversion, and increased socioeconomic differentiation (Davison 1989; Castro 1995). Unlike peoples in other parts of Kikuyu country, the people of Kirinyaga did not lose land to white settlers. However, the Ndia and Gichugu did not escape large-scale colonial land appropriation.

In other publications I have described in detail conflicts between the central government and the people of Kirinyaga over control of local forest and woodland resources (see Castro 1988, 1990, 1991b, 1991c, 1995). Briefly, the colonial government appropriated the forested and upland slopes of southern Mount Kenya in 1910. An extensive area was involved: roughly one-fourth of present-day Kirinyaga is still taken up by the Mount Kenya reserve. The neighboring peoples of Nyeri, Embu, and Meru also lost large tracts of forest. The measure was carried out in the name of conservation: to protect the forest from supposedly destructive indigenous land-use practices and to prevent white land speculators from obtaining private ownership.

The reserve boundary consisted of a cleared strip 3 meters wide, with a line of fast-growing exotic eucalyptus trees planted alongside. Although they furnished the labor for marking the border, the Kikuyu were not consulted about demarcation. On the contrary, they later complained about being deceived by officials about the purpose of the "road" they were ordered to make. As one Kikuyu group wrote in 1932, "[We] were not told, nor did we know, that the boundary lines were to be like walls of a cattle byre" (Kenya Land Commission 1934:96). Hundreds of Ndia and Gichugu families were evicted from foraging grounds, gardens, and homesteads. (A small tract called Nyagithuci was returned in the mid-1930s as belated compensation.)

The Forest Department took control over the area, restricting local use of the reserve through regulations and guards posted on the perimeter. The usurpation of Mount Kenya caused much tension between colonial authorities and the Africans, generating a long history of local resistance. Fear of colonial land grabbing permeated subsequent dealings about local forestry matters.

Demonstration Plots and Communal Woodlots

With the Kikuyu increasingly squeezed by colonial land usurpation the problem of deforestation intensified in their territory during the first years of colonial year. By World War I wood scarcity had reached crisis proportions in some areas. A report from Nyeri (including Ndia) in 1914 noted that the greater part of the district was "destitute of any timber supplies." In addition, fuelwood scarcity was "becoming yearly more and more serious." Two years later Nyeri was described by the provincial commissioner as on the verge of a fuel famine. The situation was not quite as bad in Gichugu, but the Embu district commissioner warned, "Care will have to be taken that [they are] not soon as denuded of timber as are other parts of Kikuyu country." The 1912 Native Authority Ordinance empowered chiefs to prohibit the destruction of trees. British officials saw this measure as insufficient for addressing the widespread need for trees near African settlements.

Nonforestry personnel in the colonial administration initiated the reafforestation by African households in Kirinyaga and environs. This situation reflected the political economy of early Kenyan colonialism. The Forest Department was chiefly concerned with running the government reserves, its duties limited by staff shortages and financial constraints. The ideology of "state forestry" (a combination of Western silviculture and industrial development) also led the department to concentrate on commercially oriented activities and policing functions (Castro 1995). The foresters provided some planting stock and advice to administrators engaged in African tree planting, but the department proved unable or unwilling to become directly involved.

In the early years of colonial rule the central government did not promote African agrarian advancement, because it was regarded as detrimental to the interests of white settlers (Stichter 1982). Some individual field administrators, such as Charles Dundas in Nyeri and John Ainsworth in Nyanza, disagreed with the policy (Mungeam 1966). They disdained the notion that their main task was to serve as a labor recruiter for European farmers. Dundas (1913:1) argued, for example, that colonial officers had a duty to promote improved cultivation. These men embarked on their own efforts to foster local development. Not surprisingly, their activities often included reafforestation.

Dundas became aware of the forestry situation in Kikuyuland while stationed near Nairobi. He summarized it succinctly in his memoirs: "There was constant warfare between the forest department and the natives, but the Kikuyu had no other source of timber for hut-building and firewood than the forest. For partial remedy we bethought ourselves of reafforestation in the native reserve" (Dundas 1955:66).

Transferred to Nyeri in 1913, Dundas decided to promote tree planting on a large scale. He knew that the Kikuyu grew muu (*Markhamia hildebrandtii*) and other shrubs in their *shambas* (cultivated plots). Dundas realized that such plantings occurred on a very small scale and for specific purposes. As mentioned previously, the notion of reforestation through tree planting was foreign to the Kikuyu. As Dundas recalled, "The idea of planting trees was altogether strange to the raw native [sic]—God planted trees. It had to be demonstrated that man could also do it" (1955:66–67). By establishing small woodlots in the countryside Dundas intended to show the possibilities and benefits of tree growing. Because of their agroforestry experience Dundas argued that the Kikuyu could be "easily induced" to cultivate trees.

His plan involved using influential local authorities to plant and maintain the woodlots: "On every tour I took with me numbers of tree seedlings and arranged a little planting ceremony. I myself planted the first tree, next the local chief or headmen and then each elder planted a tree which he bound himself to tend until it was well established" (Dundas 1955:67).

The trees were to be planted in 1-acre blocks, as it was regarded as easier to protect woodlots than trees scattered on farmland. The Forest Department was asked to provide enough trees the first year to establish 10 acres of woodlots.

Rapid turnovers in administrative staff were a common feature of colonial Kenya (Trench 1993), and Dundas left Nyeri in 1914 for another posting before the scheme could be implemented fully. Looking back decades later Dundas regarded his reafforestation effort as a successful first step. His memoirs recorded: "When I revisited Nyeri some thirty years later I was gratified to see many a tall grove dotted about the countryside" (Dundas 1955:67). Correspondence in the Embu District Archives reveals that some of Dundas's contemporaries disputed the success of his plan. In 1916 Edward Battiscombe, the chief conservator of forests, complained that the plantings had accomplished little. Battiscombe pointed out that Dundas's plans had not been followed up by succeeding administrators.

Battiscombe urged provincial officials to organize a formal multiyear scheme that could be followed despite the staff turnovers. However, colonial records in Embu reveal that high-ranking officials promoted communal tree plantations on a sporadic basis for local wood supply and demonstration purposes. For example, in 1920 the governor instructed African authorities to plant annually "a certain number of trees . . . suitable to the locality, raised under the supervision of the District Officers and the advice of the Forest Department."

Not all officials embraced the task of promoting communal woodlots. The Embu district commissioner complained in 1916 that his staff already had enough to do. He suggested that foresters take full responsibility for such work. Instead, when white agricultural officers were posted to Kikuyuland beginning in the 1920s, they were assigned responsibility for distributing tree seedlings. Many administrators still remained skeptical about the demonstration plots. This view was expressed concisely by the Embu district commissioner in 1924: "I do not think that merely planting a few copses of exotic trees is of any lasting benefit."

Indigenous Versus Exotic Species

A debate soon emerged about the type of trees to use in demonstration plots. Dundas wrote in a 1914 report that native trees and shrubs were either "excessively slow growing" or yielded "very inferior material." His views were shared by many foresters, including Sir David Hutchins, who had headed the Forest Department from 1907 to 1911 (Logie and Dyson 1962:17). Hutchins believed that the indigenous forest was largely incapable of regenerating itself. He advocated planting fast-growing, hardy, and commercially oriented exotic species, including trees from Australia, Central America, and India. Dundas used Australian eucalyptus in his woodlots. It was quick growing and produced good poles and firewood, with some varieties capable of multiple harvests through coppicing.

In contrast, Battiscombe, Hutchins's successor as conservator of forests, recommended using only indigenous trees. He argued in a 1916 memorandum that "the natives know them and . . . know the best species for specific purposes." Battiscombe warned that Africans were conservative and likely to have "very strong objectives to using any exotic trees either as firewood or as poles for building." Although such an approach might take longer, indigenous trees offered the greatest chance of success, Battiscombe argued.

In 1924 G. V. Maxwell, the chief native commissioner, recommended a twofold strategy for species selection based on intended end use: "For the quick production of firewood and poles the planting of Eucalyptus and Black Wattle (*Acacia mearnsii*) is generally indicated but where it is desired to plant for the production of timber and permanent shade and shelter indigenous trees should be selected and especially those of the district in which it is proposed to plant." In fact, most woodlots appeared to combine indigenous and exotic trees, although the latter were favored.

The question of what species to use proved to be complex. The Africans readily accepted some exotics. Although John Ainsworth, the field administrator in Nyanza, favored using indigenous trees, he noted in 1916 the widespread popularity of *Grevillea robusta*, an Australian species. Observation in contemporary Kirinyaga confirms *Grevillea robusta*'s continued success as an agroforestry species. Known locally as *mukema* and *mubariti*, it provides good timber and poles, as well as fodder from its leaves. The multi-use tree can be pollarded a number of times and can be intercropped with maize, beans, and other foodstuffs. And it reproduces through profuse seeding and quick germination. In short, *Grevillea robusta* fit easily into traditional agroforestry practices.

In contrast, eucalyptus (locally called *mubau*) became controversial. Two varieties were especially promoted: *Eucalyptus saligna* (blue gum) and *E. camaldulensis* (red gum). Households generally disliked eucalyptus because it strongly competed with food crops for water and nutrients. A contemporary Ndia villager reported: "People came to hate them, saying that they were consuming a lot of water in the gardens. We called them *minyua mai*, 'the ones that drink water,' because we believed that it takes all the water" from the ground.

The "tree planting" file in Embu District shows that British officials were divided over the usefulness of eucalyptus, at least for certain situations. As early as 1920 the provincial commissioner advised that it ought to be grown only near water. He also suggested that newly planted eucalyptus be intercropped with beans and maize no more than two years. A later official opposed growing eucalyptus on any arable land, suggesting that it be relegated to areas unsuited for cultivation. Despite mixed feelings, many households eventually adopted eucalyptus, growing it in small copses and along borders away from food crops.

Early Resistance to Communal Woodlots

The early reafforestation schemes involved communal or compulsory labor. The practice of compelling adult men and women to provide unpaid labor for public works originated in the first days of colonial rule (Clayton and Savage 1974). In 1916, as part of their communal labor obligations, people had to establish woodlots near chief camps and other sites. British officials assumed that Kikuyu households were willing to provide their land and labor "for the benefit of the members of their community." Although the Kikuyu were accustomed to informal labor exchanges and communal management of woodland, they had no traditions of forced labor and group tree planting.

The advent of large-scale worker migration intensified local concern about supplying labor for communal projects. The imposition of taxation had forced the Kikuyu to seek money in the White Highlands and urban areas. By the early 1920s perhaps one-third to one-half the men in Central Province were away seeking employment at any time (Stichter 1982). The responsibility for communal tasks often fell to women, who faced other demands for their labor for agricultural and household tasks. Their situation was recognized as early as 1920 by some Embu District officials, who advised that only hired labor should be used in communal tree planting. However, people were not paid regularly for such work until the 1930s.

Taking land for communal blocks of trees also provoked local concern. Officials assumed that they could easily obtain land for growing trees, given the scope and magnitude of wood shortages. On the few occasions when tenure was considered, the British presumed that the African rightholders would sacrifice their land for the common good. For example, H. R. Tate, the provincial commissioner in 1916, wrote, "[The land] need not be alienated from the owner but the usufruct must be vested in trustees pro bono publico." This fitted British perceptions of native land tenure: "It had always been assumed that among Africans there was no such thing as private land ownership; all land was believed to be held in community and allotted for cultivation by some tribal authority" (Dundas 1955:62).

Tate and others were more concerned about the location of the plots: they wanted woodlots established near easily supervised sites such as the homesteads of headmen or the overnight camps used by administrators. A few officials voiced reservations about converting farmland to woodlots. In 1916 the Embu district commissioner urged that tree planting take place only where "the owner of such land has sufficient additional land for cultivation."

Strong resistance to communal tree planting developed in Ndia, the most densely populated part of Kirinyaga. This opposition formed in response to a new campaign for establishing communal woodlots in the early 1920s. Several elderly Ndia informants recalled the colonial government's efforts to establish communal wattle plantations. An old man from Inoi location said, "People were ordered to grow it." Another remembered, "Wattle was not the will of the people. The government forced them to plant it." An elder from northern Kiine recalled that people in his area ignored the order to plant, because they disliked being told what to do. "They are big headed," he said. "If told to do something they don't [do it]. They take it easy." Wattle's early unpopularity also derived from the realization that it was "bad for crops." Wattle is a voracious surface feeder, heavily competing with other crops for water and nutrients (Tignor 1976:306). Unlike trees such as *Markhamia hildebrandtii* and *Grevillea robusta*, it cannot be interplanted with food crops. People tried a variety of methods to undermine the woodlots: absenteeism, poor planting techniques, and allowing livestock to feed on seedlings. Ironically, wattle later became a major cash crop in Kirinyaga.

Land and Deceit

Kikuyu concerns about land greatly increased in the 1920s, an era distinguished by rising racial consciousness and cultural nationalism (Rosberg and Nottingham 1970; Clough 1990). The vast amount of land appropriated by the colonialists fostered long-standing grievances throughout Kikuyuland. The Kikuyu feared further encroachment, an apprehension heightened by a 1921 Kenyan High Court decision regarding native land rights. The court ruled that the Crown Land Ordinance of 1915 and the annexation of Kenya as a colony in 1920 nullified aboriginal land rights (Bennett 1963:54). Although the colonial government officially acknowledged the existence of "native reserves" in 1926, the Africans legally remained "tenants at will of the Crown." Their land could be taken away if the government so desired.

Much of the resistance to government-sponsored tree planting was based on the belief that reafforestation might be a colonial ploy for acquiring their land. This perception was influenced by Kikuyu traditions regarding land tenure and tree planting. As mentioned earlier, one of the few reasons for planting trees in precolonial times was to demarcate property boundaries. Taking another's land by removing boundary trees, or planting new ones, had long been sources of disputes (Maxwell, Fazan, and Leakey 1929a:3). Tree planting itself conveyed long-term rights to use land. That the colonialists might use deceit to appropriate land seemed possible in the context of local experience. For many Kikuyu the eucalyptus trees along the perimeter of the Mount Kenya reserve served as reminders of an earlier deception.

By the mid-1920s intense resistance to compulsory planting and communal "blocks" had spread to other parts of Nyeri. By 1926 the district commissioner was writing in his annual report, "[Local authorities] showed themselves very averse to any schemes for reafforestation of their tribal lands except by the individuals who according to tribal land laws were considered entitled to the land or the cultivation thereof." People opposed not only the communal woodlots but also any tree planting by missionaries and schools. Intense resistance to colonial reafforestation had spread to Embu by the end of the decade.

Communal tree planting declined during the turbulent late 1920s and early 1930s. However, the practice continued through the end of the colonial era. Embu land usage bylaws passed in the late 1930s included the power to compel rightholders to plant trees on their land. Group tree planting was mainly used to protect watershed on ridges and hillsides. In contrast to the past, the Embu council or the Agriculture Department often paid these workers. Whatever their limitations, the early demonstration plots and woodlots introduced the notion of extensive tree planting to the Kikuyu. By the 1930s many households were ready to embrace the idea on their own farms.

Household Tree Planting and Decentralized Nurseries

The local native councils of South Nyeri and Embu proved instrumental in promoting tree planting by Kirinyaga households. The British government established the councils as part of its policy of indirect rule (see Kitching 1980; Maxon 1993). The councils were districtwide bodies with jurisdiction over a wide range of local matters among the African population, including land use. Limited taxation power gave them the ability to provide some public services. Each council was composed of appointed and elected members, with the district commissioner serving as its president. Kirinyaga was part of the South Nyeri council until 1933, when it was transferred to Embu.

Each council set up seed farms and tree nurseries to furnish planting stock to local families. These facilities were financed through local taxes, with technical support from agricultural and forestry officers. An experimental farm opened in Kerugoya in 1927, and two tree nurseries were operating before the end of the decade. Seedlings and seeds were available for multipurpose agroforestry species and exotic fruit trees. Agricultural reports indicated that offerings included several locally valued indigenous trees, particularly *Cordia abyssinica* (*muringa* for timber, beehives, and utensils), *Pygeum africanum* (*muiri*, or *mweria*, also prized for its timber and other uses), *Markhamia hildebrandtii*, and *Bridelia micanthra* (*mukoigo* for timber, poles and fuel). The main exotic agroforestry species were wattle, grevillea, cypress (*Cupressus lusitanica*, for timber, fuel, and as a living fence row), eucalyptus, and in the early years *Jacaranda acutifolia* (for poles). A number of exotic fruits were provided, including mango, papaya, avocado, lime, lemon, orange, and custard apple. The fruits were introduced to improve local diets, rather than for cash cropping. Indeed only wattle growing was pushed as a commercial undertaking. Trees were issued freely or for a small charge.

As early as 1929 the South Nyeri district commissioner reported that "tree nurseries are most popular with natives, unlike demonstration plots." Offering a range of low-cost trees to meet locally diverse needs was a major reason for the popularity. The most important factor had to do with power and incentive. Households decided about what to plant, how much, where to grow it, and so on. People who planted trees on their own property knew that they would be the beneficiaries, whereas the ownership rights to the communal woodlots were murky.

The fears of colonial land grabbing affected the early nursery operations. Local distrust actually altered the distribution of trees and seedlings in Embu. The Forest Department nursery used to issue seedlings at no charge to anyone interested in planting them. In the late 1920s a rumor spread to the Embu Local Native Council that the government would attempt to claim these trees and the land on which they were planted. The Embu council denied the rumors, but it decided that the people ought to pay for all seedlings in the future. Within a year the council set up its own nursery and resumed the practice of distributing free seedlings, but new rumors brought it to a halt. Once again people feared that such trees might give the government some right to the land, and the council instituted fees for all planting stock. Ironically, British officials approved of ending free issues, contending that tree survival would increase if people had to pay for the seedlings.

Wattle growing expanded spectacularly during the 1930s. What is often overlooked is the important role of the council nurseries from the 1930s to the 1950s in providing other planting stock for local agroforestry. Kirinyaga and Embu had five nurseries by 1941. District annual reports indicate that British administrators regarded the nurseries as fairly successful in meeting local demand for trees. Encouraged by the interest, the Embu council added five more nurseries by 1945. That year the council issued 92,000 tree seedlings, almost two-fifths of them indigenous species. Local households received nearly 120,000 seedlings in 1947.

Some provincial agricultural officials argued that figures about seedlings issued were misleading because the survival rates were unknown. They claimed that people ignored the seedlings after transplanting, causing a high mortality. In contrast, district administrators were impressed by the results of local farm forestry. They felt that Kirinyaga and Embu's experience differed from other Kikuyu areas. The district commissioner's 1946 annual report notes that "the natives of Embu [including Kirinyaga] are much more tree conscious than their neighbours. Parts of the district have rather too many trees for good agriculture, yet there is a strong dislike of removing any even in these areas. The number of exotic trees to be seen, too, is much greater than in neighbouring districts." The next year he reported to provincial officials, "There is a vigorous demand for more nurseries, which are being supplied."

The traditional landholding system based on land rights vested in descent groups (instead of private owners) was not a barrier to tree planting. Households had usufruct rights to their clan lands. However, tree planting generated much litigation. Conflict revolved around the issue of trees planted by temporary land occupants (*ahoi*), including whether they maintained any rights to such trees after they left the property. As early as 1924 local authorities in the South Nyeri Advisory Council drafted rules governing the planting of trees on "hired" or borrowed land. Ahoi were supposed to obtain permission from the original rightholder before planting. They had exclusive rights to such trees, plus any regenerative growth through coppicing. If the rightholder wanted to reclaim the land, the local elders were to adjudicate each rightholder's share of the trees. Rights to any subsequent regrowth were to revert to the landowner. Despite this policy, arguments about tree ownership surfaced repeatedly in Kirinyaga, especially concerning wattle and its reestablishment through profuse seeding (as many as three or four crops can be grown in succession). The Embu council dealt with the issue a number of times between the 1930s and 1950s. Some colonial observers argued that planting trees stimulated individualization of land tenure (see Maxwell, Fazan, and Leakey 1929b; Humphrey, Lambert, and Harris 1945). Certainly, it was one of many forces contributing to that trend (see Kershaw 1972).

Wattle: A Success Story

The British planted *Acacia mearnsii* on railroad fuelwood plantations, but its wood burned too quickly to be used economically by steam locomotives (Tignor 1976:296). Another commercial value was soon recognized for this Australian tree: its bark was rich in tannins, which could be extracted for use in processing leather (see figure 5.3). Kenya already was supplying Europe with tannin-laden barks from mangrove trees on the coast. However, unrestricted cutting was rapidly reducing the supply of preferred mangrove species "to the vanishing point" (*East African Standard* 1925:160). Therefore the colonial administration promoted wattle growing among white settlers in the highlands as an alternative source of bark for European tanners. A collapse in wattle bark prices at about the time of World War I caused many colonists to abandon the crop. They preferred to grow more lucrative commodities such as wheat, coffee, and tea (Africans were denied the right to plant the latter two crops).

The British introduced wattle in Kikuyuland because it was quick growing and furnished good poles and firewood. In 1925 the South Nyeri district commissioner wrote, "The natives have sufficient wattle plantation now to supply their fuel requirements and attention can be devoted to tim-

Figure 5.3   These stacks of bark from wattle trees (*Acacia mearnsii*) await pickup along a road in northern Ndia. A factory will extract from the bark tannins for tanning leather.

ber trees." What this official did not foresee was the rapid adoption of the tree by local households. Kiambu and Fort Hall families were already planting it in large numbers, and it spread eventually to the rest of the province. African smallholders in Central Province were growing more than 40,000 hectares of wattle by 1936.

Several factors contributed to the rise of Kikuyu wattle production (see Kitching 1980). Forestal, a British manufacturer of tannin extract, entered the Kenyan market in the early 1930s in search of inexpensive raw materials. Colonial officials believed that Kikuyu households using unpaid family labor could provide bark at a low cost. The administration promoted it as the major cash crop for Kikuyu areas in altitudes above 1,600 meters. White settlers did not oppose African wattle cultivation, as they did attempts to allow Africans to grow coffee or tea. In addition, agricultural officers initially regarded wattle as an almost ideal tree for maintaining soil quality: it was fast growing, easily established on steep slopes, and leguminous, thus fixing nitrogen in the soil. The tree also regenerated easily through profuse seedlings.

Agricultural officials promoted the crop by holding public meetings, operating demonstration plots, and distributing seeds (Maher 1938:57). They also directly supervised planting in some areas. This was done to ensure that wattle would be grown along the contour of ridges. Households frequently viewed the department's activities as an unwarranted intrusion. Still, wattle growing gained momentum among local households. Officials estimated that Kirinyaga and Embu had 2,400 hectares of wattle by the end of 1937 (Maher 1938:57–58). That year alone 422 hectares were planted on 1,691 holdings in Ndia, and 301 hectares were established on 996 holdings in Gichugu.

Wattle's rapid spread underscored the value of multipurpose trees. People were frequently willing to grow it, even when they were suspicious of the colonial government, because of the tree's combination of commercial and subsistence values. Elders recalled in interviews that the ability to generate cash income was especially crucial. An Ndia elder from Thaita sublocation stated, "It was the first time I realized the soil made money." Another Ndia informant (from Nduine) recalled the colonial wattle trade in Kirinyaga:

> Wattle trees were liked by the colonial rulers because of the bark from which they made leather-working fluid. Wattle bark was also used for medicine, like quinine. [After removing the bark], the trees were used for building and charcoal. People were selling it to people in Nairobi and to those who had hotels [small local inns]. We didn't know the use of charcoal because we had lots of firewood. We didn't need charcoal. People used bicycles, oxcarts, [and their own] backs to transport charcoal to hotels, or to where a lorry would pick it up.

Of course poles and firewood could also be used on homesteads. Gavin Kitching (1980) reports that wattle's total output per acre greatly surpassed the yields from available alternative crops.

Contributing to wattle's popularity was its low labor requirement. Except for thinning and harvesting, the crop needed no labor after planting. This was attractive for families engaged in wage-labor migration or such nonfarm pursuits as trading or government employment. The returns per unit of labor time were probably much higher for wattle than for any other crop available to Kikuyu cultivators at that time (Kitching 1980). It was an ideal crop for households experiencing labor scarcity.

In the 1940s agricultural officers increasingly complained about wattle husbandry standards. They contended that crowded and neglected stands produced bark of lower quality (Humphrey, Lambert, and Harris 1945:25). They also viewed poorly tended wattle as an agent of erosion. Thick stands choked off the undergrowth, leaving the bare ground more susceptible to erosion when the tree cover was removed at harvesting.

Wattle production in Kirinyaga and Embu flourished during the late 1930s and 1940s. District reports show that bark sales increased from 36.6 tons in 1935 to almost 3,000 tons in 1949. Kikuyu producers initially brought their dried or green bark to nearby shops or markets, where Asian or European merchants purchased it. The government centralized wattle trading in the mid-1930s, restricting its sale to certain sites, so that it could inspect and tax the bark. By the early 1940s the British had introduced a quota system for bark production for each district. Kirinyaga and Embu producers complained that other districts unfairly received higher quotas. Discontent also arose in 1948 over inequity in the allocation of local wattle bark sale permits. Permits had been distributed in secret, favoring influential and prosperous producers. The Embu council decided that the allocation of bark permits must take place at public meetings.

Ironically, some colonial officials feared that Kikuyu farm forestry was too successful. It was claimed that seedlings obtained from nurseries were "not being used for reafforestation in the true meaning of the word. Indeed many of [the trees] may be usurping land that should be under food crops, whilst the land that is crying out for trees remains untouched" (Humphrey, Lambert, and Harris 1945:25).

E. B. Hoskins, the chief native commissioner, claimed in his 1945 annual report that tree growing as a commercial activity threatened to become a public embarrassment because such crops were planted "on agricultural land on which there is already great population pressures." Embu administrators in the late 1940s were not as critical of farm forestry, but they acknowledged that tree planting was unevenly distributed. The fertile and moist upland ridges had plenty of trees, whereas much of the drier and less fertile southern lowlands remained treeless. Communal tree planting continued to be used in such places.

The Decline of Colonial Farm Forestry

The wattle industry collapsed as rapidly and dramatically as it had originally developed two decades earlier. Several factors led to the fall. The growth of rubber and plastic substitutes for leather depressed the international demand for tannin extract, causing bark prices to fall in the late 1940s and 1950s. Meanwhile, the governor declared a state of emergency in response to the anticolonial Mau Mau movement in October 1952. In its wake British authorities ordered large tracts of wattle to be cleared to remove cover for the insurgents. The forced relocation of Kirinyaga's entire population to guarded villages during 1954 also resulted in much deforestation, as people sought building materials (Castro and Ettenger 1994).

Large-scale wattle clearing flooded the marketplace with bark. District annual reports show that sales peaked in 1955 at more than 4,000 tons of bark, then declined to less than 1,000 tons by 1957. Prices were so low by 1958 that many growers no longer bothered to strip the bark from their harvested trees. The opening of coffee and tea production to Kikuyu farmers in the 1950s ultimately doomed wattle, because these crops offered greater opportunity to make money. In 1962 officials reported that wattle covered only about 400 hectares in Kirinyaga and Embu, compared to an estimated 4,000 hectares of coffee. Not until the 1980s would wattle, through the sale of its charcoal, regain some of its former popularity.

The disruption of government services during the emergency eventually had a strong effect on the local tree nurseries. Embu District had twenty-nine nurseries in the early 1950s. The level of maintenance slipped greatly because the council lacked staff and resources. By the late 1950s agricultural officials were absorbed in the process of land tenure reform, consolidating and privatizing holdings. This process required quick-set hedges instead of trees. A 1957 report by the forester in charge of the Embu and Fort Hall council nurseries observed that the "nurseries . . . are generally disappointing. There are too many, they are mostly too small and receive quite inadequate supervision. The result tends to be poor, expensive stock raised to the wrong sizes at the wrong seasons of the year."

Most nurseries soon closed. By 1961 only two nurseries operated in Kirinyaga (Njukiine and Kamweti), furnishing fewer than eight thousand seedlings to the public—a fraction of what had been planted two decades earlier by households. Not until the late 1970s would farm forestry regain its earlier momentum.

Many positive lessons can be learned from the five decades of colonial farm forestry interventions in Kirinyaga. A major theme is the importance of power and incentive in tree planting. Kikuyu households transformed the countryside through their voluntary growing of wattle and other species in a manner that could not have been achieved through communal or compulsory tree planting. A key difference is that households controlled decision making about what to plant, how much, where, when, and so on. People who planted trees on their own land also knew that they would be the beneficiary of their own labor, in contrast to the murky ownership rights to communal woodlots.

Tree planting was made attractive to the households in the first place through decentralized nurseries offering low-cost multipurpose trees, especially those such as wattle that combined subsistence and commercial uses. New species that fit into preexisting local agroforestry practices, as well as household land and labor constraints, are more readily accepted. Traditional landholding systems based on descent-group ownership are not necessarily a barrier to tree planting by individual households. These lessons have been learned or relearned in many contemporary projects, including Gerald Murray's celebrated case study (1987) of agroforestry in Haiti (also see FAO 1985).

There are also negative lessons. Tree planting can be counterproductive when it is imposed without consideration for whose land is being used, what species are being grown, and who is responsible for their maintenance.

Communal tree planting is especially difficult, given the lack of individual incentive, but experience elsewhere demonstrates that communal tree planting can work where a favorable sociocultural foundation exists (see FAO 1985). The collapse of farm forestry during the Mau Mau uprising also underscores the disruption of local production systems by civil conflicts.

In general, farm forestry works best when it is based on the needs and capabilities of rural people. They often have knowledge and land-use practices regarding trees that can serve as the basis for reafforestation. By definition in social forestry, households must be integral members of the planning team, because they are the land-use managers. Thus foresters and others need to continue their efforts to break down the institutional and professional barriers that have long separated them from rural communities.

Acknowledgments   Support for this research was obtained from the National Science Foundation, the University of California, the Intercultural Studies Foundation, and the Appleby-Mosher Fund of Syracuse University. I am indebted to the Kenya National Archives in Nairobi and the county council and the district commissioner's office in Embu. I alone am responsible for the views expressed in this essay. My involvement in community forestry owes much to David Brokensha and Bernard Riley. I would like to dedicate this essay to Bernard Riley on the occasion of his retirement from the University of California, Santa Barbara.

Note
*References to specific colonial documents were eliminated at the publisher's request. Those interested in detailed citations are invited to write to me at Syracuse University, Maxwell School of Citizenship and Public Affairs, Department of Anthropology, 209 Maxwell Hall, Syracuse, N.Y. 13244-1090.

References Cited

Ambler, Charles H. 1988. *Kenyan Communities in the Age of Imperialism.* New Haven, Conn.: Yale University Press.
Anderson, Robert S. and Walter Huber 1988. *The Hour of the Fox.* Seattle: University of Washington Press.
Arnold, Mike. 1992. *Community Forestry.* Rome: Food and Agriculture Organization.
Bennett, George. 1963. *Kenya.* Nairobi: Oxford University Press.
Castro, Alfonso Peter. 1983. *Household Energy Use and Tree Planting in Kirinyaga.* Nairobi: University of Nairobi, Institute of Development Studies, Working Paper No. 397.

———. 1988. Southern Mount Kenya and colonial forest conflicts. In *World Deforestation in the Twentieth Century*, John Richards and Richard Tucker, eds., pp. 33–55. Durham, N.C.: Duke University Press.

———. 1990. Sacred groves and social change in Kirinyaga, Kenya. In *Social Change and Applied Anthropology*, Miriam Chaiken and Anne Fleuret, eds., pp. 277–289. Boulder, Colo.: Westview.

———. 1991a. Indigenous Kikuyu agroforestry: A case study of Kirinyaga, Kenya. *Human Ecology* 19 (1): 1–18.

———. 1991b. Njukiine Forest: Transformation of a common-property resource. *Forest and Conservation History* 35 (4): 160–168.

———. 1991c. The southern Mount Kenya forest since independence: A social analysis of resource competition. *World Development* 19 (12): 1695–1704.

———. 1992. Social forestry: A cross-cultural analysis. In *Ecosystem Rehabilitation*, Vol. 1, Mohan Wali, ed., pp. 63–78. The Hague: SPB Academic Publishing.

———. 1995. *Facing Kirinyaga*. London: Intermediate Technology Publications.

Castro, Alfonso Peter and Kreg Ettenger. 1994. Counterinsurgency and socioeconomic change: The Mau Mau War in Kirinyaga, Kenya. *Research in Economic Anthropology* 15: 63–101.

Cernea, Michael. 1992. A sociological framework: Policy, environment, and the social actors for tree planting. In *Managing the World's Forests*, Narendra Sharma, ed., pp. 301–335. Dubuque, Iowa: Kendall/Hunt.

Clayton, Anthony and Donald C. Savage. 1974. *Government and Labor in Kenya, 1895–1963*. London: Cass.

Clough, M. 1990. *Fighting Two Sides*. Niwot: University of Colorado Press.

Colchester, Marcus and Larry Lohmann, eds. 1993. *The Struggle for Land and the Fate of the Forests*. London: Zed.

Davison, Jean. 1989. *Voices from Mutira*. Boulder, Colo.: Rienner.

Dundas, Sir Charles. 1913. *Memorandum on the Subject of Native Cultivation in Nyeri District*. Nairobi: Kenya National Archives.

———. 1955. *African Crossroads*. London: Macmillan.

East African Standard. 1925. *Kenya Manual*. Nairobi: *East African Standard*.

Food and Agriculture Organization (FAO). 1985. *Tree Growing by Rural People*. Rome: Food and Agriculture Organization.

Fortmann, Louise P. and Sally K. Fairfax. 1989 American forestry professionalism in the Third World. *Economic and Political Weekly*, August 12, pp. 1839–1844.

Gedge, Ernest 1982. A recent exploration of the River Tana to Mount Kenya. *Geographical Society Proceedings* 14: 513–533.

Guha, Ramachandra. 1990. *The Unquiet Woods*. Berkeley: University of California Press.

Hecht, Susanna and Alexander Cockburn. 1989. *The Fate of the Forest*. London: Verso.

Humphrey, N., H. E. Lambert, and P. W. Harris. 1945. *The Kikuyu Lands*. Nairobi: Government Printers.

Kenya Land Commission. 1934. *Kenya Land Commission, Evidence*, Vol. 1. Nairobi: Government Printers.

Kershaw, Gretha 1972. The Land Is the People. Ph.D. diss., University of Chicago.

Kitching, Gavin. 1980. *Class and Economic Change in Kenya.* New Haven, Conn.: Yale University Press.

Lambert, H. E. 1949. *The Systems of Land Tenure in the Kikuyu Land Unit.* Cape Town, South Africa: School of African Studies.

Leakey, Louis S. B. 1977. *The Southern Kikuyu Before 1903,* Vol. 1. New York: Academic Press.

Logie, J. and W. Dyson. 1962. *Forestry in Kenya.* Nairobi: Government Printers.

Lonsdale, John. 1992a. The conquest state in Kenya. In *Unhappy Valley,* Bruce Berman and John Lonsdale, eds., pp. 13–44. London: Currey.

——. 1992b. The moral economy of Mau Mau: Wealth, poverty, and civic virtue in Kikuyu political thought. In *Unhappy Valley,* Bruce Berman and John Lonsdale, eds., pp. 315–504. London: Currey.

Maher, Colin. 1938. *Soil Erosion and Land Utilization in the Embu Reserve, Part I.* Nairobi: Kenya Colony, Department of Agriculture.

Martin, Claude. 1991. *The Rainforests of West Africa.* Berlin: Birkhauser Verlag.

Maxon, Robert. 1993. *Struggle for Kenya.* Rutherford, N.J.: Fairleigh Dickinson University Press.

Maxwell, G. V., S. H. Fazan, and L. S. B. Leakey. 1929a. Record of evidence, *baraza* of South Nyeri District held at Karatina, September 25. London: Public Records Office.

——. 1929b. *Native Land Tenure in Kikuyu Province.* Nairobi: Government Printers.

Middleton, John and Greet Kershaw. 1965. *The Kikuyu and Kamba.* London: International African Institute.

Ministry of Finance and Planning. 1984. *Kirinyaga District, Development Plan 1984-88.* Nairobi: Republic of Kenya, Ministry of Finance and Planning.

Mungeam, G. 1966. *British Rule in Kenya, 1895–1912.* Oxford, England: Clarendon.

Muriuki, Geoffrey. 1974. *A History of the Kikuyu, 1500–1900.* Nairobi: Oxford University Press.

Murray, Gerald F. 1987. The domestication of wood in Haiti: A case study in applied evolution. In *Anthropological Praxis,* Robert M. Wulff and Shirley J. Fiske, eds., pp. 223–240. Boulder, Colo.: Westview.

Peluso, Nancy Lee. 1992. *Rich Forests, Poor People.* Berkeley: University of California Press.

Rosberg, Carl G. and John Nottingham. 1970. *The Myth of "Mau Mau."* New York: Meridian Books.

Sharma, Narendra, ed. 1992. *Managing the World's Forests.* Dubuque. Iowa: Kendall/Hunt.

Stichter, Susan. 1982. *Migrant Labour in Kenya.* London: Longman.

Stigand, Chauncey H. 1913. *In the Land of Zinj.* London: Constable.

Tignor, Robert. 1976. *The Colonial Transformation of Kenya.* Princeton: Princeton University Press.

Trench, Chenevix. 1993. *Men Who Ruled Kenya.* London: Radcliffe Press.

IV

Development as Degradation

Part 4, Development as Degradation, continues the anthropological analysis of the political economy of forests in the context of environmental history but with a focus on recent decades. Here too the analyses reveal that national governments can be important causal factors in deforestation, whereas local communities, although they have some effect on the forests, are more likely to be conservers of the forest, because it is usually in their own interest to use the natural resource sustainably.

However, in some situations the locals are forced to degrade or even destroy the forests simply to get food and other necessities for immediate survival, even though they may have viable ecological knowledge, practices, and values from their traditional culture. The authors describe how governments encourage deforestation through economic development programs and financial incentives, how gross economic inequities in the national society, such as in the distribution of land ownership, contribute to deforestation, and how the policy and assistance programs of foreign governments, economic development agencies, and international conservation organizations can also actually contribute to deforestation.

The Amazon encompasses about half the tropical forest in the world, and about half the Amazon is in Brazil. In chapter 6 Emilio Moran surveys the causes and consequences of and solutions to deforestation in Brazil and provides some specific recommendations for policy interventions. He notes that deforestation in Brazil is fairly recent in most of the Amazon, having started mainly in the 1970s, whereas deforestation in Africa and Asia started after World War II. In Brazil deforestation is caused not so much by small farmers but by economic development activities that include cattle ranching, hydroelectric dams, roads, mining, and logging. International agencies such as the World Bank—and financial incentives from the Brazilian government—have encouraged most of these development activities. Moran also points out that the institutional capacity of the government to monitor and police compliance with conservation is almost nonexistent. He predicts an increase in resource and land competition with ensuing violent conflict, migration to the frontier, diseases such as malaria, socioeconomic inequity, and long-term impoverishment.

In chapter 7 Eduardo Bedoya and Lorien Klein present a case study of the world's largest area of coca cultivation, the upper Huallaga Valley in Peru, where a complex combination of factors has contributed to deforestation. They note that whereas some 64% of Peru is forest, resources from the forest accounted for less than 1% of the country's gross national product (GNP) early in this century. To tap some of this wealth, relieve socioeconomic problems elsewhere, and enhance national security the government constructed roads into this area of the Amazon and encouraged massive migration of colonists from the highlands in the 1940s; the Inter-American Development Bank provided most of the financing. Since the 1970s farmers have concentrated land, labor, and capital in coca production, because it is three to four times more profitable than legal crops. By the 1980s the rate of deforestation had increased threefold as a result of government economic policies, regional political relations, the government drug-eradication program in conjunction with the United States, and the Shining Path guerrilla movement. The authors suggest that the only solution to deforestation in this region is to resolve the national economic crisis, reduce coca production, and control the Shining Path. They conclude that shifting cultivation is not simply a technique but a factor that must be analyzed in terms of the political economy, including the relations of production and power.

In chapter 8 Susan Stonich and Billie DeWalt examine the ecology of the social, economic, and political factors of deforestation as a social process in Honduras, emphasizing the interactions of these factors. They argue that development policies create extremes of wealth and poverty that contribute to deforestation. Poverty and inequitable land distribution lead poor rural farmers to convert forest to farms to satisfy their basic needs for survival. Although these farmers have a knowledge of forest ecology and sustainable practices as well as appropriate environmental values, the necessity to meet short-term subsistence needs overrides all other considerations. Their nutritional and health status is nevertheless deteriorating. Eventually, some resort to mangrove conversion for salt production and shrimp farming, whereas others migrate to the slums of cities as environmental refugees. In part this is a response to demands from the United States, the major export market for Honduras, and the U.S. shift in dietary habits from beef to fish, shellfish, and other animal protein. The deforestation crisis in Honduras also reflects the failure of the national government's policy for the nationalizing and managing the forests.

To some degree Honduras is representative of other countries in Central America, but there are differences. In chapter 9 James Vandemeer provides an insightful comparison of two other countries in Central America—Costa Rica and Nicaragua. In particular, he contrasts the different effects on deforestation and forest conservation of the U.S. government during the Reagan era in its cold war policies toward Costa Rica and Nicaragua. Costa Rica emphasized the proximate causes of and solutions to deforestation by developing a system of protected areas with assistance from foreign governments and international conservation organizations, especially from the United States; it nonetheless had an alarmingly high rate of deforestation. Nicaragua, which had a lower rate of deforestation, emphasized the ultimate causes of and solutions to deforestation in terms of social justice and agrarian reform with more equitable land distribution. In Costa Rica the result is islands of forests surrounded by larger areas of modern agriculture with extensive use of chemical fertilizer and pesticides and the resultant contamination of water, soil, biota, and people. In Nicaragua, in sharp contrast, the landscape ecology is evolving toward an integrated mosaic of managed plots for farming and logging, which are small enough to reflect the disturbance dynamics of natural forest. Vandemeer, a biologist, concludes that traditional conservation strategies are insufficient and that saving the forests requires greater attention to fundamental social, economic, and political concerns. This is more testimony to the relevance of anthropology and other social sciences in understanding and resolving the deforestation crisis.

In chapter 10 David Wilkie assesses the effects of logging on local foragers and farmers in the forests of the Congo basin. Although the government has advertised logging as yielding long-term employment and social services, actually it is only a short-term industry. Logging concessions are used for only about fifteen to twenty years until selective logging depletes the desired species. Thereafter the local economy collapses and other problems arise. The sale of wild game and fish is the other main source of cash income for the people of the area, but this is depleting some faunal species. Wilkie asserts that neither logging nor commercial hunting and fishing are sustainable as practiced today. He sees land rights as a crucial factor in improving this situation, but government policy and its management of forests have thus far negated this option.

6

Deforestation in the Brazilian Amazon

Emilio F. Moran

Large-scale deforestation in the Brazilian Amazon began with the decision to relocate the capital and to construct the Belém-Brasília Highway (see figure 6.1). Brasília was built so that the capital would be in a more central location that would encourage Brazilians to look away from the coastline that they had been hugging for 350 years and begin to use the vast interior of the country. The highway, construction of which began in 1958, was the first of a series of highways to be built to "integrate" the northern and western states with the rest of the country. Shipping the products of the Amazon to Europe and the United States had been easier than shipping them to southern Brazil before these north-south roads were built (Mahar 1979).

Settling of land along the Belém-Brasília was slow at first, and the road cut through a broad array of vegetation, only a small part of which was tropical moist and rain forests. Most vegetation was savannas, scrub forests, and tropical deciduous forests. In the first twenty years more than two million people settled along this dirt road, which was paved only in 1973. As this essay will show, it was land along this road that first attracted attention to the destruction of forest, especially through low-quality cattle ranches (Hecht 1980). The cattle population increased from near zero to more than five million in the same twenty-year period (Mahar 1988:12). Even feeder roads near the Belém-Brasília Highway were quickly occupied and vast areas deforested. Along a 47,000-square-kilometer stretch of the PA-150 road, the amount of cleared area jumped from 300 square kilometers in 1972 to 1,700 in 1977 to 8,200 in 1985; the leading cause was conversion from forest to pasture (Mahar 1988:13, 14).

Road building became even more important with the announcement in

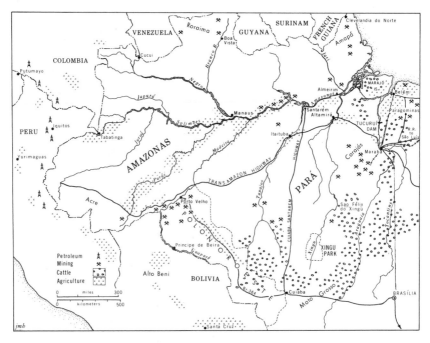

Figure 6.1   Current Development Areas in the Amazon

1971 by President Emílio Médici of the Program of National Integration (PIN), which would build north-south (Cuiabá-Santarém) and east-west (Transamazon) highways connecting locales in the the Amazon basin and the basin with the southern part of the country. Because settlement of the Belém-Brasília area was seen as "disorderly," the military governments that ruled Brazil after 1964 planned a "directed" and highly orderly occupation of the land alongside the highways with settlers from the northeast and from southern Brazil. Their goal was to move 100,000 families in the first five years and to use a hierarchical network of settlements and service communities to serve them (Moran 1981).

    The planned Northern Perimeter Road would run along the borders of Brazil. This road was more clearly geopolitical and military in nature and was intended to match the Marginal Forest Highway (Carretera Marginal de la Selva) built by the Andean countries to encourage movement of excess population from the Andean highlands to the lowlands of the Amazon (see figure 6.2). Because the 1973 oil crisis hit Brazil especially hard, only a small segment of Brazil's Perimeter Road was built in the 1970s. Brazil at the time imported nearly 80% of its oil needs. In the late 1980s the military again began to insist on the need for this road. The area

Andean Highways and Amazon Colonization

Figure 6.2 Andean Highways and Amazon Colonization

is important because it has gold, uranium, and other strategic mineral deposits—and because it cuts through Yanomamo territory. The oil crisis also affected the settlement and road building plans of the Transamazon and Cuiabá-Santarém highways. After 1973 the costs of transportation changed so radically that construction of the planned communities nearly halted, and only the main trunk of the highway was built, not the service roads to the farms. This left most farmers stranded on their landholdings, unable to get their rice, corn, and other produce to market (Moran 1976, 1981).

The changed costs of occupying the Amazon with small farmers and a change in presidents led to a policy shift in 1974, barely three years after the inception of PIN. President Ernesto Geisel announced that large-scale entrepreneurs would be more effective at developing the Amazon, and the government began to parcel out land in large units to individuals and corporations. Instead of the 100,000 families foreseen by the Transamazon Settlement Scheme in three years, only 6,000 had arrived; small farmers accounted for less than 4% of total deforestation in the 1970s (Browder 1988).

Giving priority to large-scale operators was not a new idea. In 1966 a plan to encourage Brazilians to occupy the Amazon created the Amazon Development Agency (SUDAM) and the Amazon Development Bank (BASA) through which individuals or companies could invest in projects within the Amazon region. It is important to note that in Brazil the "Amazon region" is not isomorphic with the drainage basin. Rather, it is legally defined for the purposes of development (i.e., the Legal Amazon, consisting of 5.4 million square kilometers versus 3.6 million of forests in the basin) and includes large areas of savannas and other kinds of vegetation and even areas outside the river drainage system. Individuals and companies that worked through SUDAM and BASA could invest in approved projects as much as 50% of their personal or corporate income tax liability. Participants thus paid no tax to the federal government and received three cruzeiros for every one they invested. Moreover they were able to keep all four dollars and the capital gain—tax free. Such incentives were too great to pass up. Most deforestation in the southeastern portion of the Amazon basin is traceable to this policy (Hecht 1980; Fearnside 1987a,b).

Most projects that SUDAM approved were extensive cattle ranches. Forest was converted to pasture at a rate of approximately 8,000 to 10,000 square kilometers per year in the 1970s (Mahar 1988:8), increasing to 35,000 square kilometers in the late 1980s (Fearnside 1989). By late 1985 SUDAM had approved about 950 projects, of which 631 were cattle ranches (García Vásquez and Yokomizo 1986:51 in Mahar 1988:15). Today cattle ranches cover at least 8.4 million hectares and average 24,000 hectares each (Mahar 1988:16); some are as large as 560,000 hectares. These ranches employ few people, averaging 1 cowboy (vaqueiro) for every 300 hectares.

A recent simulation of a typical 20,000-hectare ranch receiving a 75%

subsidy demonstrates that livestock activities are profitable only when they receive the full array of tax benefits. Without them they are not profitable and can achieve positive internal rates of return only through overgrazing. Although overgrazing destroys their long-term viability, the incentives to convert forest to pasture mean that ranchers clear new forested areas instead of investing in recovery of areas already cleared. Clearly, deforestation rates would have been much lower without the subsidies. In a study of a particular small farmer settlement project (Moran 1987) the correlation between bank subsidy and deforestation rate was 0.74 ($p = 0.0001$). It is probably higher still for cattle ranches.

After it changed its policy toward small-scale owners, the government diverted migrant flows to Rondônia. Construction in 1968 of the Cuiabá–Pôrto Velho Road, which happened to be cut through some of the better soils in the region— made it easier to settle this area. Rondônia is today the state with the most deforested land (24%). The average area deforested each year in the 1980s equaled the total area deforested before 1980 (Mahar 1988:34). Deforestation in Rondônia is more clearly related to the increased pace of in-migration, which averaged 160,000 people per year for the period 1984 to 1988, as compared with 65,000 in the period of 1980 to 1983, and 28,500 from 1968 to 1978 (the total population was 70,000 before 1968). Efforts to promote tree crops as a more environmentally sound way to sustain land use have faltered. Favorable treatment of cattle ranching by banks and SUDAM has led to an increase in the total amount of land in pasture—25.6% of the 1985 total—whereas land in perennial crops has remained stagnant at about 3.5% of the total (see table 6.1).

Table 6.1
Agricultural Land Use in Rondônia, 1970–85 (in km²)

Crops

Year	Annual[1]	Perennial	Pasture	Forest[2]	Total[3]
1970	323.7	127.2	410.1	15,031.1	15,892.1
1975	1,503.9	457.6	1,645.2	26,681.4	30,288.1
1980	2,425.8	1,701.8	5,101.8	41,461.1	50,690.5
1985	3,153.3	2,238.0	15,611.5[4]	39,903.7	60,906.6

1. Includes fallow land
2. Includes natural pastures
3. Area under farms at time of census; includes land unsuitable
   for agricultural use
4. Estimated

Source: Estimates made by Mahar (1988:35)

Land speculation is largely responsible for the loss of forests of Rondônia today. Because cleared forest is considered land that has been "improved," and whoever clears it gains the right to sell "the improvements," a brisk land market has developed. Average land prices have skyrocketed since 1983, leading to a frenzy of deforestation and real estate speculation. It is possible to net the equivalent of U.S.$9,000 from clearing 14 hectares of forest, planting pasture and a few crops for two years, and selling the "improvements" to a new settler (Mahar 1988:38). This is more than four times what a farm laborer can hope to make under ideal conditions. Because the government rarely collects the 25% capital gains tax on land sales,the incentive to engage in this sort of activity is strong. If 14 hectares can net a small operator that kind of return in two years, the attractiveness to average ranchers—who hold 10,000 to 20,000 hectares is obvious.

The two other sources of deforestation in the Brazilian Amazon are mining and timber activities. Mining activities do not seem to have had a major effect on the total area of forest cleared, although it is largely responsible for the exponential rise of malaria in the basin. Brazil today accounts for 1 million of the 5 million official cases of malaria worldwide, and this is probably underreported. All reports from mining areas suggest that nearly 100% of the cases are confirmed to be malaria. Malaria affects not only miners but farmers and other inhabitants. One recent large-scale mining project has begun to have a potentially devastating effect. The government offered tax benefits for the production of pig iron in the Great Carajás region, and manufacturers designed plants that could run on locally produced charcoal. By the mid-1980s the demand from the approved projects already was requiring 1.1 million metric tons of charcoal annually, a figure that was expected to double when SUDAM approved the projects it was evaluating (Fearnside 1987b). Hydroelectric projects have flooded extensive forested areas and constitute a constant threat.

The importance of timber exploitation in deforestation began to be notable only in the 1980s. Figure 6.3 shows the increasingly common sight of logging throughout previously settled parts of the Amazon. The most recent statistics available show that four of the six states in the region depend on wood products for more than 25% of their industrial production (Browder 1986:65). In Rondônia and Roraima they account for 60% of the output. The number of licensed mills increased more than eightfold since 1965, and the annual production per mill doubled during the same period. In 1984 the Amazon region of Brazil accounted for 43.6% of national roundwood production, as compared with only 14.3% ten years earlier (Browder 1988:249). The declining contribution of Asian forests to the world's demand for tropical wood products will lead to further increases in these activities in the 1990s.

Figure 6.3   Logging mahogany from farms in Amazonia.

In no small part this increase is the result of this sector's being the second-largest recipient (after cattle ranches) of income tax breaks granted to SUDAM projects. Only 18% of cattle ranches marketed any wood. Based on a conservative estimate of average 1985 stumpage values for commonly extracted Amazonian timber species, the social opportunity cost (the cost of burning the wood instead of selling it) of forest destruction reaches $1 to $2 billion (U.S.) on SUDAM-subsidized ranches alone.

The net result of road building and the associated activities of farming, ranching, mining, and logging have been devastating to the tropical forests of the Amazon. Little deforestation had taken place in Brazil's Amazon before the 1970s. As of 1975 30,000 square kilometers, or about 0.6% of Amazonia, had been cleared; much of the area was in southeastern Amazonia, an area with less forest cover than most of the Amazon. However, since 1975 the pace of deforestation has steadily accelerated. Table 6.2 shows that the deforested area increased fourfold, to 125,000 square kilometers by 1980, and twentyfold, to 600,000 square kilometers by 1988. This last figure is equivalent to approximately 12% of the Amazon and is an area the size of France. David Skole and Compton Tucker (1993) recently revised these figures; according to them, Amazonian deforestation increased from 78,000 square kilometers in 1978 to 230,000 square kilometers in 1988, an area equivalent to 6% of the closed canopy forest. These authors are quick to point out, however, that if the calculations include the effects of habitat isolation and edge effects (changes in vegetation from exposure to a different light and temperature regime), the area affected biologically by deforestation is actually 15% of the Brazilian Amazon, or 588,000 square kilometers (Skole and Tucker 1993). There was considerable controversy in Brazil during 1989 over the exact percentage and whether the estimates included areas that are not properly "tropical forests." One thing is not subject to dispute: deforestation has been concentrated in areas within easy access to roads and feeder roads, suggesting the importance of access in accelerating deforestation.

Table 6.2
Landsat Surveys of Forest Clearing in Legal Amazonia

Area Cleared

State or Territory	Area (in km^2)	By 1975 (in km^2)	By 1978 (in km^2)	By 1980 (in km^2)	By 1988 (in km^2)
Acre	152,589	1,165.5	2,464.5	4,626.8	19,500.0
Amapá	140,276	152.5	170.5	183.7	571.5
Amazonas	1,567,125	779.5	1,785.8	3,102.2	105,790.0
Goiás	285,793	3,507.3	10,288.5	11,458.5	33,120.0
Maranhão	257,451	2,940.8	7,334.0	10,671.1	50,670.0
Mato Grosso	881,001	10,124.3	28,355.0	53,299.3	208,000.0
Pará	1,248	8,654.0	22,445.3	33,913.8	120,000.0
Rondônia	243,044	1,216.5	4,184.5	7,579.3	58,000.0
Roraima	230,104	55.0	143.8	273.1	3,270.0
Totals	3,758,631	28,595.4	77,171.9	125,107.8	598,921.5

Source: Adapted from Mahar (1988:6)

How Deforestation in the Brazilian Amazon Became a Problem

Ecological Aspects of Development in the Humid Tropics, an important book published by the National Research Council in the United States in 1982, showed absolutely no sign of deep concern over Brazilian deforestation. The book recommends protective action to permit both conservation and development of the humid tropics but sounded no alarms. This is not sur-prising— until 1975 the deforested area was small and on the periphery of the Amazon basin. Nor had some political alliances been forged that would begin to give this issue prominence.

Concern about deforestation began in the late 1970s and focused pri-marily on deforestation in the Asian and African tropics, which started soon after World War II. In those areas much of the forest cover was gone by the 1970s, and in the few remaining countries with significant areas of forest the rates of deforestation were alarming. In most Asian and African countries only 2% to 15% of their forests stand today.

In terms of scale, however, the Amazon basin accounts for a much greater proportion of the total area of tropical forests in the world. One percent of the Brazilian Amazon equals 40,000 square kilometers. When the rate of deforestation increased in the 1980s, environmental organiza-tions, individuals in the scientific community, and agencies within national governments began to speak simultaneously of the devastation of tropical forests. Organizations like the National Council of Rubber Tappers, organized in 1985, sought to form alliances with national and international conservation organizations. Chico Mendes's activity within this organization brought him to the World Bank to talk to staff there, and he used that opportunity to start a dialogue with U.S. senators. The denunciation of violations of agreements led the World Bank to tem-porarily cancel the paving of the road in Rondônia. After Mendes was killed, more conservation groups joined the cause and achieved some modest victories. Now it is common for local people, working through organizations like the rubber tappers' council, to consult on strategy with nongovernmental organizations (NGOs), multilateral development banks, and Brazilian conservation organizations when they are planning protests, and conservation efforts have received favorable media cover-age. This loosely organized coalition began to call attention to the numer-ous global consequences of deforestation:

• A German scientist described the Amazon forests as the "lungs of the world," contributing large net amounts of oxygen and removing carbon dioxide. The statement, made in 1980, was incorrect, but it was highly effective in mobilizing public concern.

• More accurate evidence comes from Eneas Salati and others at Brazil's National Institute of Space Research (INPE). They have shown that the Amazon recycles 50% of its precipitation through evaporation and evapo-transpiration (cf. Salati 1985). The role of the basin in the hydrological cycle and in large-scale movement of water vapor is believed to have important consequences outside the region. Many claim that the tropical forests play an important role in stabilizing global climate and in correcting for the pollu-tants of the industrial world. INPE has played an important role in using Landsat images to estimate deforestation. Despite considerable pressure from the government, INPE has regularly released such data and provided evidence used by national and international organizations to pressure Brazil.

• Forest destruction has gained a human face: anthropologists and environmentalists have been able to show the devastating consequences of deforestation, land grabbing, mining, and other development activities on the Amazon's indigenous and peasant populations (Moran 1976, 1981, 1983; Bunker 1980; Hemming 1985; Fearnside 1987a; Browder 1986, 1988; Mahar 1988; Sponsel 1986).

In some cases, like the Kayapó, articulate and effective leaders willing and able to travel worldwide in behalf of their people's rights to those forests have galvanized public opinion. Rock stars and media personalities have also lent their names and efforts to the cause. Militancy increased from 1984 on, beginning with a Kayapó demonstration at the residence of the Brazilian president against radioactive dumping on indigenous lands and was followed by a trip by two Kayapó to Washington, D.C., and their arrest upon their return to Brazil. Indigenous people, and especially the Kayapó, were highly visible during the Brazilian Constitutional Convention in 1988 in the struggle over environmental issues and Indians' rights to their territories.

• The failure of many agricultural development projects in Amazonia has been connected to the poverty of the soils and has given impetus to efforts to stop investments in resettlement. Despite government planning, the process became disorderly, with roads to farms unbuilt, health and education services poorly provided, and the distribution of seeds unsuit-able for local ecological conditions. Brazil's deforested areas have been large, and the number of people resettled in them has been small. Outsiders increasingly see the colonization of the Amazon as enriching only the already wealthy through the government tax incentives and ben-efits (Browder 1988; Mahar 1988; Fearnside 1989). Now that it is clear that Brazil has misused its borrowed capital, will the international banks to which Brazil is indebted force the country to change its priorities?

• Scientists also are concerned about the loss of biological diversity. Norman Myers, one of the more effective advocacy scientists, has called the Amazon the "single, richest region of the tropical biome" (1984:50). People are said to be destroying the as-yet undiscovered medicinal plants and pest-resistant genetic materials that the world will need in the future (Mahar 1988:5). Current interest in molecular biology and in biotechnology suggests that protecting such genetic material is important.

• "Nuclear winter" simulations appeared in the mid-1980s, raising concern about global warming, the greenhouse effect, and the effects of clouds of smoke (from the fires touched off by the nuclear blast) on productivity of the biosphere. At about the same time the first press reports emerged about how land clearing had led to thousands of simultaneous fires in the Amazon basin. Still, Robert Dickinson, a prominent climatologist, wrote in 1981 that the global climate changes from even complete tropical deforestation "are expected to be no larger than either natural climate fluctuation or the changes that will result from past combustion of fossil fuels" (p. 436). He adds that future carbon contributions from tropical deforestation will simply exacerbate the pollution problem caused primarily by fossil fuels. Few environmental activists and atmospheric scientists share Dickinson's moderate view.

People active in conservation and in the protection of indigenous rights have learned to move quickly and have become more sophisticated in articulating their concerns to a sympathetic media. The politics of deforestation pit diverse and increasingly well-organized and well-funded nongovernmental conservation organizations and a diverse group of scientists in the biological and social sciences against the highly nationalistic Brazilian government, which prefers to frame the confrontation in terms of territorial sovereignty. It is difficult to predict which direction deforestation will take in the Brazilian Amazon. The rates had been moving steadily upward, but because of hyperinflation and economic recession in Brazil they have steadily declined since 1988 to less than half of the peak reached in 1987 (Moran et al. 1994).

Under pressure from international NGOs (e.g., World Wildlife Fund, Nature Conservancy, Greenpeace, Environmental Defense Fund, Cultural Survival, and Survival International) Brazil has recently created a program it calls Our Nature, which purports to deal with these concerns internally. This agency has shown that it has the power to punish violators, but it applies its power selectively, and violators remain largely free in the frontier setting. The appointment of José Lutzenberger to the Secretariat for the Environment (Secretaría do Meio Ambiente) was meant to suggest that the presidency of Fernando Collor de Mello wanted to be perceived as truly concerned. Lutzenberger claims that in 1989 the number of burns was down by 60% over 1987's peak figures. However, Lutzenberger proved to be largely ineffective, and he was removed from office right before the Earth Summit in Rio in 1992. Since then IBAMA (the Brazilian equivalent of the Environmental Protection Administration) has been trying to be more effective, supported by a World Bank—sponsored program called the Pilot Program to Conserve the Brazilian Rainforests (Batmanian 1994).

What Policy Interventions Might Be Effective?

Brazil is more likely to take corrective action if its critics recognize that territorial rights in the Amazon are a raw nerve. Recent economic analyses of deforestation have suggested that several corrective steps are available:

1. Eliminate tax benefits outright. They constitute an income transfer to the wealthy and have promoted wasteful deforestation of vast areas. Politically, this could be popular with the masses but difficult to implement, given the power of the beneficiaries.
2. Eliminate fiscal incentives for cattle ranching. Cattle ranching is an Iberian cultural tradition that already enjoys considerable cultural preference. To give it further encouragement distorts the economic behavior of individuals. Ranching incentives should be limited to transfers of technical expertise and more intensive production methods.
3. Collect the 25% capital gains tax on land sales. This can pay for increasing the staff needed to collect it. Collecting the tax is likely to have the most direct effect on the current wave of land speculation.
4. Institute and collect an annual progressive property tax on landholdings. In areas zoned for conservation— especially where soils are poor or where biodiversity is high—those who keep the forested land intact could be exempted from this tax. This would reverse current trends that see deforestation as an improvement.
5. Institute favorable tax rates for those who buy and reclaim degraded areas for farming, ranching, or forestry activities. This would not require tax breaks but simply a lower rate than would be assessed of those clearing still-forested areas.
6. Extend the terms for timber licenses and tie them to regular monitoring of reforestation (Repetto and Gillis 1988:386). Violators should be assessed stiff penalties, and logging companies and mills should be charged for the costs of road maintenance and regrading. Today logging trucks make the largely dirt roads impassable almost as soon as they have been regraded. This cost should be charged to those responsible for causing most of the transportation problem.
7. Stop constructing roads that fail to present a clear environmental and social impact statement. Some observers call for stopping all road construction (cf. Fearnside 1989), but this seems impractical. Road plans that guarantee adequate policing of resource use in the area cut by the road should be given favorable consideration. Roads should not be initiated until indigenous land

rights and biological reserves have been demarcated; this would decrease the number of roads that cut through protected areas, thereby opening them to illegal entry and devastation by outsiders. Road building in already opened areas that have productive agriculture may in fact lead to greater stewardship by farmers. This would need to be tied to zoning of the Brazilian Amazon, so that large areas remain off-limits to development. Farmers would welcome this because it would probably would increase the value of their open land.

8. Beef up research on intensive agriculture with low input approaches (crops that require little in the way of fertilizer, insecticides, water, and capital) so that population growth becomes less of a factor in deforestation. Brazil's population, and that of Amazonia in particular, will continue to grow at a fast rate. Even if many of these actions are taken, population pressure alone will continue to affect the forests unless researchers find economical ways to increase food production on less land (Nicholaides et al. 1983; Moran 1989; Fearnside 1989; Sánchez et al. 1982; Serro and Homma 1993).

9. Expand Indian reservations: indigenous peoples have often proved to be sophisticated stewards of the forest (Balée and Gély 1989; Balée 1994). Moreover they have an inherent right to land on both historical and humanistic grounds. These rights extend to intellectual property rights over the resources they have husbanded. Further, the Brazilian constitution protects these rights. Indigenous territories need more forceful demarcation and more effective protection, which should not be tied to criteria for stewardship imposed by outsiders. Indigenous people, like ourselves, are looking to develop viable ecological and economic systems that meet their needs. Many important biological areas are probably important biotically because of the actions of indigenous people. Their removal could mean a loss of biodiversity. The most conservative strategy is probably to allocate to indigenous peoples land they already are managing, even if that land is biotically precious. A parks model might protect some areas that are too far away for indigenous people to protect.

Brazilian government policies clearly have encouraged the deforestation activities of both ranchers and farmers in the Amazon. Some policies could be changed at the stroke of a pen, but their implementation is less than likely. The government does not have the institutional capacity to monitor and police compliance with conservation. In areas like Peru, where an antigovernment guerrilla effectively terrorized the population, it is difficult to imagine how the government could even enter some areas safely, much less be able to impose its will. In Brazil the arm of the government reaches unevenly and ineffectively into the Amazon.

Lack of action is likely to lead to even greater violence over access to resources, growing rates of migration to the frontier, an exponential increase not only in malaria but in many other diseases, greater socioeconomic inequality, and the long-term impoverishment of the Brazilian people. NGOs estimate that more than eight hundred people were killed in 1989 in land-related conflicts. Land redistribution and agrarian reform throughout Brazil are the only methods likely to stop the violence, and one of the easiest ways to effect these reforms might be to collect the existing tax levies. The military might be charged with its collection to prevent well-armed farmers and ranchers from harming tax collectors.

Debt-for-nature swaps are unlikely to work in Brazil because of their concern with the "internationalization" of the Amazon (Ferreira Reis 1968). More likely to succeed are technical and scientific assistance in demarcating areas as Brazilian biological reserves, assistance in training forest rangers, and training in using economical monitoring techniques. The rapidly growing conservation organizations within Brazil need help so that they can fight their own battles internally. Anthropologists and other social scientists investigating native systems of forest use and conservation need research support (cf. Balée and Gély 1989; Moran 1989, 1993a). This work is consistent with technical recommendations that promote agroforestry approaches to forest management and recognize the importance of local ecosystems' variability. Current fiscal incentives place the small farmer at a great disadvantage to the large operators, whose operations receive copious capital from the government. Policies that would encourage greater production of staples would enhance food security internally as well as permit food exports. The subsidies of mostly export crops, like soy beans and orange juice concentrate, need to be eliminated to reduce the distortions they create. The subsidies help these industries to compete globally, but their profits are paid by most of the population of Brazil, which is thereby further impoverished.

The forces leading to deforestation vary from place to place within the Amazon. Global approaches that fail to take into account this intraregional diversity will miss the mark. Global effects have local causes. Future research must be more sensitive to these internal differences in climate, soils, forest type, economic activities, and socioeconomic and political forces that interact in the various parts of the Amazon. Ecologically oriented anthropologists are in a particularly good position to contribute basic and applied research on local variations and their significance for larger-scale processes.

Acknowledgments   This essay was originally presented at the annual meeting of the American Anthropological Association, November 28–December 2,

1990, New Orleans. Considerably revised versions have appeared in the journal *Human Ecology* (Moran 1993b) and as a chapter in a volume published in Göteborg, Sweden, in 1992. The author thanks the editors and the volume reviewers for their constructive suggestions, the result of which is this version.

References Cited

Balée, William. 1994. *Footprints of the Forest: Ka'apor Ethnobotany—The Historical Ecology of Plant Utilization by an Amazonian People.* New York: Columbia University Press.
Balée, William and Anne Gély. 1989. Managed forest succession in Amazonia: The Ka'apor case. *Advances in Economic Botany* 7: 129–158.
Batmanian, Garo. 1994. The pilot program to conserve the Brazilian rainforests. *International Environmental Affairs* 6 (1): 3–13.
Browder, John. 1986. Logging the Rainforest: A Political Economy of Timber Extraction and Unequal Exchange in the Brazilian Amazon. Ph.D. diss., University of Pennsylvania.
———. 1988. Public policy and deforestation in the Brazilian Amazon. In *Public Policies and the Misuse of Forest Resources*, R. Repetto and M. Gillis, eds., pp. 247–297. Washington, D.C.: World Resources Institute and Cambridge University Press.
Bunker, Stephen. 1980. The impact of deforestation on peasant communities in the Medio Amazonas of Brazil. *Studies in Third World Societies* 13: 45–60.
Dickinson, Robert. 1981. Effects of tropical deforestation on climate. *Studies in Third World Societies.* 14: 411–441.
Fearnside, Philip. 1987a. Causes of deforestation in the Brazilian Amazon. In *The Geophysiology of Amazonia: Vegetation and Climate Interactions*, R. E. Dickinson, ed., pp. 37–61. New York: Wiley and United Nations University.
———. 1987b. Deforestation and international economic development projects in Brazilian Amazonia. *Conservation Biology* 1 (3): 214–221.
———. 1989. A prescription for slowing deforestation in Amazonia. *Environment* 31 (4): 17–20, 39–40.
Ferreira Reis, Arthur. 1968. *A Amazônia e a cobiça internacional* (The Amazon and international covetousness). Rio de Janeiro: Gráfica Record Editôra.
García Vásquez, J. and C. Yokomizo. 1986. Resultados de 20 anos de incentivos fiscals na agropecuaria na Amazônia (Results of 20 years of fiscal incentives to Amazon agropastoralism). Fourteenth meeting of the Economics Association of Brazil. *Encontro Nacional de Economía* 2: 47–84.
Hecht, Susanna. 1980. Deforestation in the Amazon basin: Magnitude, dynamics, and soil resource effects. *Studies in Third World Societies* 13: 61–108.
Hemming, John, ed. 1985. *Change in the Amazon Basin.* 2 vols. Manchester, England: Manchester University Press.

Mahar, Dennis. 1979. *Frontier Development Policy in Brazil: A Study of Amazonia.* New York: Praeger.

———. 1988. *Government Policies and Deforestation in Brazil's Amazon Region.* Washington, D.C.: World Bank.

Moran, Emilio. 1976. *Agricultural Development along the Transamazon Highway.* Bloomington, Ind.: Center for Latin American Studies, Monograph Series.

———. 1981. *Developing the Amazon.* Bloomington: Indiana University Press.

———. 1987. Socioeconomic considerations in acid tropical soils research. In *Management of Acid Tropical Soils for Sustainable Agriculture*, P. Sánchez, E. Pushparajah, and E. Stoner, eds., pp. 227–244. Bangkok: International Board for Soil Research and Management.

———. 1989. Models of native and folk adaptation in the Amazon. *Advances in Economic Botany* 7: 22–29.

———. 1993a. *Through Amazonian Eyes: The Human Ecology of Amazonian Populations.* Iowa City: University of Iowa Press.

———. 1993b. Land use and deforestation in the Brazilian Amazon. *Human Ecology* 21: 1–21.

Moran, Emilio, ed. 1983. *The Dilemma of Amazonian Development.* Boulder, Colo.: Westview.

Moran, Emilio, E. Broindizio, P. Mausel, and Y. Wu. 1994. Integrating Amazonian vegetation, land use, and satellite data. *BioScience* 44 (5): 329–339.

Myers, Norman. 1984. *The Primary Source: Tropical Forests and Our Future.* New York: W. W. Norton.

National Research Council. 1982. *Ecological Aspects of Development in the Humid Tropics.* Washington, D.C.: National Academy Press.

Nicholaides, John, P. Sanchez, D. Bandy, J. Villachica, A. Coutu, and C. Valverde. 1983. Crop production systems in the Amazon basin. In *The Dilemma of Amazonian Development*, E. F. Moran, ed., pp. 101–153. Boulder, Colo.: Westview.

Repetto, Robert and Malcolm Gillis. eds. 1988. *Public Policies and the Misuse of Forest Resources.* New York: Cambridge University Press and World Resources Institute.

Salati, Eneas. 1985. The climatology and hydrology of Amazonia. In *Key Environments: Amazonia*, G. Prance and T. Lovejoy, eds., pp. 18–48. London: Pergamon.

Sánchez, Pedro, D. Bandy, J. Villachica, and J. Nicholaides. 1982. Amazon basin soils: Management for continuous crop production. *Science* 216: 821–827.

Serrão, Emmanuel Adilson and Alfredo Homma. 1993. Brazil. In *Sustainable Agriculture and the Environment in the Humid Tropics*, Committee on Sustainable Agriculture of the National Research Council, pp. 263–351. Washington, D.C.: National Academy Press.

Skole, David and Compton Tucker. 1993. Tropical deforestation and habitat fragmentation in the Amazon: Satellite data from 1978 to 1988. *Science* 260: 1905–1910.

Sponsel, Leslie. 1986. Amazon ecology and adaptation. *Annual Review of Anthropology* 15: 67–97.

7

Forty Years of Political Ecology in the Peruvian Upper Forest: The Case of Upper Huallaga

Eduardo Bedoya and Lorien Klein

Deforestation in Peru's Amazon basin is directly related to several government policies instituted in this century. By the 1980s government economic policies, regional political relations, the government drug-eradication program in conjunction with the United States, and the Shining Path guerrilla movement had multiplied the rate of deforestation by a factor of three. Although two-thirds of the country is forest, forest products represented a minuscule percentage of Peru's gross national product early in this century. The government encouraged massive migration to the forests with a road-building program during the 1940s in order to tap some of this wealth, relieve socioeconomic problems elsewhere, and enhance national security. Since the mid-1970s farmers have concentrated on coca production, because it is three to four times more profitable than legal crops, and their rights to do so have been championed by the terrorist organization Shining Path.

In this essay we examine these problems from the perspective of agricultural systems, which constitute specific technological methods by which human societies obtain edible crops. However, those systems and certain forms of social organizations correspond in significant ways, as do agriculture systems and specific modes of production. Likewise, agricultural systems tend to associate with different levels of local and regional demographic densities. Equally important is the relationship of one spe-

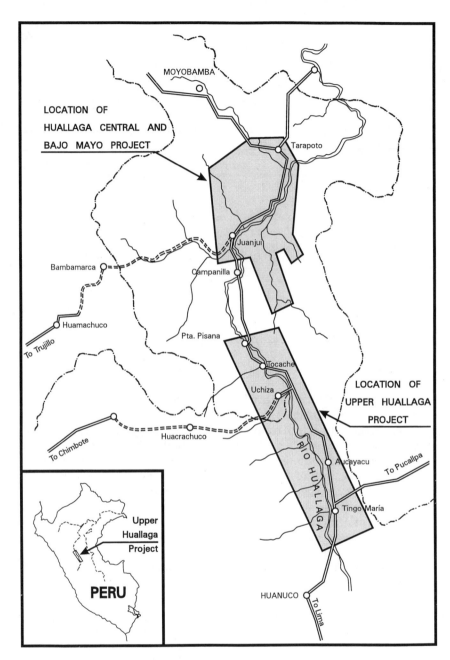

Figure 7.1   Huallaga Valley

cific agricultural system, the political structure, and the effect of specific economic policies on the peasant population. Expressed differently, the analysis of diverse agricultural systems and rural environmental processes cannot be restricted to the study of their technological characteristics (Schmink and Wood 1987; Blaikie and Brookfield 1987; Peluso 1990).

If the massive process of deforestation in the Peruvian Amazon is a direct consequence of the agricultural practices of peasant Andean immigrants, especially in the upper forest region (*selva alta*), it is necessary to examine the social and economic causes of deforestation.[1] In an effort to understand the deforestation process in the Peruvian Amazon this essay analyzes the expansion of the agricultural frontier in the upper Huallaga, located in the central upper forest (see figure 7.1) and representing the largest area of coca (*Erythroxylon coca*) cultivation in the world.

Deforestation in the Peruvian Amazon

The greatest renewable resource in Peru is the abundant forest, 96% of which is located in the Amazon watershed, covering some 93% of that area (Bueno 1984). There are 72.8 million hectares of natural rain forest, 170,000 hectares of cultivated forest (forest managed for the purpose of selecting certain species), and 10.3 million hectares of land suitable for forestation; together these comprise 64% of Peru. However, because of a series of historical, economic, and cultural factors, forest resources account for less than 1% of the gross national product (Bueno 1984).

In the late 1970s a series of investigators issued extremely pessimistic and alarming projections regarding the rates of deforestation of the Peruvian Amazon. The analyses of Jorge Malleux (1975) and Marc Dourojeanni (1983) stressed the immense size of the territory being deforested and how it resulted from the migratory agriculture practiced by the Andean peasants in the high and low regions of the Peruvian Amazon watershed. The investigators emphasized that in fifty years the peasants had cleared approximately 6 million hectares. It is estimated that the area will have lost 20 million hectares between 1980 and 1999. The growth of indiscriminate deforestation caused by the highland agriculturists has an exponential increment such that each year the number of cleared hectares rises.

Between 1925 and 1974 a total of 4.5 million hectares were cleared, an average annual deforestation of approximately 90,000 hectares. Most of the deforestation occurred in the upper forest. However, the figure of 90,000 hectares per year is slightly misleading, because the principal migratory movements to the high forest did not start until the 1940s, when major roads to the tropical uplands were constructed (Aramburú 1982). Road construction likely is why the annual rate of deforestation was less than 90,000 hectares from 1925 to 1940 and more than 90,000 hectares after 1940.

Until 1940 the expansion of the economic frontier in the tropical forest during the cycles of rubber, quinine, and timber extraction did not provoke the expansion of the demographic frontier. The "rubber boom" (1880–1914) did result in an increase of the population of Iquitos from ten thousand to fifteen thousand, but the city returned to its normal population level once the boom was over (Werlich 1968). Only recently—since 1940—has massive settlement occurred in various basins of the upper forest; it has followed road construction, which penetrates the area and connects it with the coastal Andean regions. Nevertheless the expansion of large landholdings in certain areas of the upper forest occurred well before 1940. The haciendas of Cuzco date from the early colonial period (sixteenth century), and the coffee plantations of Chanchamayo were established at the beginning of the twentieth century. The tea and coffee plantations of upper Huallaga date from the late 1940s.

According to the confidential correspondence of the haciendas, or plantations, that operated during the 1940s in valleys such as La Convención in Cuzco (Fioravanti 1969) and the region of Satipo-Chanchamayo, the haciendas did not cultivate more than 5% of their territory during the first three decades of this century. Large plantations, which represented the dominant form of tenancy during this period, do not appear to have caused massive deforestation because labor was extremely scarce (Bedoya 1981).

According to data published in the mid-1980s, the rate of deforestation between 1981 and 1985 was 270,000 hectares per year (World Resources Institute 1986:73). This unfortunately supports the projections established in the 1970s. Furthermore, with coca expansion throughout the high Peruvian forest this figure could easily surpass 300,000 hectares annually. However, Dourojeanni (1984) projects that by the year 2000 only 20% of the deforested hectares in the Peruvian Amazon will actually be used for agriculture or grazing, the rest lying fallow or abandoned. He bases this calculation on current land use in some regions of the forest, where only 1 cleared hectare in 4 is actually used for agriculture or grazing. It is important that the total number of unproductive hectares not be considered as territory at rest or regrowth. Approximately 2 in 5 hectares cleared do not even possess the potential for forest cultivation. Some of these areas include local forests on grades sharper than 40 degrees that are useful only for forest cover in the face of strong rains and the prevention of erosion (Agreda 1984).

One principal reason that protected forests and land with forestry potential is occupied is that roads into the high forest were built through land that is inadequate for agriculture (Dourojeanni 1983). During his first administration Peruvian president Fernando Belaúnde (1963–1968) addressed agrarian underdevelopment by promoting frontier settlement as an alternative to agrarian reform and land redistribution. Belaúnde was thus pressured into tropical colonization projects. In his first administration he constantly confronted a conservative alliance of both traditional landowners and the political party APRA (Cotler 1979). This alliance in the Peruvian Congress blocked all reformist measures proposed for the rural highlands. In this political context the Belaúnde administration concentrated all its efforts on promoting human settlement in the upper regions of the Peruvian Amazon basin.

The government promoted high forestry development by building a network of roads on the eastern slopes that would run parallel to the Peruvian highlands, connecting each tropical valley with the national highway system (Stocks 1988:2). In the past the government had sought the shortest and most economical routes to connect specific regions in the highlands with particular destinations in the upper forest. In contrast, Belaúnde sought a road system that would open up the maximum amount of land to settlement (Belaúnde 1965; Werlich 1968:461). He proposed the construction of the Marginal Highway of the Forest, which would use existing and newly constructed roads to link the highlands and the coast (Werlich 1968:462). To promote economic and agricultural expansion along with road construction the government initiated a massive program of colonization in the valley of the upper Huallaga (Bedoya 1981). Loans from the Interamerican Development Bank financed much of this tropical colonization.

During his second administration Belaúnde promoted the expansion of the Proyectos Especiales de Desarrollo (Special Development Projects), which had been started by the preceding military governments. These projects were designed to furnish the techno-agropastural, social, and infrastructural services that the local colonists in important regions of the high forest required. However, the projects focused on the expansion of the agrarian frontier, which relied on building roads to the tropical regions. Road construction was not accompanied by adequate consideration of marketing, price structure, and technical assistance, as clearly seen in the analysis of the budgeted costs of the projects. In some cases, as for example in Huallaga Central–Bajo Mayo and Pichis-Palcazú, building access roads ate up more than 50% of the budget. For the remaining cases the cost was most important in proportional terms. Only 7% of the total amount was budgeted for forest development and management of the environment (Salazar 1984).

Francisco Verdera (1984:186) rightly indicates that, in a context of relative abundance along the highway infrastructure, the individual adjudication of lands without proper marketing and technical assistance leads to extensive land use. He also states that, according to the plans of the Proyectos Especiales, 55% of the total lands to be incorporated in the high forest were classified as "new lands" and the rest as "improved lands." In addition, the activities related to intensifying the use of the land, such as building irrigation works, were concentrated in two projects, upper Mayo and Jaén–San Ignacio–Bagua (Salazar 1984). In the regions of central Huallaga–lower Mayo and the upper Huallaga the projects did not even consider the areas of forest development and environmental control in their budgets and programming. It is logical to argue that a policy of this kind leads to deforestation and an unproductive use of the land.

The unproductive land use reflects the low intensity of the agricultural systems of the tropical forest. This is expressed in the low mean productivity of the principal crops of the region. For example, according to the data gathered in 1984 and published in the *Encuesta Nacional de Hogares Rurales* (National Survey of Rural Households, Instituto Nacional de Estadística [INE] 1987), both hard yellow corn and rice harvested in the Amazon watershed of Peru represented about 67% and 48%, respectively, of the national hectarage. Thus analyzing the yields per hectare shows that hard yellow corn and rice fell 20% and 32%, respectively, below the national average (INE 1987; see also Verdera 1984). Most of the forest area planted with corn and two-thirds of the rice area are cultivated using the traditional system of slash and burn, with prolonged periods of fallow and an extremely low use of modern technology (INE 1987; see also Verdera 1984). Finally, all this recently accumulated data indicate how important Amazonian agriculture has become to Peru during the last two decades. However, this has occurred at great ecological and economic cost.

Much of the responsibility for the inadequate land management falls on the Ministry of Agriculture, which grants certificates of possession and land titles in protected forests. After cutting lumber, private forest enterprises usually refuse to reforest, arguing that the Ministry of Agriculture grants the farmers titles of ownership or certificates of possession in those areas (Dourojeanni 1983). Additionally, the ministry issues the titles or certificates of possession for lands unsuitable for cultivation; the possibility that these producers will develop a stable and intensive agriculture is quite remote. Most often the peasants cultivate for a short period and then migrate to other areas, resulting in the continued practice of shifting agriculture.

The contradictory results obtained by various land studies in different regions of the high forest have contributed to the irrational management of natural resources. The general tendency of the studies carried out during the 1960s and 1970s was to overestimate the quantity of land with agricultural potential and underestimate the amount of land for forestry and conservation purposes (Dourojeanni 1984; Bedoya 1987). Studies of agrarian Peru have always emphasized the mountains and the coast. The Amazon basin has been perceived, implicitly or explicitly, as an immense free territory to be conquered, with great agricultural potential.

Although more precise data on the differential effects of the large landholder and the migratory agriculturalist in Peru are necessary, the agriculture developed by the small producers is an important element in the expansion of the agricultural frontier and the deforestation of the Peruvian Amazon. Today's economic crisis, which has provoked a surprising resurgence in the peasant economy throughout the country, clearly reaffirms this trend. In addition, according to the national survey (INE 1987), 39% of the national agricultural area is located in the Amazon basin. The total number of agricultural hectares contracted by associated enterprises—agrarian production cooperatives created during the second agrarian reform of the 1970s—shows that the forest encompasses 31.8% of the national area in cultivation. The relatively small difference between both percentages reflects the fact that the expansion of the agricultural frontier in the Peruvian forest has been produced through the small- and medium-sized agropastoral family units; 64% of these units are farms of 5 to 50 hectares (Bedoya 1991).

Although the interest in cattle raising in the Peruvian Amazon never reached the level or speculative characteristics of Brazil's, in some areas large- and medium-size cattle ranches have been established that have also converted agricultural lands to pasture (Bidegaray and Rhoades 1988). According to William Loker (1989), writing about small- and medium-sized producers in the Andean part of the Amazon watershed, the areas cultivated as pastures fulfill important functions for the colonists. First, they constitute physical proof of land occupation in a context in which property titles are often missing and conflicts over land are part of daily life. Second, the colonists gain income by renting their pastures to neighbors who keep cattle. The colonists are interested in cattle raising because it represents capital investment that maintains its value in a highly inflationary context (Loker 1989).

The enactment of agrarian reform in colonization zones meant a reorientation from agriculture to meat-cattle production (Ministry of Agriculture 1974:18).[2] In 1971 four cooperatives of agrarian production and five service cooperatives in the upper Huallaga region began to use lines of credit stemming from Interamerican Development Bank funds primarily dedicated to the purchase of sixty-five hundred head of cattle from Panama and Central America. By 1973 the loans drawn for cattle rose to 59% of the total debt incurred by functioning cooperative enterprises. The turn toward cattle was partially based on soil studies carried out between 1964 and 1970 by the Ministry of Agriculture and the National Office of Natural Resources. This study concluded that 38% of the land in the colonization zone was appropriate for raising cattle (Ministry of Agriculture 1974:54). According to this source, 51% of the soil was considered to have agricultural potential, whereas 11% was suitable for forests, and no land was considered to require protective forest cover. Recent information on the effect of introducing cattle raising into a tropical moist forest region suggests that the study may have been wrong in promoting this form of land use for 38% of the land. Still, the expansion of cattle raising was carried out despite soil studies that showed agricultural activity to be the most suitable use for 51% of the land. Regional development of cattle raising responded to the increase in urban demand for meat and to the industry's relatively low labor requirements.

In the upper Huallaga Valley government cooperatives acquired a considerable number of cattle, and the relative importance of coffee declined. Likewise, new crops such as rice and corn were cultivated in response to new policies of the military government. The cooperatives introduced heavy machinery to clear virgin forests or secondary forest. The region's labor scarcity prompted the decision to use heavy machinery for clearing and made it impossible to quickly switch any of the lands to cultivation. Using machinery on tropical soils was catastrophic, and within a few years the government cooperatives could no longer pay off their loans on the heavy machinery. Moreover the heavy machinery compressed the soil, producing extremely low yields of both corn and rice and diminishing the nutritive quality of pasture and grazing lands (Ministry of Agriculture 1976:8–10). Low agricultural productivity meant low revenues. In short, labor scarcity influenced the type of technology used, and the technology was highly destructive. Ecological imbalance resulting from the technology caused the government cooperatives to fail.

Coca Production and Deforestation

Coca Plantations, Labor Scarcity, and Deforestation

Coca has been cultivated in the Andes for thousands of years. During the period of Inca expansion in the fifteenth century the Chupaychu ethnic group, centered in the highlands, extended its territorial control toward tropical and subtropical regions to gain access to such resources as wood, feathers, and coca (Santos 1985). During the colonial period the Chupaychu retained access to these regions, although some resources, including coca, were absorbed by Spanish operations based in the tropical region of Chinchao, north of upper Huallaga (Santos 1985). Unfortunately, little is known about the technology used in the cultivation of coca in pre-Colombian and colonial Peru. Studies by W. Guerra (1961), J. Paez (1937), and C. Bües (1937) do indicate that traditional coca cultivation used irrigation and flat terrains or well-constructed terraces.

Directly or indirectly, the cultivation of coca has been one of the most important factors provoking massive peasant migrations to the upper Huallaga region. The contemporary expansion of coca production in upper Huallaga is an expression of deep-rooted cultural traditions as well as an economic necessity. Several factors make coca an important product for highland peasants. Roderick Burchard, in a 1974 study of coca and the exchange of food in that region, described the many social functions, rituals, and economic needs that coca satisfies. First, coca is required in many rituals that form a system of reciprocal exchange between humans and the supernatural. Members of the peasant community, or *comuneros*, believe that if a peasant loses an animal to robbery or death, chewing coca leaves is the only way the supernatural hills will offer him a new animal (Burchard 1974:242). Second, wage workers contracted for work on the family plots are paid in coca (Dewind 1987:303), as is any peasant who assists in community agricultural activity. When a family receives the assistance of another—*ayni*—for one or two days of work on the plot, the family is obliged to give that person coca. In reality, all the adults chew—*chaqchan*—coca during ayni (Burchard 1974:242). Coca is also an essential component of social consumption for marking changes in community political authorities, festivals, and many other occasions. Third, one of coca's most important functions is to establish a basic means of exchange for all other agricultural products. Because some communities do not possess land that can produce everything they need, they obtain these goods through interregional exchange. For example, in the community studied by Burchard (1974:246–247) not all peasants had land at altitudes higher than 3,500 meters. Therefore they had a permanent shortage of potatoes and other root crops. To obtain these products the comuneros engaged in an exchange system in which coca was the main currency. In Tingo María, the community Burchard studied, the rate of exchange was 25 pounds of coca for 1 full sack of potatoes. The corresponding exchange rate in Puquio Pampán was 3 pounds of coca for 1 sack of potatoes. According to Burchard (1974:248), a simple calculation shows that it was possible to convert one sack of potatoes into eight of the same product, entirely through this mechanism of reciprocal exchange among peasants located in different ecological zones. In other words, the peasants used coca to maximize their limited production capacity. In this sense coca functioned as "special purpose money" (Bohannan 1959).[3]

Coca became a substitute for currency. On one hand, it served to increase the scanty revenues provided by daily wages from the plantations and mines. On the other hand, it partially protected the peasants from inflation. This combination of factors explains the immense interest the comuneros had in cultivating coca in the tropical regions. This also explains the peasants' general lack of interest in working for the regional coffee and tea plantations.

Coca generated prosperity for many large, medium, and small plantations in upper Huallaga. These plantations were the most important focal point of migrations from the highlands. In Tingo María 82% of the peasants—men and women—had worked on the coca plantations, Burchard (1974) found. A third of the comuneros had coca plots in that tropical zone. In 1962 coca and coffee were the plants yielding the highest economic profit in the area of Tingo María (Servicio de Investigación 1962:17). In this region 75% of the coca estates were cultivating less than 1 hectare, and 24% were working 1 to 10 hectares, whereas 1% were plantations of more than 10 hectares (Burchard 1974:219). The largest estates had an average of 20 cultivated hectares. The biggest three were Shapajilla, Porvenir, and Santa Rita with 67, 26, and 24 hectares, respectively (Centro Nacional [CENCIRA] 1973:248). In all three cases coca production was based on a diversified and complex system of social relations of production. The owners possessed and managed one section of the estate, whereas other plots were managed by the permanent and temporary peasants and their families.

During the 1950s and 1960s the expansion of the coca crop caused a small but prosperous labor market to develop in upper Huallaga. Unlike the situation on the coffee and tea plantations, free laborers, or *huayreros* (a Quechua word), made up a relatively large percentage of the labor force on coca estates (CENCIRA 1973:244–245). The workers were obliged to attend to the daily tasks of the plantation, directly supervised by the owner. For this they received wages, which occasionally were lower than the tea and coffee estates paid, a ration of coca, and some food (CENCIRA 1973:245–246). The permanent and temporary workers in turn managed plots of staple crops within and outside the estate.

Those peasants who managed coca plantations within the private estates did so under the system of *mejoreros*.[4] The daily ration of coca, or the possibility of planting it as mejoreros, was a powerful labor magnet. Wages were lower than what the coffee and tea estates paid, but coca could be substituted for money in the exchange mechanism. As a whole these estates not only offered the possibility of planting staple crops but also of obtaining coca through diverse means, in addition to money they received for wages.

The development of a labor market for coca crops harmed the coffee and tea plantations, which could not offer rations or coca plots along with the wages. For all these reasons highland peasants preferred to migrate freely to the coca plantations, which aggravated the regional labor scarcity. To compete effectively with coca, coffee and tea plantations would have had to raise wages to a level at which workers could recoup the equivalent of coca rations or personal crop cultivation. Indeed plantations could not solve the problem; they simply could not compete with the coca crop.

The coca plantations, and their subsequent development of a labor market, is one explanation for the extensive cultivation among the coffee and tea plantations. Because of the labor shortage the coffee and tea plantations could not implement an intensive agriculture system. Coffee and tea planters or their administrators never had sufficient labor for weeding and fertilizing their plantations (Bedoya 1993). They could not use modern techniques to avoid diminishing returns. The coffee or tea enterprises thus constantly required the clearing of new plots in order to maintain a certain level of agricultural productivity. In other words, large-scale planters also practiced shifting agriculture.

Coca Farmers and Deforestation

The first period of the military government (1968–1975) saw the continuation of the planned settlement projects, although they were adapted to the agrarian reform policy (Bedoya 1981). The agrarian reform established agricultural production cooperatives but did not properly address the problems of production and productivity experienced by family farms located within and outside the boundaries of the newly formed cooperatives (Bedoya 1981). As a result at the end of the 1970s the upper Huallaga region continued to be characterized by low productivity of legal food crops and minimal use of modern techniques such as fertilizers. The average yields per hectare of rice, corn, manioc, and beans were much lower than those of upper Mayo, lower Mayo, and central Huallaga. The overall differences in productivity placed the upper Huallaga farmers at a disadvantage in comparison with agriculturalists from other tropical valleys.

The economic stabilization rules imposed by the World Bank and the International Monetary Fund since 1976, which removed subsidies and let the market reflect real scarcities, have reduced demand for foodstuffs. Moreover the terms of trade between agricultural prices and production costs have shown an increasingly unfavorable trend for the producer, especially since 1973 (Alvarez 1983). In the economic crisis of 1981 and 1983 the terms of trade deteriorated even further. In upper Huallaga overall production costs increased 2.7 times more than agricultural prices did (Aramburú, Alvarado, and Bedoya 1985). In this context coca production burgeoned. Furthermore, during the first half of the 1980s the international demand for coca rose by nearly 50%, increasing the levels of coca production dramatically. This was especially apparent in the upper Huallaga region where the amount of land devoted to coca production was conservatively estimated at 24,000 hectares in 1984 and 70,000 hectares in 1986. In other zones increases in the rate of production have not been as great but continue to climb steadily (ECONSULT 1986). In 1986 coca production could have contributed to the deforestation of 150,000 hectares. Because of the deep economic crisis that has gripped Peru since late 1987—with negative indexes in nearly every indicator of economic growth—coca production has expanded considerably. In the region of San Martín, where the largest number of coca plantations and the Upper Huallaga Special Project (PEAH) are located, the figure may approach 300,000 hectares.[5] This is five times the amount of land planted in corn, which in 1988 constituted the most important legal crop in this region (Loker 1989).

The Tingo María region of the upper Huallaga Valley is officially considered a national park. Similarly, other national parks, like that in Abiseo located in the Department of San Martín, and national forests, like that of Von Humboldt located in Ucayali and Huánuco, or Biabo in San Martín and Ucayali, have also been invaded by coca-producing peasants. As the economic crisis in Peru continues to worsen, and the prices for legal agricultural products remain low, the expansion of coca production will continue at its current rate. It seems likely that the annual rate of destruction will reach higher levels, and the gravity of the ecological problem will acquire larger dimensions.

Coca farmers both inside and outside the boundaries of the project area have chosen locations far from the main highways to avoid police repression. As in other upper forest regions coca is cultivated not only on land appropriate for agriculture but also in areas suitable for forestry. The hillsides are the safest alternative, but although it is a permanent crop, coca has the harmful effects of an annual crop when it is cultivated on deforested steep slopes, especially because it is typically cultivated in vertical furrows rather than in terraces that mitigate the damaging effects of rain (Dourojeanni 1989; Bedoya 1987).

In the 1980s fallow agriculture covered most of the area delimited by the PEAH. This extensive land use is also one significant effect of coca production on legal agriculture in upper Huallaga. Farmers have directed land, capital, and labor to coca production since the late 1970s, because they found it more profitable than other crops. In the interior of the PEAH area such crops as yellow maize and rice, cultivated using a fallow system, have been slighted by increasing production costs brought about by the expansion of coca production. As table 7.1 shows, the small plot of coca is surrounded by large areas of extensively cultivated annual food crops and permanent cash crops. An important issue is the destination of noncoca production, particularly annual crops. Annual crops are consumed by farming families in addition to being sold, although prices are normally low. Food is also provided to hired workers as part of most labor arrangements.

Competition by coca production led to a chronic labor shortage for legal agricultural crops. The expansion of coca has also deterred the development of technology specifically related to the production of legal crops. In upper Huallaga most of the fertilizers and pesticides are used on coca plantations (Aramburú, Alvarado, and Bedoya 1985). Almost all yellow corn and rice is cultivated by using slash-and-burn agriculture, incorporating little modern technology. The 1980s saw no significant expansion in total hectares under cultivation or in the productivity of the most important legal crops within PEAH boundaries. The main goal of the farmers was to intensify coca production on hillsides and underuse the plots with better soils because of the low value of most legal crops. This strategy of extensive land management—although the coca plots are actually intensively cultivated—results in continued deforestation.

Deforestation caused by coca production in upper Huallaga does not seem to be as serious when analyzed plot by plot, until we consider the large number of agriculturalists, both inside and outside the borders of the PEAH, who have cleared steep slopes to cultivate coca. Using the information gathered in 1981 from the National Foundation for Development (FDN), we found that among 348 farmers from upper Huallaga, 136 were coca producers. Our study analyzed agricultural systems and the rate of deforestation among the coca farmers. The annual deforestation rate is clearly determined by extensive soil use and not by the number of coca hectares. The extensive agricultural system means that the bigger the plot, the higher the annual deforestation rate and the higher the annual increase in fallow areas in relation to a relatively smaller increase in the cultivated area (see columns 8, 9, and 10 in table 7.1).

Table 7.1

Coca Hectares, Annual Rate of Deforestation, Annual Increase of Agricultural and Fallow Hectares, According to Plot Size Among Coca Producers in Upper Huallaga in 1981

Plot size	Percentage of cases	(1) Average plot size (ha.)	(2) Average cocoa (ha.)	(3) Average of cultivated hectares (annual + permanent crops)	(4) Average of natural and cultivated pastures (ha.)	(5) Average of fallow hectares	(6) Total cleared area (ha.) (3 + 4 + 5)	(7) Number of years occupying plot	(8) Annual increase of agricultural hectares*	(9) Annual increase of fallow area†	(10) Annual rate of deforestation‡
0.1 <10	13.0	8.59	0.91	4.11	0.19	2.50	6.80	11.13	0.37	0.22	0.61
10.1 <20	45.0	15.71	1.14	5.93	1.48	4.54	11.95	10.91	0.54	0.42	1.10
20.1 <30	21.0	24.20	0.97	6.06	2.44	8.84	17.34	13.88	0.44	0.64	1.25
30.1 +	21.0	45.61	1.02	6.92	8.49	12.67	28.08	10.07	0.69	1.26	2.79
Total	100.0	22.89	1.05	5.93	3.00	6.89	15.82	11.18	0.53	0.62	1.42

Source: Elaborated from National Foundation for Development survey (1981)

* Col. 3 divided by col. 7
† Col. 5 divided by col. 7
‡ Col. 6 divided by col. 7

The information that we pulled from the National Development Foundation figures reveals an average of 1.05 hectares of coca per family plot. As in the Chapare region of Bolivia, the colonists in upper Huallaga are not seeking to maximize coca production (Painter and Bedoya 1991). Table 7.1 shows that the size of the family plot and the number of coca hectares have no relationship. The average number of coca hectares ranges from 0.91 to 1.14. As in Chapare the risks associated with a crop that is subject to price instability and police repression are the main reasons that coca producers are cultivating smaller coca plots.

The annual rate of deforestation from coca production is less than that from holdings cultivated only with legal crops (compare table 7.1, column 10 with table 7.2, column 10). If we compare the information related to deforestation rates of both groups, we find that for each range of plot size the coca producers have a lower rate of deforestation than the legal producers. The noncoca producers also have a higher annual increase in fallow area (compare table 7.1, column 9 with table 7.2, column 9). Furthermore, for 51% of the land area of the coca producers, the cultivated area is larger than the fallow area. On the land of the noncoca farmers 66% of those farmers have fallow areas larger than their cultivated areas. In other words, the legal producers use more area of land in their type of cultivation than do the coca farmers. This is provoked by the extremely high profitability of the illegal coca economy compared with the legal agricultural economy. When coca prices are high, the profitability is three or four times more than other crops such as cacao (*Theobroma cacao*), which is the next most lucrative legal crop.

This economic disparity has led to a relocation of human and economic regional resources toward the production of coca. As in the 1950s and 1960s, when coca plantations attracted the majority of wage laborers, in the 1980s small coca producers generated the most dynamic sector of the labor market, thus causing a regional labor shortage. The legal farmers suffered labor scarcity because of the elevated wages paid to the hired workers by the coca producers (Bedoya and Verdera 1987). Of all the farmers who had labor shortages, 74% were legal agriculturalists and 26% were illegal producers. Labor scarcity did not allow legitimate farmers to implement intensive agriculture. Similarly, the relatively low economic productivity of legal farming led to the extremely low use of modern farming techniques. Only 1% of the legal farmers used fertilizers, whereas 18% of the coca producers did. It is important to note that coca prices were at a very low level when the FDN survey was taken. During periods of higher prices most coca producers use modern techniques such as fertilizers. Although coca cultivation increased at the expense of legal crops, it was the hillside forests that were affected most negatively. The intensification of land use was concentrated in the hillsides outside the boundaries of the Special Project, where the greatest number of monocrop coca farmers exist. The size of the coca

Table 7.2

Annual Rate of Deforestation, Annual Increase of Agricultural and Fallow Hectares,
According to Plot Size Among Noncoca Producers in Upper Huallaga in 1981

Plot size (ha.)	(1) Percentage of cases	(2) Average of plot size (ha.)	(3) Average of cultivated hectares (annual + permanent crops)	(4) Average of natural and cultivated pastures	(5) Average of fallow hectares	(6) Total cleared area (ha.) $(3 + 4 + 5)$	(7) Number of years occupying the plot	(8) Annual increase of agricultural hectares[*]	(9) Annual increase of fallow area[†]	(10) Annual rate of deforestation[‡]
0.1 < 10	15.0	7.61	4.47	0.68	2.29	7.44	9.83	0.45	0.23	0.76
10.1 < 20	34.0	16.34	6.18	2.40	6.18	14.76	7.38	0.84	0.84	2
20.1 < 30	21.0	29.94	5.24	2.83	9.90	17.97	10.73	0.49	0.92	1.67
30.1 +	30.0	57.19	8.04	13.28	15.16	36.48	10.86	0.74	1.40	3.36
Total	100.0	29.18	6.30	5.52	9.09	20.91	9.48	0.68	0.94	2.16

Source: Elaborated from National Foundation for Development survey (1981)

[*] Col. 3 divided by col. 7
[†] Col. 5 divided by col. 7
[‡] Col. 6 divided by col. 7

plantations in this zone varies from 2 to 3 hectares. These plantations are intensively managed and use many modern farming techniques. Inside the boundaries of the Special Project the strategy of extensive land management results in continued deforestation. On the hillsides negative ecological consequences result from the large amounts of chemical compounds used to produce cocaine paste (Dourojeanni 1989) as well as from the process of deforestation. Expanding coca production in the upper Huallaga region has produced a curious pattern in which the lands under extensive cultivation are located closer to main highways. Plots that are monocropped in coca and under intensive cultivation are located in regions more remote from the Marginal Highway and other routes of communication. This trend is the opposite of that predicted by modern theories of agricultural intensification.

Coca Eradication and Shining Path

Another important factor in the destruction of the rain forest is the program of eradication aimed at reducing the amount of coca production. The eradication campaign being carried out in the upper Huallaga region offers a clear example of the problem. Eradication has been initiated in the southern part of the valley along the eastern margin in the vicinity of Tingo María. The agriculturalists moved north and opened new fields for their coca plantations when they learned about the program.

As the eradication program spread to northern regions, the peasants resumed their migratory patterns and moved to more remote areas of the valley or to other areas, such as the central Huallaga and the Aguaytía zone of Pucalpa on the Ucayali River. The dynamics of the coca eradication program and of peasant migration have led to a general dispersion of coca plantations throughout the entire forest. This also explains why most coca producers cultivate several dispersed plots of the illegal crop. This has also contributed to the massive deforestation of areas designated as protected forest zones, or national parks, with all the negative consequences mentioned previously.

The problem of deforestation exacerbated by expanding coca production and corresponding policies of eradication is not confined to the national economy or to the institutional decisions related to the eradication campaign. The dilemma of coca and deforestation has important political dimensions. The agrarian reform eliminated the political and social elite constituted by the regional owners of the coffee and tea plantations in upper Huallaga and failed in its attempt to create a new political elite through the regional government bureaucracy and the cooperative leaders. This political vacuum, plus the weakness of the government's hold on tropical demographic frontiers such as the Peruvian Amazon basin, eased the expansion of coca cultivation and political organizations such as the Shining Path—*Sendero Luminoso.*

The political terrorist organization Sendero Luminoso maintains a significant presence in the upper Huallaga region. It is the only organization that openly defends the rights of the peasant coca producers, which guarantees its members legitimacy within the region. Sendero does not limit itself to organizing and mobilizing the peasant population through terrorist methods; it also encourages them to undertake continuous migrations in order to avoid the eradication program and police repression. Sendero Luminoso has thus pressured peasant agriculturalists into relocating to the most distant and fragile zones of the upper Huallaga.

Another aspect of the problem is the type of war that the Peruvian government has organized against the narcotraffickers and organized terrorists. Despite evidence of ties between these sectors, government institutions have treated them as two distinct battle fronts (ECONSULT 1986). The government strategy has been to concentrate all its efforts for a given period on one sector, leaving the other relatively free to operate. When the government decided that Sendero was becoming too powerful in the region, it abandoned its struggle against the narcotraffickers and withdrew the U.S. Drug Enforcement Administration (DEA), Mobile Rural Patrol Unit of the Peruvian Civil Guard (UMOPAR), and the eradication program and shifted to the military suppression of the guerrilla terrorists. Control over narcotraffickers virtually disappeared, leading to an increase in coca production in the area. In ecological terms this signaled an accelerated rate of deforestation.

Once the Peruvian government determined that Sendero was relatively under control, the military withdrew, police forces returned to fight the narcotraffickers, and the eradication program was reinstated. Sendero then reorganized and mobilized the peasant population to migrate to other regions. The deforestation resulting from the planting of new coca fields occurred repeatedly. It has become clear that both government strategies of repression carry weighty ecological implications.

One final aspect of the problem is ideology. According to our experience in the region, Sendero typically seeks to control a specific portion of the coca producers. The guerrillas try to persuade these peasants not to grow cash crops and to instead produce crops for family consumption while cutting back on coca hectares. Its antimarket ideology explains Sendero's actions with respect to the coca producers of the Huallaga region. The guerrillas require the peasants to pay tributes, and coca becomes a necessary exception to their rigid and dogmatic ideological scheme. Here again the ecological problem arises. Because Sendero promotes the development of economic self-sufficiency, which means extensive cultivation with slash-and-burn agriculture,the inevitable result is deforestation. Limiting the amount of deforestation in the upper Huallaga requires restrictions on the expansion of the coca plantations and reductions in coca production. The government must control both the national economy and Sendero Luminoso.

The tremendous amount of deforestation in the upper Huallaga region is not only a result of coca expansion but also of several decades of inadequate government economic policies. The extensive soil use in the Peruvian upper Amazon begins with the development of a government policy that promotes the expansion of the agricultural and demographic frontier through the construction of roads in fragile ecosystems, altering the productive conditions of the impoverished colonists. This political and economic policy was maintained throughout the 1960s, 1970s, and 1980s. Several colonization projects did not improve the peasant productive organization, and no attempts to formulate a stable ecological policy were made. Emphasis on road construction led to extensive soil use that was highly destructive to the environment. Likewise, regional political relations and labor scarcity forestalled any possibility of developing an intensive pattern of agriculture. Deforestation caused by coca producers is a consequence of traditional methods of shifting agriculture, practiced by impoverished farmers and never modified in several regions of upper Huallaga. The coca economy consumes most of the region's productive resources and deters the technological development of legal economy. As in the past, this has led to expansion of shifting agriculture and an increased rate of deforestation.

Notes

1. 'Upper forest' is the translation of the Spanish *selva alta*, a recognized ecological zone of the Peruvian tropical forest that lies between the eastern Andean slopes and the Amazon basin at an altitude of 400 to 800 meters.

2. From the time of the military coup on October 3, 1968, the military regime led by General Juan Velasco implemented and carried out the most extensive plan for structural reform in the republican history of the country. In general terms this was a government revolving around the redistribution of national revenue, the nationalization of petroleum, the banking industry, and other important sectors of the economy. Ideologically speaking, the military government tried to distance itself from both socialism and capitalism (Pease 1977). The military government implemented one of the most radical programs of agrarian reform, which affected the large agroindustrial complexes of the northern coast, the traditional haciendas of the Andean highlands, and the modern plantations of the upper forest. The reform generally was an attempt to substitute the former land tenure system and rural economic structure with a more efficient capitalist pattern (Caballero 1981). Rural reform meant transforming private agrarian companies into agrarian production cooperatives.

3. The only economic sphere in which I verified that coca was not used as a

means of exchange was in the acquisition of land. For this reason we do not refer to it as general purpose money.

4. José María Caballero describes the social organization of the forest plantations, differentiating them from the highland haciendas: "The typical colonist—called *mejorero*—was a worker who was paid by the job, in charge of putting wilderness areas into production. He was paid in money, through a system of advances, based on the amount of land improved. He was authorized to grow crops for his own subsistence on small marginal plots" (1981:263). Care of the established cultivated crops was covered, at least on most plantations, by the bonded workers who were partially paid with money and partially through the right to clear land and grow crops for their own consumption.

5. To tackle the problem of coca expansion in upper Huallaga, the United States and the Peruvian government in 1981 designed Project Paper to carry out PEAH. The design included research, extension, highway maintenance, and credit development. Furthermore, Project Paper excluded work in steeply sloping areas, where a significant number of coca producers are settled. Because PEAH's work was limited to farmers in flat areas, the effect of the project on coca producers was limited from the beginning (ECONSULT 1986). Unfortunately, forestry was not even considered. This was a serious shortcoming in a region in which a considerable proportion of the land is classified for forestry use (Bedoya 1987).

References Cited

Agreda, Víctor. 1984. *Frontera agrícola y demográfica en la selva alta* (Agricultural and demographic frontier in the forest). Lima: Instituto Nacional de Planificación.

Alvarez, Elena. 1983. *Política económica y agricultura en el Peru* (Political economy and agriculture in Peru). Lima: Instituto de Estudios Peruanos.

Aramburú, Carlos. 1982. Expansión de la frontera agraria y demográfica en la selva alta Peruana (Agricultural and demographic frontier expansion in the upper Peruvian forest). In *Colonización en la Amazonía* (Colonization of the Amazon), pp. 1–46. Lima: Centro de Investigación y Promoción Amazónica.

Aramburú, Carlos, Javier Alvarado, and Eduardo Bedoya. 1985. *La situación actual del crédito en el alto Huallaga* (The current credit situation in the upper Huallaga). Lima: U.S. Agency for International Development.

Bedoya, Eduardo. 1981. *La destrucción del equilibrio ecológico en las cooperativas del alto Huallaga* (Ecological destruction in the upper Huallaga cooperatives). Working paper No. 1. Lima: Centro de Investigación y Promoción Amazónica.

———. 1987. Intensification and degradation in the agricultural systems of the Peruvian upper forest. In *Lands at Risk in the Third World*, P. D. Little and M. Horowitz, eds., pp. 290–315. Boulder, Colo.: Westview.

184   Eduardo Bedoya and Lorien Klein

——. 1991. *The Social and Economic Causes of Deforestation in the Peruvian Amazon Basin: Natives and Colonists.* Working paper No. 60. Binghamton, N.Y.: Institute for Development Anthropology.

——. 1993. Bonded Labor in Peru: The Upper Huallaga Case. Ph.D. diss., State University of New York, Binghamton.

Bedoya, E., and F. Verdera. 1987. *Estudio sobre mano de obra en el alto Huallaga* (Labor force study in the upper Huallaga). Lima: Ronco.

Belaúnde, Fernando. 1965. *Peru's Own Conquest.* Lima: American Studies Press.

Bües, C. 1911. *La coca.* Lima: Ministerio de Fomento.

Bidegaray, P. and R. Rhoades. 1988. *Los agricultores de Yurimaguas: Uso de la tierra y estrategias de cultivo en la selva Peruana* (Farmers in Yurimaguas: Land use and crop strategies in the Peruvian forest). Working paper No. 10, Lima: Centro de Investigación y Promoción Amazónica.

Blaikie, Piers, and A. Brookfield. 1987. *Land Degradation and Society.* Methuen: London.

Bohannan, Paul. 1959. The impact of money on an African subsistence economy. *Journal of Economic History* 19: 491–503.

Bueno, Jorge. 1984. *Evaluación del desarrollo forestal en el ambito de los proyectos especiales de selva: Análisis comparativo* (Forest development evaluation in the special projects of the forest: Comparative analysis). Working paper No. 4. Lima: Instituto Nacional de Desarrollo/Apoyo a la Política de Desarrollo en la Selva Alta.

Burchard, Roderik. 1974. Coca y trueque de alimentos (Coca and food barter). In *Reciprocidad e intercambio en los Andes Peruanos* (Reciprocity and exchange in the Peruvian Andes), Giorgio Alberti y Enrique Mayer, ed., pp. 209–251. Lima: Instituto de Estudios Peruanos.

Caballero, José María. 1981. *Economía agraria de la sierra Peruana y pobreza campesina* (Agrarian economy in the Peruvian highlands and peasant poverty). Lima: Instituto de Estudios Peruanos.

Centro Nacional de Capacitación e Investigación para la Reforma Agraria (CENCIRA). 1973. *Diagnóstico socio-económico de la colonización Tingo María—Tocache y Campanilla* (Socioeconomic situation of the Tingo María, Tocache, and Campanilla colonization). Lima, Peru.

Cotler, Julio. 1979. *Clases, estado y nación en el Peru* (Classes, state, and nation in Peru). Lima: Instituto de Estudios Peruanos.

Dewind, Josh. 1987. Peasants become miners. In *The Evolution of Industrial Mining Systems in Peru, 1902–1974.* New York: Garland.

Dourojeanni, Marc. 1983. Bosques Amazónicos: Ecología y desarrollo rural (Amazon forest: Ecology and rural development). In *Socialismo y participación* (Socialism and participation), pp. 69–81. Lima: Centro de Estudios para el Desarrollo y la Participación.

——. 1984. Potencial y uso de los recursos naturales: Consideraciones metodológicas (Potential and natural resource use: Methodological consdierations). In *Población y colonización en la alta Amazonía Peruana* (Population and colonization in the Peruvian upper Huallaga), pp. 110–121. Lima: Consejo Nacional de Población y Centro de Investigación y Promoción Amazónica.

———. 1989. Impactos ambientales de la coca y la producción de cocaína en la Amazonía Peruana (Environmental impact of coca and cocaine production in the Peruvian Amazon). In *Coca y cocaína* (Coca and Cocaine), León and Castro de la Mata, eds., pp. 281–299. Lima: Centro de Información y Educación para la Prevención del Abuso de Drogas.

ECONSULT. 1986. *Informe final de la evaluación del proyecto* (Final report of the evaluation of the project Alto Huallaga). Report No. 527–0244 prepared for the U.S. Agency for International Development. Desarrollo Rural del Area del Alto Huallaga. ECONSULT.

Fioravanti, E. 1969. *Latifundio y sindicalismo agrario en el Peru: el caso de los valles de la Convención y Lares* (Large estates and peasant unions in Peru: The case of the Convención and Lares). Lima: Instituto de Estudios Peruanos.

Guerra, W. 1961. *El cultivo de la coca y su estudio en relación al indígena* (Coca cultivation and its study in relation to the indigenous population). Lima: Universidad Agraria.

Instituto Nacional de Estadística (INE). 1987. *Encuesta nacional de hogares rurales* (National suervey of rural households). Lima, Peru.

Loker, William. 1989. *Environment and Agriculture in the Peruvian Amazon: A Methodological Experiment.* Paper prepared for Rural Development in the Peruvian Amazon: Ecological, Socioeconomic and Technological Factors, workshop sponsored by Centro Internacional de Agricultura Tropical, November, Lima.

Malleux, Jorge. 1975. *Mapa forestal del Peru* (Forest map of Peru). Lima: Universidad Nacional Agraria.

Ministry of Agriculture. Peru. 1974. *Análisis de la situación de la colonización Tingo María—Tocache—Campanilla* (Analysis of the colonization of Tingo María—Tocache—Campanilla). Agrarian Zone no. 9. Aucayacu, San Martín: Ministry of Agriculture.

———. 1976. *Características de los suelos de la colonización Tingo María, Tocache y Campanilla, y los problemas derivados de su mal manejo* (Soil characteristics of the colonization of Tingo María, Tocache, and Campanilla and the problems caused by mismanagement). Aucayacu, San Martín: Ministry of Agriculture.

National Foundation for Development (FDN, Fundación para el Desarrollo Nacional). 1981. *Plan de ejecución del proyecto de desarrollo rural integral del alto Huallaga. diagnóstico social* (Plan of action of the rural development project in upper Huallaga: A social analysis). Lima: Fundación para el Desarrollo Nacional

Paez, J. 1937. Coca. Lima: *Agronomía* 2 (2): 3–44.

Painter, Michael and E. Bedoya. 1991. *Socioeconomic issues in agricultural settlement and production in Bolivia's Chapare region.* Working Paper No. 70. Cooperative Agreement on Settlement and Natural Resource Systems Analysis. Binghamton, N.Y.: Institute for Development Anthropology.

Pease, Henry. 1977. *El ocaso del poder oligárquico, lucha política en la escena oficial* (The demise of the oligarchic power, political struggle in the offical arena). Lima: Desarrollo Comunal (DESCO).

Peluso, Nancy Lee. 1990. A history of state forest management in Java. In *The Keepers of the Forest: Land Management Alternatives in Southeast Asia*, Mark Poffenberger, ed., pp. 27–55. West Hartford: Conn.: Kumarian Press.

Salazar, Alvaro. 1984. Situación actual de los proyectos especiales de selva (Actual situation of the special projects of the forest). In *Población y colonización en la Amazonía alta* (Population and colonization in the upper Amazon), pp. 245–274. Lima: Centro de Investigación y Promoción Amazónica—Instituto Andino de Estios en Población y Desarrollo.

Santos, Fernando. 1985. Crónica breve de un etnocidio o la génesis del mito del gran vacío Amazónico (Short history of the genesis of the legend). *Amazonía Peruana* 6 (11): 9–38.

Schmink, Marianne and Charles H. Wood. 1987. The "political ecology" of Amazonia. In *Lands at Risk in the Third Word*, Peter Little and Michael Horowitz, eds., pp. 38–57. Boulder, Colo.: Westview.

Servicio de Investigación y Promoción Agraria. 1962. *La actividad cafetalera en Tingo María* (Coffee activity in Tingo María). Lima: Ministry of Labor.

Stocks, Anthony. 1988. *Fragile Lands Development and the Palcazu Project in Eastern Peru*. Working Paper No. 34. Binghamton, N.Y.: Institute for Development Anthropology.

Verdera, Francisco. 1984. Estructura productiva y ocupacional en la selva alta (Productive and occupational structure in the upper forest). In *Población y colonización en la alta Amazonía Peruana* (Population and colonization in the upper Amazon of Peru), pp. 169–186. Lima: Centro de Investigación y Promoción Amazónica—Instituto Andino de Estudios en Población y Desarrollo.

Werlich, D. P. 1968. The Conquest and Settlement of the Peruvian Montana. Ph.D. diss., University of Minnesota.

World Resources Institute. 1986. *World Resources 1986*. Alexandria, Va.: Forte Group.

8

The Political Ecology of Deforestation in Honduras

Susan C. Stonich and Billie R. DeWalt

> I can only expect destruction for my family because I am provoking it with my own hands. This is what happens when the peasant doesn't receive help from the government and the banks—he looks for the obvious way out which is to farm the mountain slopes and cut down the mountain vegetation. Otherwise how are we going to survive? We're not in a financial position to say, "Here I am!—I would like a loan to plant so many hectares!" I put in my request but the banks don't want to give me credit because I cannot guarantee to cover the loan. I know what I am doing—as a person I know. I am destroying the land.
>
> —*Honduran peasant, 1990*

In 1972 the United Nations Conference on the Human Environment met in Stockholm to consider how human activities alter the global environment. At that time few people appreciatedthe extent to which economic development schemes affected the natural environment, and even less empirical research was aimed at understanding the complex relationships involved.

In June 1992 the United Nations Conference on Environment and Development (also known as the Earth Summit) convened in Rio de Janeiro, marking the twentieth anniversary of the Stockholm conference as well as the increased global awareness of the connections between development initiatives and the state of the natural environment. In addition, the conclusions of many international councils, which they reached during that twenty-year interval, point to the growing recognition of the complex interconnections among the economic, social, and demographic, as well as environmental, consequences of development efforts.

Especially important was the United Nations Commission on Environment and Development—the so-called Bruntland Commission—established in 1983 with the mandate to examine the causes rather than the effects of global environmental deterioration. The final report of the commission included the following recommendations: (1) to significantly change the way development projects are conceived and implemented, (2) to integrate environmental concerns in international and national programs of economic development, and (3) to recognize that such changes can only come about as a result of political action because of the lack of genuine consensus about the environment (Bruntland 1987).

Consequently, ameliorating global resource abuse will require what we term a *political ecology* of development.[1] Political economic perspectives traditionally have focused on understanding the tension between the government and the market, or on the interaction of the pursuit of wealth and the pursuit of power, as means of organizing human society (e.g., Gilpin 1987:11). In these conceptions the ecological effects of these processes have not been of much concern (Redclift 1984, 1987). In contrast, the political ecology approach looks at how the government and market interact to transform the environment and pursues questions of how political means may be applied to ensure that humans develop symbiotic, rather than destructive, relationships with the natural environment. By assuming that natural environments or ecosystems are in large part social constructs, political ecology also significantly expands much ecological analysis.

In other words, in terms of political ecological analysis it is insufficient to take the attitude toward the environment of former assistant secretary of state Elliott Abrams. When *Time* magazine reported that he was involved in a scheme to exploit Honduran timber using cargo-carrying blimps (and dubbed him Mahogany Man), he was quoted as saying, "I'm making lots of money. It's great" (June 5, 1989). Michael Redclift has asked whether it is possible "to undertake environmental planning and management in a way that does minimum damage to ecological processes without putting a brake on human aspirations for economic and social improvement" (1987:33). Our response is that our survival depends on pursuing economic and social improvement in ways that do minimum damage to ecological processes.

This essay uses a political ecology approach to examine the problem of deforestation and other abuses of natural resources in Honduras. The political ecological analysis includes an examination of the interconnections among the dominant export-led development model, the ongoing economic crisis, the policies and actions of the state, the competition among various classes and interest groups, and the survival strategies of an increasingly impoverished rural population. An examination of the Honduran case indicates that deforestation cannot be understood apart from the associated social processes and suggests that what is happening in Honduras is representative of processes occurring throughout the Central American isthmus. Analysis begins with southern Honduras, one of the most densely populated regions of the country and an area in which natural resources are most threatened (figure 8.1).

We will show that

> 1. Although deforestation in Honduras has many immediate causes, the roots lie in misdirected development strategies that have emphasized export-led growth.

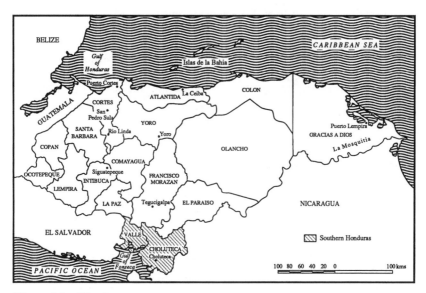

Figure 8.1   Honduras

2. Development in the region has in fact exacerbated structural inequalities and extremes of wealth and poverty that have intensified resource abuse throughout the country.
3. Governments (especially the United States in collaboration with the government of Honduras) and bilateral and multilateral aid and lending organizations are exacerbating resource destruction by focusing solely on short-term needs to generate foreign exchange and so-called development, defined only in terms of economic growth
4. Reversing deforestation and other resource abuse will require an altered development agenda that directly addresses extremes of wealth and poverty and other issues of social and environmental justice.

Although this essay focuses on Honduras, the patterns that we discuss relate to other areas of Central America, as people like Robert Williams (1986), H. Jeffrey Leonard (1987), and others (e.g., USAID 1989a; Faber 1992) are demonstrating. Extensive evidence shows that because of economic and population pressures, governments and individuals alike are overexploiting the natural resources that they control in order to generate income to satisfy immediate needs—whether those requirements are to generate foreign exchange at the national level or to increase income at the level of the household. The costs of these short-term strategies are already apparent in the accelerated deterioration of forests, soils, fisheries, and other crucial resources and in probable longer-term declines in food security, economic growth, and human well-being (Williams 1986; Leonard 1987; USAID 1989a). These processes have to be reversed, or the continued destruction of Central American natural resource systems will doubtless exacerbate the existing widespread problems of economic stagnation, rural impoverishment, and social and political instability (USAID 1989a).

Development Trends in Honduras

Except for the banana industry established at the turn of the century along the relatively isolated north coast, extensive agrarian capitalism in Honduras did not arise until after World War II during a period of temporarily high prices on the world market for primary commodities like cotton, coffee, and cattle. At that time the industrialized countries promoted capitalist enterprises through increased foreign investment, and national security interests prompted the U.S. government to expand programs of economic and military assistance. The Honduran government became an active agent of development, creating a variety of institutions and agencies to expand government services, modernizing the country's financial system, and undertaking a number of infrastructural projects (Stonich 1993). With the infrastructural improvement, landowners and investors in the southern part of the country found it profitable to expand production for the global market, and southern Honduras was firmly integrated into national and international markets for the first time. Since then diversification and growth of agricultural production for export have characterized the southern Honduran economy. With financial assistance from multilateral and bilateral development and lending institutions (most important: the United States Agency for International Development [USAID], the World Bank, and the International Monetary Fund [IMF]), cotton, then sugar and livestock were the primary commodities first promoted in the south. By the mid-1970s these products were supplemented by sesame and melons and later by a wider variety of so-called nontraditionals, especially cultivated shrimp (Stonich 1991a, 1992, 1993).[2]

The Honduran government's continued efforts to expand export agriculture are more understandable, given Honduras's extreme economic dependence on agriculture and its continued economic crisis. Honduras remains predominantly an agricultural country; in 1990 agriculture generated about 30% of its gross domestic product, 75% of export earnings, and 55% of employment (Comisión Nacional 1992:67). Indications of the international economic crisis emerged in Honduras in 1981 and intensified through the end of the decade. Productive activity declined drastically, unemployment intensified, and inflation deepened. The balance of payments and the national treasury suffered imbalances, and the real income of a large proportion of the population declined. Honduras was significantly constrained in supplying imported materials, and private investment dropped as a result of the region's political and social problems and disturbances in exchange and monetary systems. This situation was aggravated by the economy's vulnerability to external fluctuations, which affected the demand and price of its most important traditional export products such as bananas and coffee (Stonich 1993).

By 1989 the Honduran external debt of U.S.$3.3 billion was 120% of the annual gross domestic product—larger than the per capita debt of either Brazil or Argentina (Daniels 1990). By late 1989 all the major financial lending institutions (the World Bank, International Monetary Fund, and Inter-American Development Bank) had placed Honduras on the list of countries that were ineligible for new loans because of overdue payments on earlier credits, as well as because of the Liberal government's reluctance to continue its economic adjustment program. Also in 1989, for lack of what it perceived as a sound economic reform program, USAID did not release U.S.$70 million that had been approved to support Honduras's balance of payments.[3]

Economic liberalization was a central component of the platform of the National party, which came to power in early 1990. One of President Rafael Callejas's first actions was to declare the nation bankrupt. Barely a month after taking office Callejas, with the support of his new legislative majority, passed a major reform of the Honduran economy that was both in line with the demands of major creditors and designed to make Honduras more attractive for investors and hence promote exports: the national currency (the lempira) was devalued by 100%, and a crawling peg rate of exchange was adopted; protective import tariffs were slashed from 135% to 20%, and investment regulations—both for foreigners and national entrepreneurs—were simplified.

Fiscal deficit reduction actions included decreased public spending (achieved in part by laying off approximately ten thousand government workers, about 20% of the government's employees, in January 1991), elimination of subsidies, increased water and energy tariffs, and modification of prices to actual market values. The exchange rate of the lempira (per U.S. dollar) rose from 2.0 before the devaluation to 3.5 immediately afterward, to 4.9 by July 1990, and to 5.5 by July 1991 (*Latin American Regional Report* 1991a). Inflation during the twelve-month period of May 1990 to May 1991 was 38.7% (*Latin American Regional Report* 1991a). The ensuing rise in the cost of living further hurt the economic circumstances of the most vulnerable sectors of Honduran society, whose minimum wages remained unchanged and who were also most affected by the sharp rise in unemployment. Despite the apparent effects of the severe structural adjustment program on the poor and the presidential election of 1993, which returned control to the Liberal party, the ongoing economic crisis makes it highly unlikely that the national government will direct its policies away from attempting to expand export production (Stonich 1993).

In this critical time the natural resource base of the country has come under severe pressure. Honduras is highly dependent upon renewable natural resources to generate income from agriculture, forestry, and fisheries. Natural resource-based commodities were the principal means of earning foreign exchange, providing more than 80% of export earnings throughout the 1980s (World Bank 1984–94). During the fiscal crisis grappling with the repayment of growing external debt has been more important to the Honduran government than conserving natural resources. Raising cotton, cattle, melons, and shrimp draws international financial assistance and helps meet foreign exchange requirements—whatever their social and environmental costs.

The Status of Honduran Forests

During the 1980s Latin America's average annual rate of deforestation was the highest in the world (approximately 1.3% of existing forests were lost annually). This overall rate was exceeded within Central America, which underwent estimated annual losses of 1.6% during the period (World Resources Institute [WRI] 1990:42). During that same period average forest loss in Honduras was appraised at 2.3% annually (WRI 1990:42). Table 8.1 compares the results of an inventory of Honduran forests compiled by the Food and Agriculture Organization of the United Nations in 1964 with a similar inventory completed in 1986 by the parastatal Honduran Forestry Corporation (Corporación Hondureña de Desarrollo Forestal (COHDEFOR) in charge of forest management. It reveals a total loss of forests of 26% (approximately 1.76 million hectares) from 1964 to 1986 and shows that the greatest loss was in broadleaf forests (34.8%) compared to pine forests (12.5%).

In general, rapid rates of deforestation of broadleaf forests first occurred in the southern part of the country in what were primarily tropical dry deciduous forests but more recently have accelerated in the tropical humid forests located in northeasterly portions of the country. The recent *Environmental Profile of Honduras—1989* identifies the principal causes of deforestation (in upland and noncoastal zones) as

1. Rapid population growth, which led to cultivation of increased marginal land and to an expansion of the agricultural frontier
2. Inefficient and wasteful lumbering practices
3. Lack of supervision and control by COHDEFOR

Table 8.1
Forest Loss in Honduras, 1964–86 (in Thousands of Hectares)

Type of forest	FAO in 1964	COHDEFOR in 1986	Forest loss	Percentage in 22 years	Annual deforestation
Pine forest	2,739	2,397	(342)	12.5%	16
Broadleaf forest	4,072	2,654	(1,418)	34.8%	64
Totals	6,811	5,051	(1,760)	25.8%	80

Source: Corporación Hondureña de Desarrollo Forestal (COHDEFOR 1988); FAO is the Food and Agriculture Organization of the United Nations.

4. No local incentives for protecting and conserving forests, which translates into indifference on the part of the population
5. Unequal and insecure land tenure
6. No clear national forestry policy
7. Entrepreneurs unaware of the need to manage forest resources in an orderly and sustainable manner
8. Failure by the government to implement a systematic and persuasive education campaign to create public awareness of the necessity to protect and use the forest resources rationally
9. Instability in the group of public administrators that decides forestry policy
10. Lack of an agrarian reform law that takes into account forest management and the rational use of forest resources (Secretaría de Planificación [SECPLAN] and USAID 1989).

There also has been increasing concern about degradation of coastal zones, especially the significant loss of ecologically vital mangrove forests and associated ecosystems in areas surrounding the Gulf of Fonseca (SECPLAN and USAID 1989; Stonich 1991a, 1992, 1993; Foer and Olsen 1992; International Union [IUCN] 1992; Vergne, Hardin, and DeWalt 1993). According to the recent *Environmental Study of the Gulf of Fonseca* (Vergne, Hardin, and DeWalt 1993), the area in high-quality mangrove stands declined by about 6,760 hectares (22%) since 1973. Of this total, approximately 2,132 hectares (32% of the total area lost) was the direct result of the construction of shrimp farms. An undetermined amount of loss can also be indirectly attributed to the expansion of the shrimp industry because road building and pond construction lead to changes in hydrology.

The remaining mangroves are lost to a combination of factors, including the construction of salt-making ponds, the cutting of trees for fuelwood and construction materials, and the gathering of bark from red mangroves for the tanning industry (SECPLAN and USAID 1989; IUCN 1992). For example, approximately 46,300 cubic meters of mangrove fuelwood, equivalent to the loss of 250 to 350 hectares of forest, are used annually (Flores and Reiche 1990). An undetermined but probably significant amount of mangrove destruction can also be attributed to the increased sediment loads carried by freshwater runoff from mountainous watersheds and deposited in coastal zones. Highland deforestation and intensive agriculture on steep hillsides have produced extremely high rates of soil erosion and excessive sedimentation.[4] The destruction of mangrove areas, along with the disappearance of seasonal lagoons, deteriorating water quality, and a declining gulf fishery have precipitated widespread social conflict and placed southern Honduras in the center of increasingly violent confrontations between opposing interest groups (Stonich 1991a, 1993; Vergne, Hardin, and DeWalt 1993; Stonich, Murray, and Rosset 1994).

Government Policy Regarding Forestry Management

In part because of increased concern over the clear-cutting of upland forests by foreign lumber companies, the Honduran government began to assume a greater role in forestry resource management in the early 1970s. The principal laws governing forest management were enacted: Decree 85, the Forest Law, which outlined national forest conservation and management requirements, and Decree 103, which created COHDEFOR as manager of the nation's forests (USAID 1982). The specific mandate of COHDEFOR was to halt clear-cutting by foreign companies and to regulate the extraction and marketing of forest products in order to generate income to finance various government development programs. To accomplish this the Honduran government in effect nationalized the forests.

Although the government was given exclusive ownership of Honduran forests, new or existing groups of farmers living in the forest were considered (at least on paper) the chief means of executing programs to conserve and regenerate the forests. Established within COHDEFOR was the national Social Forestry System (Sistema Social Forestal), the goal of which was to promote the formation of farmer cooperatives or other groups to protect forests by preventing fires, overgrazing, illegal cutting, and the expansion of pasture and shifting agriculture. In addition to supporting cooperatives, COHDEFOR created government-sponsored forest-management zones (areas of integrated management—AMIS) on large forest tracts that were allocated to specific community level groups. The government provided technical advice, materials, and markets, and rural people were to supply the labor. Although by 1987 fifty AMIS had been established, in reality neither the forestry cooperatives nor the AMIs ever received much financial or technical assistance from COHDEFOR (SECPLAN and USAID 1989).

In the wake of passage of laws 85 and 103 a number of serious problems arose, especially regarding enforcement. Among the most crucial were lack of clearly defined forestry policies, regulations, and guidelines, lack of coordination and communication both within COHDEFOR and between COHDEFOR and the many other institutions that affect the management of forestry resources (including several government agencies and ministries as well as organizations of farmers and ranchers), and inadequate execution of plans and decisions. These difficulties resulted in making COHDEFOR a vast, unwieldy, and indecisive bureaucracy and contributed to the uncontrolled and ecologically unsound exploitation of Honduran forests (USAID 1982; SECPLAN and USAID 1989).

SECPLAN and USAID (1989) identified the failures of the national Social Forestry System, as well as the far-reaching powers and inadequate management of COHDEFOR, as among the principal causes of deforestation. In response, the government of Rafael Callejas significantly revised its natural resource policy (Johnston et al. 1990). Preliminary measures included ending COHDEFOR's monopoly on wood exports and doubling stumpage fees in order to discourage overexploitation of forests. Although the government maintained that it was committed to conserving Honduras forests, in 1992 it attempted to enter into a preliminary forty- year contract with the Stone Container Corporation of Chicago to establish a pine plantation and chip mill in La Mosquitia, the last remaining large area of tropical humid forest in the country (Honduran Popular Action Group 1992). Only widespread public resistance by national and international environmental groups thwarted that effort.

Later the same year, however, Honduras passed the Law for the Modernization and Development of the Agricultural Sector (Decree 31–92), which included controversial forestry provisions (passed in 1993). The law stripped COHDEFOR of all authority except its supervisory and enforcement powers (which remain important) and gave the right to cutting and commercial forest production only to private persons or entities. In addition, companies engaged in various facets of commercial forestry could include foreign owners, partners, and investors and could use foreign capital without limitation (Fandell 1994). Thus shortly after rejecting Stone Container's proposal in response to national and international environmental protests, the government enacted legislation that opened Honduran forests to forestry corporations all over the world. Nor were Stone Container Corporation's efforts to establish a new plantation and mill in Central America blocked. After failing to reach agreement with the Hondurans, the company began negotiations to transfer the operation to the Punta Estrella rain forest in Costa Rica (Scanlan 1994).

Protection and management of mangrove ecosystems received legal status in Honduras through the articles of the Fisheries Law of 1959, which prohibit clearing of mangroves on shorelines, and the Forestry Law of 1958, which declared mangroves protected forestry zones. Although modified by subsequent forestry laws (most important was the creation of COHDE-FOR in 1974), the effectiveness of national forestry legislation has suffered from the lack of clear operational directives and shortages of trained staff (Vega 1989). With regard to aquaculture development and mangrove areas, the Honduran government has administrative authority over lands that lie between high tide and a point 2 kilometers inland. Until recently the government exercised this mandate through the Honduran Institute of Tourism, but it has been assumed by the Ministry of Natural Resources. Despite this chance to directly influence the effects of shrimp-farm expansion on mangrove zones, the agency has established no direct link between the granting of concessions for farm construction and requirements for mangrove protection (in part because of the lack of clear procedures governing concessions) (Vergne, Hardin, and DeWalt 1993:22–23).

Southern Honduras: Environment and Demography

Southern Honduras is located in tropical dry and subtropical moist forest zones between the borders of El Salvador and Nicaragua (Holdridge 1962). The zone includes the departments of Choluteca and Valle and has a total surface area of about 5,757 square kilometers, about 5.2% of the national territory. Three major geomorphic areas can be defined within the region: the coastal zone, the plains, and the highland (mountains). The coastal area of the south that lies adjacent to the Gulf of Fonseca provides Honduras with its only access to the Pacific Ocean. This is an area rich in biodiversity—extensive stands of mangroves, seasonal lagoons, estuaries, mud flats, and enclaves of dry tropical forests. The coastal mangrove forests, estuarine waters, and wetlands generally have a high biological productivity and serve as nursery areas for many species of finfish, shellfish, and crustaceans.

Beyond the mangrove forests lies one of the few extensive plains on the Pacific coast of Central America. The plains can be divided into two zones, an alluvial sedimentary shelf that stretches from the coastal area to 15 meters above the mean high-tide mark and a higher shelf that continues as much as 200 meters above the high-tide mark. This savanna gives way to steep foothills, which quickly become the jagged mountain ranges that form a broad base to the northeast and comprise the majority of the region. Although these volcanic mountains rarely reach altitudes of more than 1,600 meters, they are exceedingly rugged and form myriad isolated valleys.

Remnants of tropical dry forest occur inland from the coastal zone. Such tropical deciduous forests are found in areas where marked seasonality of precipitation predominates and were once prevalent along the entire Pacific coastal plain of Central America. Although deciduous forest once represented the dominant vegetation type in the lowlands of the Pacific coastal region of southern Honduras as well, agriculture (crops and cattle) has almost completely eliminated it. Only a few fragments remain, mostly as scattered gallery forests along streams and rivers.

Pine and oak associations, corresponding to Leslie R. Holdridge's subtropical moist forest (1962), occur at altitudes of 600 to 1,800 meters. Predominant species are oak (*Quercus*) and pine (*Pinus oocarpa*) at lower elevations and pine (*Pinus psuedostrobus*) at higher elevations of the zone. Understory varies from grassy cover to low shrubs and tall grasses. Slash-and-burn agriculture, cattle grazing, cutting of trees for fuelwood and construction, and commercial logging of pine for export have greatly modified this habitat.

Islands of cloud (montane rain) forest are found at elevations of 1,350 to 2,300 meters; the almost daily cloud build-up and the lower evaporation rates on mountain peaks provide moisture for the lush plant growth. These highland broadleaf forests generally are surrounded at lower elevations by pine and oak forest. Cloud forests are important in the regulation of surface and groundwater supplies for drinking, irrigation, and hydroelectric power production. Because of their rugged terrain many of these cloud forests remained fairly intact until the 1980s. However, they are being seriously degraded as increasing populations of desperately poor farmers expand slash-and-burn cultivation to these formerly remote areas.

Adding to these environmental concerns has been the considerable climatic instability of the last few decades (Stonich 1993:36). In a region characterized by erratic precipitation the 1980s were marked by the worst drought in fifty years and accompanied by an increase in median ambient temperature of 7.5 degrees centigrade (Almendares et al. 1993). The growing ecological crisis in the region has not only increased the agricultural risk, especially for small farmers, but has also altered the distribution of vector-borne diseases affecting people, crops, and other crucial species. (Comprehensive Resource 1984; Stonich 1986, 1989, 1993; SECPLAN and USAID 1989; and IUCN 1993 contain more complete discussions of the environmental context and the natural and agricultural potential of the area.)

Demographic Considerations

The rate of population growth in Honduras has been among the highest in the world, averaging 3.1% per year from 1950 to 1974 and rising to 3.4% from 1974 to 1988 (Stonich 1993:40). In 1990 the population of Honduras was estimated at 5.1 million, nearly double the 1970 population of 2.63 million (World Bank 1992:268). Although the total fertility rate for Honduras dropped from 7.4 births per woman in 1970 to 5.4 in 1989, and the annual growth rate declined to 2.96% by 1990, the country's population continues to grow rapidly, and the population is expected to reach 6.2 million by the year 2000 (SECPLAN 1991:206).

Persistently high rates of population growth have been accompanied by escalating population densities nationally: from 12.2 people per square kilometer in 1950 to 39.1 in 1988 (Stonich 1993:41). Southern Honduras is the most densely settled region of the country, comprising only 5.2% of the total national land area but approximately 9.3% of the population (Stonich 1989:277). Population density remains well above the national average, climbing from 29.8 persons per square kilometer in 1950 to 72 in 1988, with population densities near 150 people per square kilomter in some highland municipalities (Stonich 1993:41).

Although population densities continue to be significantly higher than that of the nation as a whole, since 1950 the rate of growth in the south has not been as high as in other areas of the country. This is primarily the result of extensive out-migration from the region and in part the result an infant mortality rate that is higher than the national average. Almost half of all people born in the region migrate to other parts of the country; the most popular destinations are the capital city of Tegucigalpa, the industrial center of San Pedro Sula, and the rural "agricultural frontier" areas in the northeastern part of the country. Considerable migration from rural to urban areas of the south (the cities of Choluteca and San Lorenzo) also is occurring. Despite migration to urban centers within the region, the south remains more rural than the country as a whole, with three-quarters of the population living in rural areas in contrast to 60% nationally (Stonich 1991b, 1993).

Agrarian Transformation and Ecological Consequences

The Cotton Boom

It was cotton cultivation that first transformed traditional social patterns of production in southern Honduras (Stares 1972:35; Durham 1979:119; Boyer 1982:91). Although cotton had been grown in the area since preconquest times, large-scale commercial cultivation of cotton was introduced in the late 1940s and 1950s by Salvadorans who brought seeds, chemicals, machinery, and their own labor force into the area. Salvadoran farmers secured Honduran bank loans, rented (or purchased) large tracts of land from Honduran owners, and began commercial production. They were joined by Honduran farmers who first began producing on a minor scale but who by 1960 expanded production and formed their own ginning and marketing cooperative. When the Salvadorans were expelled from the country after the Salvadoran- Honduran War in 1969, their property was confiscated and became available to the Honduran growers (Stonich 1986:118).

As in El Salvador and Nicaragua commercial cotton cultivation in Honduras involved considerable mechanization in land preparation, planting, cultivation, and aerial spraying and was dependent on the heavy use of chemicals (especially insecticides and fertilizers).

The indiscriminate use of pesticides in the cotton-growing regions remains among the most pervasive environmental contamination and human health problems throughout Central America (Central American Institute [ICAITI] 1977; Weir and Shapiro 1981; Bull 1982; Botrell 1983; Boardman 1986; Williams 1986; Leonard 1987). Water from cotton-growing areas of southern Honduras shows heavy contamination from DDT, dieldrin, toxaphene, and parathion (USAID 1982). A 1981 study of the levels of pesticide poisoning in the area around the city of Choluteca, Honduras, revealed that approximately 10% of the inhabitants had pesticide levels sufficiently high to be considered cases of intoxification (Leonard 1987:149). A number of reports show that the land and water contamination from pesticides, as well as high levels of pesticide residues in food supplies, continue to have substantial effects on human health (Williams 1986; Leonard 1987; Murray 1991).

The major social effect of the cotton boom was to increase inequalities in access to land. Large landowners revoked peasant tenancy or sharecropping rights and raised rental rates exorbitantly so that peasants would leave the land. Landowners also laid claim to many wilderness areas and forcibly evicted peasants from national land or from land of undetermined tenure (Parsons 1975; Durham 1979; Boyer 1982:94). Increased cotton cultivation thus displaced many poor farmers from the more suitable agricultural lands in the south. At the same time, however, cotton provided many seasonal jobs during the harvest season, because the long-staple cotton grown in the region was largely picked by hand.

Production of cotton in the south fluctuated considerably before the cotton boom finally ended in the late 1980s. The build-up of pesticide-resistant insect populations and the increasingly high costs of pesticides, combined with low market prices, effectively ended cotton cultivation in southern Honduras. Although attempting to resurrect cotton cultivation using integrated pest management techniques has been discussed, virtually no cotton was planted in the south through 1992.

The Cattle Boom

The expansion of the cattle industry probably had the most extensive and devastating environmental effects in the south. During the 1960s the

Alliance for Progress and the growing demand for inexpensive beef by the expanding U.S. fast-food industry helped to fuel a livestock boom throughout Central America.

Honduras increased its export quotas to the United States, implemented development initiatives that stimulated the beef trade and modernized beef production, and instituted credit programs to help expand beef production. From 1960 to 1983 57% of all loans allotted by the World Bank for agriculture and rural development in Central America financed the expansion of beef for export. During that same time Honduras received 51% of all World Bank funds disbursed in Central America, of which 34% was used for livestock projects (Stonich 1992). This assistance was funneled into the country through institutions and projects controlled by national elites as well as foreign (especially U.S.) interests (Stonich and DeWalt 1989).

In a context of declining agricultural commodity prices, high labor costs, unreliable rainfall, and international and national support for livestock, landowners reallocated their land from cotton and/or grain cultivation to pasture for cattle (Stonich 1986; Stonich and DeWalt 1989). Cattle appealed to landowners in Honduras, because they can be husbanded with little labor. With only two or three hired hands and extensive pasture a landowner can manage a herd of several hundred cattle. Ironically, land reform programs also encouraged the expansion of pasture for livestock. Landowners who feared expropriation of unused fallow and forest land fenced it and planted pasture to establish use of the land without substantially increasing their labor costs (Jarvis 1986:157; Stonich 1986, 1992).

Large landowners also exploit the growing inequalities in access to land with an inexpensive way to convert land from forest to pasture: by renting hillside land in forest to land-poor peasants (DeWalt 1983, 1985, 1986). These renters cut the forest down in order to plant maize and sorghum, their principal subsistence crops. During the second or third year of cultivation, when land fertility declined, landowners instructed the renters to sow pasture grasses among the maize and/or sorghum. This converted the land, usually permanently, into pasture for cattle. Renters recognize that they are destroying their potential source of livelihood as more fallow and forest land is converted into pasture. They are caught because they have to meet their short-term needs for survival, yet they jeopardize their long-term future by participating in the pasture conversion process. In the words of one small farmer, "Right now we have land available to rent, but each year you can see the forest disappearing. In a few years, it will all be pasture and there will be no land available to rent. How are we to produce for our families then? We see what is happening, but we have no choice because our families have to eat now."

The expansion of pasture caused extensive changes in land-use patterns in Honduras through the 1960s and 1970s. Growth took place in the lowlands and foothills, where cattle raising traditionally occurred, and in the highlands, where many of the wealthier peasant farmers augmented cattle production with income generated by agricultural production (Durham 1979; Boyer 1982; Stonich 1986). Increased livestock production in the lowlands and the highlands accelerated the expulsion of peasants from national and private lands (White 1977:126–156; Stonich 1986:139–143). From 1952 to 1974, for example, pasture in the southern region of the country increased from 41.9% of the land to 61.1% and was associated with the simultaneous and precipitous decline of land in fallow and in forest (Stonich 1989, 1993). Thus both deforestation and serious soil erosion accompanied the cattle boom. It has been estimated that Honduras is losing its forests at the rate of 10,000 hectares per year and, if current trends continue, "the forest resource will be exhausted in a generation" (USAID 1990:3). Most dry tropical forest in the south has already disappeared, and soil erosion rates are alarming.

Local and Regional Consequences of Development

The social consequences of the expansion of the cotton and cattle industries—of economic development—on rural areas of the south have been discussed in detail elsewhere (see White 1977; Durham 1979; Boyer 1982; Stonich 1986, 1989, 1993; Stonich and DeWalt 1989). Briefly, development led to ever greater socioeconomic inequalities of households in the region. Farmers with medium and large holdings sought to improve their competitive position in the world marketplace. Using the international foreign assistance that was channeled through government loans, they tried to cut their costs by investing in commodities and techniques that were labor displacing rather than labor absorbing; they tried to achieve economies of scale by acquiring more land and expanding their operations; and as material costs rose and prices fell for cotton, they increasingly switched their operations to cattle, a commodity that requires small amounts of labor and large amounts of land (DeWalt 1986; Stonich and DeWalt 1989).

The appropriation of land for commercial agriculture and for extensive livestock raising relegated resource-poor individuals to the most marginal areas of the south. Using shifting cultivation systems, peasants in the foothills and highland regions expanded production to steep slopes, interplanting maize and sorghum (their primary subsistence crops) for a few years before leaving the field in fallow to regain its fertility (Stonich 1993). The conversion of land to pasture, combined with the rapid growth of the human population, has increased the pressure on the remaining cropland. During the last several decades fallowing periods in the south have decreased. In some communities fallow periods have been eliminated entirely, whereas in others the fallowing interval has decreased, from fifteen to twenty years in the 1950s to just a few years (Stonich 1993:150–152).

This trend toward permanent cultivation has led to depletion of the soil and has exacerbated the soil erosion problems on steep slopes (Stonich 1993:150–152). Thusthe landscape of southern Honduras has been transformed in recent decades. The greatly disturbed regional ecology has been left vulnerable to the volatile weather patterns since the mid-1980s and has resulted in extensive flooding, landslides, and watershed destruction.

The concentration of agricultural land, combined with the lack of alternative economic options and growing environmental destruction, led many resource-poor families to seek opportunities elsewhere (Stonich 1991b). Between 1974 and the late 1980s out-migration from the southern region averaged 1.3% annually. Approximately half as many people left the region permanently each year as were added to the population by both its high birthrate and limited in-migration. Many poor families engaged in cyclical or permanent migration to the cities or came to depend on remittances from family members (Stonich 1991b). The urban population growth rate in Honduras was about 5.6% from 1974 to 1987, a rate much higher than the overall national population growth rate of about 3.4% for the same period (USAID 1989b). The expanding squatter settlements on the edges of Tegucigalpa and San Pedro Sula bear witness to the environmental problems caused by this rural to urban migration.

Migrants from environmentally degraded areas in the south also have extended the agricultural frontier by settling in the departments of Olancho and El Paraiso, which border the relatively unpopulated tropical humid forest region of La Mosquitia in northeastern Honduras. According to the national population census of 1974, the adjacent departments of El Paraiso and Olancho rank behind only the largest cities (Tegucigalpa and San Pedro Sula) as the predominant extraregional destinations of migrants from the south (Stonich 1991b). Community-level research shows that by the 1980s these two departments accounted for more than 50% of the total destinations of male householders from rural highland communities in the south (Stonich 1991b).

The first organized migration of people from the south to La Mosquitia began in the early 1970s, and by the 1980s communities had settled along the entire upper reaches of the Rio Patuca. The colonization of this area of tropical humid forest has extended into the Rio Platano Biosphere Reserve.[5] Replicating processes taking place throughout Latin America, deforestation has taken a heavy toll on ecosystems, as newly arriving colonizers (many using the illegal roads constructed by loggers) clear forest for crops, cattle, and fuelwood, thereby facilitating the expansion of ranching interests and encroaching on the lands inhabited by Honduras's small remaining indigenous population.

Another strategy for resource-poor households is to relocate within the southern region to the relatively sparsely populated coastal region of mangrove, mud flats, estuaries, and seasonal lagoons along the Gulf of Fonseca.

Unsuitable for large-scale cultivation of crops, pasture, or most other commercial uses, this area has become populated by increasing numbers of migrants from other municipalities in the south. From 1974 to 1988, a period of substantial out-migration from the southern region as a whole, rural populations in the six municipalities that border the Gulf of Fonseca grew faster than the country as a whole. The families settling the coastal communities survive by exploiting the resources of the coast and the estuaries. They clear the wilderness to cultivate crops but have come to depend as well on fish, shrimp, shellfish, animals, and wood gathered from the surrounding common resource areas—lagoons, mangroves, estuaries, and the Gulf of Fonseca. Until the early 1980s the only major competition for these coastal resources was from commercial salt-making operations.

The "New Nontraditionals": The Shrimp Boom

Central American countries are championing shrimp mariculture as a principal means of attacking the region's continuing economic crisis. Several international development agencies, including the United Nations Development Program and USAID predict that shellfish will be the most important primary nontraditional export commodity from the region during the 1990s. Bilateral and multilateral development assistance agencies, national elites, as well as private investors from North America, Japan, Taiwan and elsewhere, are fostering the growth of shrimp mariculture in coastal zones (Stonich 1991a, 1992).

Exports of shrimp from Central America increased significantly throughout the 1980s as more producers became involved in this nontraditional market and began shipping large quantities of frozen shrimp from the Pacific and Caribbean coasts. During the 1980s, as overfishing and destruction of habitats dramatically reduced catches from capture fisheries, the Central American shrimp industry grew increasingly dependent on mariculture to supply shrimp for export. Honduras, Panama, and Costa Rica led the region in the expansion of cultivated shrimp production, with most operations located along the Pacific coast. By 1987 Honduras's foreign exchange earnings from shrimp were exceeded only by export earnings from bananas and coffee (Stonich 1991a, 1992). Most of this growth was the result of the expansion of shrimp farms in coastal zones along the Gulf of Fonseca in southern Honduras; the expansion occurred as beef was losing its status as the most important agricultural commodity from the zone. By the mid-1980s principal investors in the industry included transnational corporations and government and military leaders, as well as consortiums of private investors.

As in the rest of Central America the growth of the shrimp industry was financed by national, international, private, and public capital (SECPLAN

and USAID 1989). This included direct financing through loans and technical assistance supplied by USAID and indirect funding in the form of incentives to foreign investors (USAID and Honduran Federation 1989).
Although USAID reports written through the mid-1980s emphasized the
importance of integrating resource-poor households in the shrimp industry, mainly by forming and supporting cooperatives, more recent reports
conclude that only the larger semi-intensive operations are profitable
(USAID 1989a), and USAID has virtually curtailed its efforts with small producers. By 1993 the south had approximately 11,500 hectares of semi-intensive shrimp operations, with sales of more than 4 million kilos of shrimp
valued at U.S.$ 40.2 million (Vergne, Hardin, and DeWalt 1993). According
to the Chamber of Commerce of the departments of Choluteca and Valle,
the shrimp industry provided employment to some 11,900 people (90%
women) through 25 commercial farms, 6 packing plants, and 6 ice-making
operations (Vergne, Hardin, and DeWalt 1993).[6]

Despite providing jobs, the expansion of the shrimp industry has raised
a number of environmental and social justice concerns that in turn have
incited widespread conflict and increasingly violent confrontations. The
major environmental issues center around the consequences of alterations
in, and the loss of, mangrove ecosystems,[7] whereas social justice issues
focus on diminished access to common property resources brought about
by the government-controlled concession process.[8]

Contributing to the conflicts is the lack of adequate and reliable information on the regional environment and ecology. In 1990 (several years
and millions of dollars into USAID's efforts to promote shrimp mariculture
in southern Honduras) the USAID-funded environmental analysis of its
Investment and Export Development Project asserted that the "pitifully little research on the natural resources of the Gulf of Fonseca's estuaries,
mangrove forests, and mudflats" made it impossible to evaluate adequately the significance of environmental changes emanating from the
ongoing expansion of shrimp farms (Castañeda and Matamoros 1990).
Because USAID financing of shrimp industry expansion was channeled
through a financial intermediary (a national development bank), USAID
was able to circumvent its own environmental regulations. More recently,
in response to internal and external pressures USAID commissioned a number of environmentally related studies, including *Environmental Study of the
Gulf of Fonseca* by Tropical Research & Development, a Gainesville, Florida,
environmental consulting firm that has done extensive work in Honduras
(Vergne, Hardin, and DeWalt 1993).

From 1973 to 1982 (before the boom in shrimp mariculture) the number
of hectares of high-quality mangrove declined from 30,697 to 28,776, representing a loss of 1,927 hectares (6%).[9] By 1992 total hectarage of high-
quality mangrove was 23,937 hectares, a drop of 4,839 hectares (17%) since
1982. During that same ten-year period the area occupied by shrimp farms

increased from 1,064 hectares to 11,515 hectares—an increase of almost 1,000%.

Mangroves lost between 1982 and 1992 as a direct result of building shrimp farms are estimated at 4,307 hectares (37%). Of this amount, 2,132 hectares (18.5%) are dense *Avicennia, Rhizophora,* and some *Laguncularia* from forested stands bordering salt and mud flats and estuaries. The remainder, 2,174 hectares (18.9%), are the lower density, stressed, young, or dwarf mangroves associated with salt and mud flats (Vergne, Hardin, and DeWalt 1993).

By early 1993 the government had approved shrimp-farm concessions occupying a total of 28,699 hectares and conversion of 2,720 hectares more was under consideration. The total—31,419 hectares—is virtually equal to the sum of the high-quality mangrove areas plus the area in mud and salt flats (where lower-quality mangroves are found) that existed in the early 1970s. Of the 31,419 hectares of lands under concession status, only about 12,000 hectares of shrimp farms have been constructed. Of these, approximately 8,000 hectares of ponds are owned by members of the National Association of Honduran Shrimp Farmers, representing the largest and most technically sophisticated (intensive) producers. The balance of about 4,000 hectares is comprised of small and artisanal farms (some integrating seasonal salt production) whose methods are more extensive. A significant portion of the remaining undeveloped concessions is located in areas with denser vegetation cover than exists in already developed areas. Therefore the amount of mangrove that could be destroyed by construction of the remaining 19,904 hectares of lands conceded would be significantly higher than what has been taken so far. Although the removal of only 2,174 hectares of low-density and stressed mangrove might not have had a serious adverse effect on the ecology of the region, the removal of some 10,000 hectares of additional dwarf, stressed, and forested stands of mangrove is significant.

Because the characteristics of areas most appropriate for shrimp farms and those of high-quality mangrove forests are mutually exclusive, large industrial shrimp-farm development thus far appears to be responsible for relatively little invasion and destruction of the higher quality (i.e., mature, young, and regenerating) mangrove categories. If possible, more capitalized and powerful investors have chosen unforested areas, especially the interiors of large mud flats, for large pond construction, although clearing of mangrove stands for roads, perimeter dikes, and pumping stations also has taken place.

However, because the most suitable sites were chosen first, since the mid-1980s construction of shrimp farms has expanded into areas of dwarf and stressed stands, and the tendency is to build new larger shrimp farms on lands occupied by dense stands of dwarf mangrove. At the same time a detectable trend is for artisanal farms to be constructed in areas of higher-quality mangrove. These farms often feature dikes that are constructed by hand from soils with mature stands of *Avicennia* (Vergne, Hardin, and DeWalt 1993:47). The width and steep slopes of these dikes suggest that their purpose is to segregate the enclosed mangrove stand from tidal inundations. The resulting alterations of tidal regimes kill the mangroves in the contained area, thereby expediting burning and the ultimate clearing of the pond site.

Thus, although these artisanal ponds are quite small (most are 5 to 50 hectares), they can have considerable negative environmental costs because of their hydrological effects as well as the important biological value of the destroyed mangroves. Estimates based on aerial photos suggest that such small artisanal farms comprised about 400 hectares in 1992. The tendency for artisanal operations to use less-desirable land (i.e., higher-quality mangrove stands) suggests that these farms have disproportionately adverse environmental consequences. These concerns also imply an eventual human tragedy for artisanal farmers, given the high probability for farm failure associated with widespread problems of pond management and mangrove soil acidity in such zones.

The loss of access to seasonal lagoons has been an especially serious point of contention, and the most serious confrontations in the region have taken place between shrimp farmers and communities that exploit the lagoons. The temporary ponds develop annually on the sparsely vegetated mud flats behind the mangrove fringe. Seasonal peaks in high tides (resulting from elevated water levels in the creeks and rivers) create brackish conditions in the pools and introduce larval and postlarval stages of fish and crustacea. At the end of the rainy season most lagoons become isolated from open water and begin to dry out. From then on, as the lagoons shrink and finally dry out, they are heavily exploited by human populations in the region (as well as by migratory bird populations and other species).

Artisanal fishers enter the lagoons as shrimp and fish are becoming concentrated in the dwindling pools. Such efforts can be highly productive (although sporadic) and represent an important economic option for poor rural people. Conflicts over the use of lagoons arise from their high suitability for conversion to shrimp farms. Dry most of the year, with sparse vegetation and easy access to seawater, they are ideal site for shrimp farms. Some communities have constructed gates and fences and act as armed guards to prevent unauthorized access to the lagoons by shrimp-farm personnel, wild larva gatherers who supply the larger farms, and other outsiders. Nevertheless a number of farms have been constructed on what were seasonal lagoons. Examination of areas now occupied by shrimp farms, and of maps of the concessions that have been granted, suggests that about one-third of the area of seasonal lagoons has already been, or will be, physically lost. Some remaining areas of lagoons could be reduced further if the shrimp farms fence and control the access roads. In addition, future shrimp-farm expansion may alter the regional hydrology, impede water flow to the winter lagoons, and thereby further reduce their productivity and jeopardize their long-term viability.

Similarities in the emerging social and ecological costs related to the boom in shrimp mariculture and the earlier promotion of export commodities in the region are striking. Many of the same international and national agencies are promoting the development. "Enclosure movements," which once removed small farmers from relatively good agricultural land, often by force and with the compliance of local authorities, are being repeated on the intertidal lands that have not been cleared. Intertidal land, once open to public use for fishing, shellfish collecting, salt production, and the cutting of firewood and tanbark, is now being converted to private use. Conflicts have arisen among the large operations, local medium-scale entrepreneurs, and campesino cooperatives and communities over land and access. Violent confrontations have also taken place between shrimp farmers and artisanal fishers (Stonich 1991a, 1992; Stonich, Murray, and Rosset 1994).

The mounting growth of shrimp farms, reminiscent of earlier peasant movements that stemmed from the loss of forest, range, and farm land, is taking place over the protests of local people dislocated by the shrimp industry. One such group, the Committee for the Defense and Development of the Flora and Fauna of the Gulf of Fonseca, has been especially successful in mobilizing local people to resist the expansion of the industry as well as in garnering national and international support for its efforts (Stonich 1991a, 1993; Stonich, Murray, and Rosset 1994). Such grassroots groups challenge the transformation of coastal resources used by many people for many purposes into private property controlled by foreigners and national elites that have the political power to obtain concessions or title to coastal lands (Stonich 1991a; Stonich 1993).

Since the end of World War II the landscape of Honduras has been transformed through deforestation, overgrazing, changes in agricultural systems, and other environmental stresses. Along with other seriously degraded areas of the world such as Haiti, the Philippines (see Eder, chapter 11 in this volume), southeastern Kenya (see Castro, chapter 5 in this volume), and Nepal's middle mountains, it has been designated a critically endangered region where basic life-sustaining systems, including water and soils, are threatened (Kasperson, Kasperson, and Turner in press). Environmental decline within the country has been most severe in the southern zone, where semidesertification and growing rural impoverishment have spurred extensive migration to other areas within and outside the zone.

The paradox is that environmental degradation is most serious in an area that has been an important target for a series of economic development initiatives. The political ecology of development in Honduras reveals the interconnections of the dominant development strategy, deforestation (and other forms of environmental destruction), and worsening rural poverty. As part of an overall strategy of export-led growth, a series of non-traditional agricultural commodities has been championed in southern Honduras since the 1950s. This prevailing development strategy has altered the agrarian structure of the region, exacerbated existing social and economic inequities, and shaped the ways in which natural resources have been exploited.

By fostering economic growth at the expense of human populations and the environment, this strategy has encouraged environmental degradation as well as political instability and violence.

An analysis of the growth of the shrimp industry in Honduras is particularly useful in showing how the latest development trend has advanced the social and ecological processes established with the cotton and cattle booms, spatially as well as temporally, to coastal zones now having greatly enhanced economic value. Diminished access to common property resources, brought about by government-sponsored privatization efforts and encouraged by international agencies, is not a new occurrence in southern Honduras. Nor are enclosure movements, supported by force, that result in rural displacement, repression, and violence. The pattern of expansion of the shrimp farms raises serious social questions about who benefits and who pays the price for growth of the industry. At the same time, although the expansion of the shrimp industry has brought some short-term economic benefits to the region, it has done so at some environmental expense. Although less than half the decline in high-quality mangroves since 1982 can be directly attributed to shrimp farm construction, an equal area of dwarf and stressed stands of mangrove and significant areas of mud flats were also destroyed. Should the remaining 20,000 hectares of shrimp-farm concessions be developed, the destruction of stress, dwarf, and mature mangroves will be more serious.

A political ecological perspective allows analysis of deforestation and other forms of environmental decline and human poverty to go beyond overly simplistic explanations that ascribe blame to particular commodities (e.g., the "hamburger connection"). According to measures of land scarcity, displacement, poverty, and environmental degradation, outcomes have been similar regardless of which commodities have been promoted. Although the specific commodities being promoted vary, the underlying social and economic relations remain the same.

The repetition of these processes through time and through space demonstrates the extent to which these dynamics are part of the structure of Honduran society and tied to the dominant development model.

Political ecological analysis also moves beyond a fixation on population growth as the only, or the most important, factor in explaining environmental degradation. The political ecological approach demonstrates that blaming the population increase for environmental degradation in the region is too facile and diverts attention from the complexity of issues facing the region and from a more comprehensive explanation.

Although the rapid increase in population growth in the region is a matter for serious concern, population growth per se cannot adequately explain the destructive land-use patterns that have emerged. Although population growth may be a part of the explanation for some environmental problems, the nature of agricultural development in the region is more responsible for most problems. Development in the region has been highly uneven, not only in terms of the distribution of economic costs and benefits but also in terms of its effects on the spatial distribution of people. Political economic factors related to the expansion of export-oriented agriculture constrain access to the most fertile lands of the region. This results in a highly unevenly distributed population in which the greatest population densities occur in the highlands—the areas most marginal for agriculture. The growing population in the highlands has few opportunities to earn a living and continues to distribute a diminishing amount of land among more and more people while intensifying agricultural production and expanding into areas more marginal for agriculture. Growing rural poverty also stimulates out-migration from the more densely packed south, thereby decreasing population pressure in highland areas and simultaneously augmenting urban populations and escalating pressure on heretofore undamaged coastal zones in the south and tropical humid forests in other parts of the country.

Within the south, in urban centers throughout Honduras, and in frontier areas being settled, the mounting evidence of ecological and human decline may portend long-term and immutable threats to human, economic, and environmental sustainability. Moreover the government appears to be rushing into the new privatization scheme for its agricultural land and forests without ensuring that it has the capacity to enforce new regulations and ameliorate social and environmental consequences. Deforestation and other grave environmental abuses in Honduras will not improve unless the basic social structural inequalities in the region are confronted and alleviated.

Deforestation will continue so long as people do not have enough land, jobs, and food. Environmental catastrophe will likely ensue unless the predominant development agenda is transformed to remedy expanding social inequalities as well as environmental ills.

Notes

1. Elsewhere, Susan Stonich (1989, 1993) has critiqued the dominant paradigms used to explain environmental degradation (including deforestation) in

tropical areas of the developing world: neo-Malthusian, neoclassical eco-
nomic/technological, and dependency. The argument is that although several
of these major paradigms identify one or more factors relevant to a comprehensive explanation of social and environmental change, no single model adequately explains poverty and environmental deterioration in areas of the
developing world such as southern Honduras. As an alternative, the overall
approach here is a more comprehensive framework that integrates political,
economic, and human ecological analysis. The political economic analysis
examines the interacting roles that social institutions (international, national,
regional, and local) play in providing constraints and possibilities that affect
human decisions that in turn affect those institutions as well as the natural
environment. Human ecological analysis allows the consideration of demographic trends, environmental concerns, and issues related to human health
and nutrition. It expands the perspective of political economy to include an
examination of the distribution and use of resources and the dynamic contradictions between society and natural resources. A more comprehensive discussion of political ecology appears in Stonich 1993, chapter 1.

2. Melons grown on irrigated land have also been an important nontraditional export promoted in southern Honduras in recent years. For discussions
of the social, economic, and environmental effects of the melon industry see
Murray 1991 and Stonich et al. 1994.

3. These funds were released to the new Honduran government that took
office in January 1990. In July 1991 the Honduran Central Bank reached an
agreement with the IMF that paved the way for an influx of American capital–$1.8 billion-worth of external finance over a three-year period: U.S.$300
million in 1991, U.S.$70 million in 1992, and U.S.$750 million in 1993
(Honduras/International Monetary Fund [Honduras/IMF] 1991a:5). In August
1991 Honduras requested from Mexico and Venezuela the rescheduling of its
U.S.$51.2 million bilateral debt and a new loan of U.S.$120 million
(Honduras/IMF 1991b:6).

4. Erosion is estimated to occur at rates as great as 13 tons per hectare per
year in the upper Choluteca watershed, and about 168 cubic meters of soil per
second are transported in the river at the bridge on the outskirts of the city of
Choluteca (Vega 1989).

5. The Rio Platano Biosphere Reserve, located within the region referred to
as La Mosquitia, was established in 1979 and became a world heritage site in
1980. It covers almost the entire watershed of the Rio Platano, approximately
525,000 hectares on the Caribbean coast of Honduras. When it was created, it
had approximately 4,450 inhabitants, mostly Miskito Indians and a few Pech,
and some *ladino* villages within the boundaries. The reserve was inaccessible by
road. By 1990 the reserve still had no conventional roads, but illegal access was
facilitated by widespread logging and gold mining. By then several hundred
Nicaraguan Miskito Indians had settled within the reserve.

6. Estimates of employment generated by the shrimp industry vary widely.
According to the environmental profile published in 1989, the shrimp industry
employs fewer than one person per hectare (SECPLAN and USAID 1989:179). In

contrast, the National Association of Shrimp Farmers of Honduras, whose members tend to be owners and operators of large farms, issued their own estimate of 1.5 jobs per hectare in 1990 (National Association 1990), or a total of 25,000 direct jobs.

7. Related environmental concerns include modifications in the regional hydrology because of obstruction of water flow and sedimentation, the capture of wild postlarvae and the associated indiscriminate introduction of hatchery-raised seedstocks, discharges of shrimp-farm effluent, diminished water quality (including hypernutrification and eutrophication), the effects on populations of migratory birds, reptiles, amphibians, and aquatic mammals from the destruction and transformations of habitats (especially seasonal lagoons) and the antipredator measures taken by farmers, and contamination by pesticides purportedly used by shrimp-farm owners (Vergne, Hardin, and DeWalt 1993; Stonich et al. 1994).

8. Other social issues include inequities in hiring, wages, and the ability to organize; the creation of a variety of economies of scale (in terms of concessions, land, credit, technical assistance, marketing, and so on) that effectively constrain small producers and cooperatives but favor highly capitalized investors, thereby increasing unequal access to resources; and blatant harassment, death threats, and killings (Stonich et al. 1994).

9. This section is based on the results of the recent environmental study conducted in 1993 (Vergne, Hardin, and DeWalt 1993). This article uses the categories of mangrove forest developed by COHDEFOR in 1987 and also used by Verge and colleagues (1993). These categories include dwarf, stress, and mature, or high quality. Dwarf mangroves occupy soils at the outermost limits of soil salinity tolerance (usually more than 100 parts per thousand) and are characterized by sparse stands less than 1 meter in height. Stress mangroves are located at the outer fringes of mangrove zones where soil salinity is limited. Generally, stands are 1 to 3 meters in height and show other signs of stress, such as poor flowering and vulnerability to insect pests. Mature, or high-quality, mangroves are found in areas of frequent tidal inundation within an ideal range of soil salinity and may range from 15 to 20 meters in height (Vergne, Hardin, and DeWalt 1993).

References Cited

Almendares, Juan, Miguel Sierra, Pamela K. Anderson, and Paul R. Epstein. 1993. Critical regions: A profile of Honduras. *Lancet* 342 (December): 1400–1402.

Boardman, Robert. 1986. *Pesticides in World Agriculture.* New York: St. Martin's Press.

Bottrell, Dale. 1983. Social problems in pest management in the tropics. *Insect Science and Applications* 4 (1–2): 179–184.

Boyer, Jefferson. 1982. Agrarian Capitalism and Peasant Praxis in Southern Honduras. Unpublished Ph.D. diss., University of North Carolina, Chapel Hill.

Bruntland Commission (United Nations Commission on Environment and Development). 1987. *Our Common Future*. Oxford, England: Oxford University Press.

Bull, David. 1982. *A Growing Problem: Pesticides and the Third World Poor*. Oxford, England: Oxfam.

Castañeda, Catherine and Z. Matamoros. 1990. *Environmental Analysis for the Investment and Export Development Project*. Tegucigalpa, Honduras: U.S. Agency for International Development–Honduras.

Central American Institute of Investigation and Industrial Technology (ICAITI). 1977. *An Environmental and Economic Study of the Consequences of Pesticide Use in Central American Cotton Production*. Washington, D.C.: U.S. Agency for International Development.

Comisión Nacional del Medio Ambiente y Desarrollo. 1992. *Honduras Environmental Agenda*. Tegucigalpa, Honduras: Comisión Nacional del Medio Ambiente y Desarrollo.

Comprehensive Resource and Inventory and Evaluation System. 1984. *Resource Assessment of the Choluteca Department*. East Lansing: Michigan State University and the U.S. Department of Agriculture.

Corporación Hondureña de Desarrollo Forestal (COHDEFOR). 1988. *Mesa redonda: La participación internacional en el desarrollo forestal de Honduras* (Round table: International participation in the forestry development of Honduras). Tegucigalpa, Honduras: COHDEFOR.

Daniels, Mitch. 1990. Encourage Honduras in its bold economic reforms. *Christian Science Monitor*, April 30, p. 18.

DeWalt, Billie R. 1983. The cattle are eating the forest. *Bulletin of the Atomic Scientist* 39: 18–23.

———. 1985. Microcosmic and macrocosmic processes of agrarian change in southern Honduras: The cattle are eating the forest. In *Micro and Macro Levels of Analysis in Anthropology: Issues in Theory and Research*, B. R. DeWalt and P. J. Pelto, eds., pp. 165–186. Boulder, Colo.: Westview.

———. 1986. Economic assistance in Central America: Development or impoverishment? *Cultural Survival Quarterly* 10: 14–18.

Durham, William. 1979. *Scarcity and Survival in Central America: The Ecological Origins of the Soccer War*. Palo Alto, Calif.: Stanford University Press.

Faber, Daniel. 1992. *Environment Under Fire: Imperialism and the Ecological Crisis in Central America*. New York: Monthly Review Press.

Fandell, Sharon. 1994. Foreign investment, logging, and environmentalism in developing countries: Implications of Stone Container Corporation's experience in Honduras. *Harvard International Law Journal* 35 (2): 499–534.

Flores, Juan and Carlow Reiche. 1990. *El consumo de leña en las industrias rurales de la zona sur de Honduras* (Fuelwood consumption by rural industries in southern Honduras). Turrialba, Costa Rica: Centro Agronómico Tropical de Investigacíon y Enseñanza.

Foer, Gordon and Stephen Olsen, eds. 1992. *Central America's Coasts.* Washington, D.C.: U.S. Agency for International Development, Research & Development, Environmental and Natural Resources.

Gilpin, Robert. 1987. *The Political Economy of International Relations.* Princeton, N.J.: Princeton University Press.

Holdridge, Leslie R. 1962. *Mapa ecologico de Honduras* (Ecological map of Honduras). San José, Costa Rica: Tropical Science Center.

Honduran Popular Action Group. 1992. Honduras signs forty-year pact with Stone Container Corp. *Occasional Newsletter,* January.

Honduras/International Monetary Fund (Honduras/IMF). 1991a. *Latin American Regional Report* RM-91–06, 5, July 18.

———. 1991b. Inflation and exchange rates in the region. *Latin American Regional Report* RM-91–07, 4, August 22.

International Union for the Conservation of Nature (IUCN). 1992. *Conservación de los ecosistemas costeros del Golfo de Fonseca* (Conservation of coastal ecossytems of the Gulf of Fonseca). Project proposal submitted to the Danish International Development Agency. Tegucigalpa, Honduras: International Union for the Conservation of Nature.

Jarvis, Lovell S. 1986. *Livestock Development in Latin America.* Washington, D.C.: World Bank.

Johnston, George et al. 1990. *Honduran Natural Resource Policy Inventory.* Draft, Technical Report No. 111. Washington, D.C.: ABT Associates for U.S. Agency for International Development.

Kasperson, Jeanne X., Robert E. Kasperson, B. L. Turner. In press. *Critical Environmental Regions: International Perspectives.* Tokyo: United Nations University Press.

Leonard, H. Jeffrey. 1987. *Natural Resources and Economic Development in Central America.* New Brunswick, N.J.: Transaction Books.

Murray, Douglas L. 1991. Export agriculture, ecological disruption, and social inequality: Some effects of pesticides in southern Honduras. *Agriculture and Human Values* 8 (4): 19–29.

National Association of Shrimp Farmers of Honduras. 1990. *Shrimp Cultivation: A Positive Support to the Development of Honduras.* Choluteca, Honduras: National Association of Shrimp Farmers of Honduras.

Parsons, Kenneth. 1975. *Agrarian Reform in Southern Honduras.* Research Paper No. 67, Land Tenure Center. Madison: University of Wisconsin.

Redclift, Michael. 1984. *Development and Environmental Crisis.* New York: Methuen.

———. 1987. *Sustainable Development: Exploring the Contradictions.* New York: Methuen.

Scanlan, David. 1994. An environmental test case. *Christian Science Monitor,* July 26, p. 11.

Secretaría de Planificación, Coordinación, y Presupuesto (SECPLAN). 1991. *Urgencias y esperanzas: Datos prioritarios para los retos del noventa* (Urgencies and hopes: Data priorities for the challenges of the nineties). Tegucigalpa, Honduras: Secretaria de Planificación, Coordinación, y Presupuesto.

214   Susan Stonich and Billie R. DeWalt

Secretaría de Planificación, Coordinación, y Presupuesto (SECPLAN) and United States Agency for International Development (USAID). 1989. *Perfil ambiental de Honduras 1989* (Environmental profile of Honduras 1989). Tegucigalpa, Honduras: U.S. Agency for International Development.

Stares, Rodney. 1972. *La economia compesina en la zona sur de Honduras: 1950–1970* (The peasant economy of southern Honduras). Paper prepared for the bishop of Choluteca, Honduras.

Stonich, Susan C. 1986. Development and Destruction: Interrelated Ecological, Socioeconomic, and Nutritional Change in Southern Honduras. Unpublished Ph.D. diss., University of Kentucky, Lexington.

———. 1989. Social processes and environmental destruction: A Central American case study. *Population and Development Review* 15 (2): 269–296.

———. 1991a. The promotion of nontraditional exports in Honduras: Issues of equity, environment, and natural resource management. *Development and Change* 22 (4): 725–755.

———. 1991b. Rural families and income from migration: Honduran households in the world economy. *Journal of Latin American Studies* 23 (1): 131–161.

———. 1992. Struggling with Honduran poverty: The environmental consequences of natural resource based development and rural transformation. *World Development* 20 (3): 385–399.

———. 1993. *I Am Destroying the Land: The Political Ecology of Poverty and Environmental Destruction in Honduras*. Boulder, Colo.: Westview.

Stonich, Susan C. and Billie R. DeWalt. 1989. The political economy of agricultural growth and rural transformation in Honduras and Mexico. In *Human Systems Ecology: Studies in the Integration of Political Economy, Adaptation, and Socionatural Regions*, S. Smith and E. Reeves, eds., pp. 202–230. Boulder, Colo.: Westview.

Stonich, Susan C., Douglas L. Murray, Peter M. Rosset. 1994. Enduring crises: The human and environmental consequences of nontraditional export growth in Central America. *Research in Economic Anthropology* 15: 239–274.

Time. 1989. Grapevine: Elliott Abrams and retired general Paul Gorman in logging business. *Time,* June 5, p. 40.

U.S. Agency for International Development (USAID). 1982. *Country Environmental Profile.* McLean, Virginia: JRB Associates.

———. 1989a. *Environmental and Natural Resource Management in Central America: A Strategy for A.I.D. Assistance.* Report prepared for the Latin America and Caribbean Bureau of USAID by the Regional Office for Central America and Panama.

———. 1989b. *Strategic Considerations for the Agricultural Sector in Honduras.* Draft copy of report by the Office of Agriculture and Rural Development of USAID, Tegucigalpa, Honduras.

———. 1990. *Agricultural Sector Strategy Paper.* Tegucigalpa, Honduras: U.S. Agency for International Development–Honduras.

U.S. Agency for International Development and Honduran Federation for the Promotion of Agricultural Exports. 1989. *Plan de desarrollo del camaron en*

Honduras (Development plan for shrimp mariculture in Honduras). Tegucigalpa: U.S. Agency for International Development–Honduras.

Vega, Alberto. 1989. *Medio ambiente y desarrollo en la zona sur de Honduras* (Environment and development in southern Honduras). Tegucigalpa: Dirección General de Planificación Territorial, Proyecto de Estrategias de Desarrollo para Tierras Frágiles (DESFIL).

Vergne, Philippe, Mark Hardin, and Billie DeWalt. 1993. *Environmental Study of the Gulf of Fonseca*. Gainesville, Fla.: Tropical Research and Development.

Weir, David and Mark Shapiro. 1981. *Circle of Poison: Pesticides and People in a Hungry World*. San Francisco: Institute for Food and Development Policy.

White, Robert. 1977. Structural Factors in Rural Development: The Church and the Peasant in Honduras. Unpublished Ph.D. diss., Cornell University.

Williams, Robert. 1986. *Export Agriculture and the Crisis in Central America*. Chapel Hill: University of North Carolina Press.

World Bank. Various editions 1984–94. *World Development Report*. New York: Oxford University Press.

World Resources Institute (WRI). 1990. *World Resources 1990–1991*. New York and Oxford, England: Oxford University Press.

9

The Human Niche and Rain Forest Preservation in Southern Central America

John Vandermeer

Dramatic events have transpired in the southern half of Central America since 1980. Buoyed by the popular revolution of 1979, Nicaragua entered a decade of radical political experimentation that included a reformation of its entire social program, a readjustment of the rural economy, and a concomitant growth in nationalist fervor. An electoral turnaround in 1990 brought that experiment to a sudden halt. Costa Rica entered the 1980s with more than thirty years of relatively stable internal politics and at least ten years of strong lobbying by foreign conservationists, which apparently resulted in social calm and national commitment to conservation and preservation programs, especially with regard to tropical forests.

A glance at major international media coverage of the two countries during the 1980s reveals a dramatic difference in their agendas. Nicaragua was the revolutionary society, praised or maligned depending on the writer's ideology, struggling to develop a new model for the Third World, emphasizing questions of social justice, Third World solidarity, national identity, and sovereignty. Right-wing ideologues frequently portrayed Nicaragua's need to defend itself as a desire to expand its territory. It was also a country at war, with a large and relatively modern military. Costa Rica's image could hardly have been more different. Thirty years of a demilitarized society and relative social peace had given the country the image (self-promoted) of the Switzerland of the Americas, and it undoubtedly was the most advertised tourist destination in Central America. Concomitantly, it was subjected to a great deal of propaganda, largely generated by U.S. citizens, regarding the use and abuse of natural areas, the importance of biological diversity in the tropics, and the importance of such notions to the nation. The result was an educated class that exhibited a strong commitment to programs of conservation and preservation, especially in forested areas. An impressive outcome of this trend was a development of the most extensive system of preserved areas in all the neotropics.

In both countries the major driving force for any sort of development or underdevelopment was (and of course still is) the influence of the government of the United States, a factor that hardly needs extensive documentation. Continuing a history of direct and indirect assertion of political power by the United States, the Reagan administration took a keen interest in the region; its chief security administrator regarded Nicaragua as the main target for the rollback of the "communist" system. What began as historically conditioned differences between the two countries was enormously exaggerated by the Reagan administration. Nicaragua fell into a state of war, whereas Costa Rica received some of the most generous nonmilitary development aid ever handed to a Central American country.

Thus the picture of Nicaragua that emerges during the 1980s is of a country besieged by war, struggling to promote a radical social justice agenda, and little concerned with the conservation or preservation of natural areas. In contrast stood Costa Rica, not only free of war but a recipient of extremely generous foreign aid, strongly promoting conservation as a national priority, but little concerned with anything that might be called a radical social justice agenda. On the surface, given the dramatic differences between the two countries, most people would predict that their rates of deforestation would quite different. Indeed, if the generally accepted strategies of conservationists are valid and workable, here is a perfect test case—Costa Rica is the experimental subject that has adopted all the right programs. Nicaragua is the control with none of them. What happened?

According to the 1990 report of the World Resources Institute, Costa Rica experienced the highest rate of deforestation in the world. This is a remarkable and truly stunning figure that demands first an explanation and second considerable reflection on the general strategy that ought to be adopted for rain forest conservation in Central America.

Deforestation in Central America follows a general pattern. Modern lumbering operations locate individual trees, or localities with relatively high concentrations of valuable species, build an access road, selectively cut, and leave. The immediate damage is thus the access road and those few trees that were cut, although substantial incidental damage is inevitable, especially if heavy machinery is used. Subsequently, land-hungry peasants use the access road, cutting out pieces of the remaining forest,

Figure 9.1   Central America   Diagonal lines approximate position of original lowland rain forest cover before European contact.

until whole areas that were crisscrossed with access roads become cleared of all forest and converted to peasant agriculture. After this stage the peasant farmers are often removed, with or without compensation, and the land is converted to intensive agriculture of one form or another (e.g., banana plantations, cattle ranches, and so on). As this essay will show, the particular form of this sequence goes a long way toward explaining the differences between Nicaragua and Costa Rica and should guide any strategies proposed for curbing the unacceptable rain forest loss that we see today (see figure 9.1).

Contrasting International Attention in the 1980s

Having had the good fortune to be located next to a country that experienced a radical political transformation, Costa Rica was positioned to receive the bountiful largesse of the United States. The goal, at least in part, was to create an alternative model that could be used to persuade other poor Third World countries that revolutionary changes are not desirable. To a great extent the goal was met (of course with the paired strategy of economic and military pressure on Nicaragua). A massive influx of U.S. dollars created the illusion of great prosperity in this tiny Central American country. The Costa Rican press, strongly influenced by U.S. covert actions (Hruska 1986), pounded a monotonous drum beat of anti—Sandinista propaganda.

Illustrative of the success of the apparent goal, in the fall of 1988 Costa Rica's president Oscar Arias, in an address to the people of Los Chiles (on the northern border with Nicaragua), noted that an examination of conditions across the border should convince all Costa Ricans that the road to development chosen by the Sandinistas was not for them. Generally, the result is that the average Costa Rican is dramatically disillusioned about reform and revolution in Nicaragua and certainly does not view Nicaragua, even in theory, as a useful model for Costa Rica. Although Costa Rica has seen some unrest, including a growing realization that its dependence on U.S. dollars has not been economically healthy, this unrest has not visibly translated into respect for or desire to imitate the Nicaraguan model in even a small way. In other words, the part of the Reagan administration's program that sought to create a distinction between Nicaragua and Costa Rica, by negative actions against Nicaragua, positive actions toward Costa Rica, and an energetic propaganda campaign, was evidently a dramatic success in Costa Rica.

The massive U.S. aid that entered Costa Rica did so against a backdrop of intense concern over environmental issues, especially among the educated classes, and especially associated with questions of conservation (note that similar concern about environmental issues is not a major agenda item in other equally important areas, such as abuse of pesticides, which is rampant in Costa Rica [Hilje et al. 1987]. Foreign tourists and academics had long been flocking to Costa Rica for the "tropical experience" and had created a worldwide image of Costa Rica as a tropical garden. The interest and attention of First World academics did not go unnoticed by educated Costa Ricans, and the full pockets of foreign ecotourists did not escape the attention of Costa Rican planners. "Conservation" thus became a big issue in Costa Rica.

Consequently, Costa Rica developed an internationally applauded environmental ethic, one with which international funding agencies seem comfortable and foreign conservationists applaud. Nicaragua's environmental ethic was different, not nearly as attractive to conservative conservation organizations, despite what might be argued was an underlying conservationist philosophy. Although Costa Rica saw environmental deterioration as a proximate problem for which direct solutions were necessary, Nicaragua saw them as indirect consequences of dominant social, political, and economic forces. Thus whereas Costa Rica was attempting to deal with environmental problems head on, Nicaragua was attempting to structure the socioeconomic system in such a way that environmental problems would not arise in the first place.

The situation in Nicaragua was, and still is, distinct from all other Central American countries,' especially that of its neighbor, Costa Rica. First, the nature of the Contra war mitigated against significant lumber operations (except in some pine forests), so extensive access roads were not built. But another factor seems to have been important. The massive agrarian reform program initiated after the 1979 revolution all but eliminated the land-hungry peasant in Nicaragua. In 1978 the fraction of land owned by farms larger than 850 acres was 36%. By 1985 it had dwindled to 11%. In 1978 Nicaragua had no production cooperatives. In 1985 the cooperative sector had 9% of the land and 50,000 families. From 1980 to 1985 more than 127,000 families received title to their own land (Kamowitz 1986). Nicaragua saw agrarian reform without precedent in the history of Latin America.

Costa Rica's agrarian reform program is entirely different, bearing the fingerprint of similar programs in El Salvador and many years ago in Vietnam. Organized by the government agency Instituto de Desarrollo Agrario (IDA), the program locates landless peasants on or near undeveloped areas, only occasionally on the lands of large landowners. The system makes no provision for expropriating large inefficient enterprises, and land titles are given with ten- or thirty-year mortgages, the repayment of which requires something more than traditional production (i.e., more than for household subsistence). Traditional farmers seeking a plot of land, if they get anything at all, get a piece of marginal undeveloped land and, probably for the first time in their life, a bank debt.

Interviews during the 1980s with small farmers in both Nicaragua and Costa Rica reflected this basic difference in agrarian reform programs. Costa Ricans emphasized the land tenure issue, voicing concern either with their lack of land title or their inability to pay their recently acquired mortgage (many were quite surprised at their new debt); they focused their economic concerns on the attainment of land security. Nicaraguans were the opposite. Although they voiced many legitimate gripes, lack of land was not one of them. Only rarely would Nicaraguan peasants talk about needing a piece of land to call their own. The expected consequences regarding pressure on rain forests are obvious.

With this model of land tenure in Nicaragua we might expect that Costa Rica's pattern of land-hungry peasants following lumber roads into the forest would not be a problem in Nicaragua. And that seemed to be true during the 1980s. However, the political changes induced by the elections of 1990 stopped—and in many ways reversed—the experiment. Nicaragua quickly abandoned the radical agrarian reform program of the Sandinistas. As expected, the new class of landless peasants is now busy following new lumber roads, clearing forests in southeastern Nicaragua.

Managing the Human Niche and Rain Forest Preservation

Certain programmatic conclusions seem obvious from the lessons of the past fifteen years. Three points seem hardly to require further discussion:

- The conventional conservation strategies, as typified in Costa Rica, do not work to preserve rain forests.
- Social programs in the more general society represent an important component of any program of rain forest conservation.
- International players, especially the United States, are major determinant factors.

But these conclusions demand further analysis. What is to be done now? That is, if we are forced to live with these three conclusions, what strategies for rain forest conservation ought to be pursued in Central America? Indeed it certainly seems that the time is ripe for a new agenda, given the surprising results of the 1980s.

The science of ecology has some insights to offer. The role of periodic disturbance has long been recognized in ecology as an important organizing force (Pickett and White 1985). Tree falls create light gaps in a forest, initiating a process of succession on a microscale (Denslow 1980), and catastrophic wind damage is a well-known determinant of certain structural features of forests (e.g., Lugo et al. 1983). Work on Nicaragua's Atlantic coast (Yih, Boucher, and Vandemeer 1989; Vandermeer, Zamora, and Boucher 1990; Boucher 1989) suggests that direct regeneration of the forest after a hurricane may be the norm.

Whether the main form of disturbance is a hurricane or simply a light gap, the implications of disturbance ecology are the same—it should be possible to design an extractive system that provides for relatively rapid recuperation of the forest. Stewart Pickett (personal communication 1990) suggests that the message of disturbance ecology might be to tune production activities to mimic as much as possible the normal disturbance events in the ecosystem. The strip cuts in the Palcazu project in Peru (Hartshorn

1989) represent just such a philosophy—strips were designed to mimic the natural dynamics of tree-fall gaps. A logging operation could be informed by the sort of damage done by a hurricane (e.g., the hurricane topples or truncates most large trees but does little damage to the understory, including the saplings that will eventually form the new tree layer;this implies that the damage the logging machinery does to the forest is far more significant than the logging operation itself [Vandermeer et al. 1990; Boucher 1989]).

The work of Andrew Johns (1983) is particularly interesting in this respect, although its location makes application to Central America somewhat questionable. Working in forests in West Malaysia, Johns found no evidence of species loss as a result of logging. He did find some dramatic differences in some population densities, and primate behavior was detectably different. But with regard to the species composition per se he could detect no difference between a logged forest and a primary forest. Indeed an auto trip through another dipterocarp forest (in the Danum Valley of Borneo) shows his results quite vividly. The logged-over forest looks much like hurricane-damaged forests in Nicaragua and Puerto Rico, clearly damaged and severely so but also clearly regenerating.

To some extent the outlines of a new program can be seen in the philosophies espoused by some foresters (espoused, not necessarily practiced), rather than the one-dimensional programs commonly promoted by traditional conservationists. D. Poore's 1989 summary of timber extraction in many tropical areas and J. Westoby's excellent 1989 history of lumbering are two of many books that could serve as primers. The former is especially interesting for its optimism regarding sustained yield forestry potential in the tropics and the latter for its acknowledgment of the role of sociopolitical forces in the question of sustainability. The point on which these and other forestry analyses agree is the role of secondary activity in the lumbering operation, the most important of which is the invasion of agriculture in logged areas.

However, the story is yet more complicated because agricultural conversion is not a homogeneous process. A small peasant farm usually creates an environment that shares some features of tropical forest structure. If abandoned, such a farm is likely to revert to forest quite rapidly. A peasant farmer who uses chemical pesticides may do further damage to the ecosystem, such that a return to forest after abandonment takes a bit longer. A small commercial farmer who uses machinery and chemical fertilizers and pesticides will do yet more damage, and a banana plantation or cattle ranch or similar modern intensive activity is likely to set the process back even further. That is, the nature of the agricultural activity is clearly an important variable.

The recent work of I. Perfecto (in press; Perfecto and Snelling in press) is illustrative. Traditional coffee plantations, with their overstory shade trees and semi-organically managed coffee, represent an ecosystem that has superficial structural similarities to a tropical forest. On the other hand, the modern coffee management system is monocultural, with no shade trees and a heavy reliance on pesticides. The number of ant species in the traditional system varies from 20 to 30, whereas the number in modern coffee is two, rarely three or four. A similar area under natural forest cover in this region might contain 30 to 40 species of ants. Two shade trees in the traditional system yielded 106 and 128 species of beetles, whereas much larger trees in a tropical rain forest in Panama had 115 to 335 species. The modern coffee system had none. These data all suggest that the difference between the natural forest and the traditional coffee system is far less than the difference between the traditional and the modern coffee systems. Where should conservation efforts be aimed if a specific concern is biological diversity?

These observations lead to the generalization that deforestation often proceeds as a three-stage process, defined by two factors: the amount of damage done to biodiversity by an extractive pattern and the time required to recuperate.These actors evidently are highly correlated. But associated with this generalization is a programmatic conclusion regarding the "human niche" and its relationship to the preservation of biological diversity.

The three-stage process is as follows:

- Timber is extracted for commercial purposes. The recovery time for this disturbance is zero to fifty years (obviously not to the full recovery of primary forest conditions but to recovery of the area such that forest cover once again dominates), and direct damage to biological diversity is minimal. (This latter point is sometimes difficult to appreciate, because the physical damage done by logging is so striking. Nevertheless evidence from several sources [e.g., Yih, Boucher, and Vandermeer 1990; Vandermeer et al. 1990; Johns 1983] suggests that despite the dramatic damage done by catastrophic events such as hurricanes and logging, if no further incursions occur, the regenerative processes of the forest are remarkable, and the actual biodiversity is virtually unchanged.
- The area may be converted to peasant or traditional agricultural activities, representing a disturbance that requires a recovery time of 50 to 150 years;direct damage to biological diversity is significantly higher than the original logging operation.
- The introduction of an intensive modern agricultural system may occur; this represents a disturbance that requires a

Table 9.1

Type of Damage and Extent Estimated for Central American Lowland Rain Forest

Damage type	Damage to biodiversity	Time to recovery (in years)
Logging		
Light selective	minimal	0–20
Selective	minimal	10–30
Clear cut	minimal	30–50
Peasant Agriculture		
Structurally diverse	intermediate	50–75
Monotonous	large	75–150
Modern Agriculture		
Cattle ranch	very large	80–175
Intensive fruit production	extreme	100 +

recovery time of one hundred to thousands of years (for example, if the soil undergoes permanent physical or chemical damage, the ecological horizon shifts dramatically from secondary succession to primary succession, with a concomitant dramatic shift in recovery time). (See table 9.1.)

I have concluded that management on a local level should be informed by this three-stage process, but that management on the landscape level ought to be concerned with the question of direct damage and the potential rate of recuperation. At a local level the concern is with a logging procedure that minimizes recovery time and controls the rate and style of conversion to peasant and modern agriculture. At the landscape level we can imagine two extremes on an obvious continuum. At one extreme are islands of primitive pristine rain forest, protected with electric fences and armed guards and set in a sea of pesticide-drenched rice fields, banana plantations, and cattle ranches, with the rural proletariat that is inevitably necessary for such intensive agricultural activities concentrated in shantytowns on the edges of the pristine rain forest. At the other extreme is a mosaic of natural forest, old logged forest, traditional agriculture, abandoned agricultural fields, and modern agriculture designed with structural features resembling a forest. Indeed in the second extreme a satellite photo might categorize the entire area as "under forest cover." As to which area truly conserves the most biological diversity, there can be no certain answer, but it seems intuitively obvious that a jaguar is more likely to regard the entire landscape as part of her territory in the second scenario but not be terribly pleased with the pesticide-drenched rice fields in the first. The second strategy seems far more likely to preserve more biological diversity than the first.

So the somewhat utopian solution I propose for rain forest preservation in Central America includes a managed landscape with a mosaic of sustainable and small-scale agriculture, interspersed with ecologically managed large-scale agriculture, interspersed with timber extractive reserves, interspersed with small islands of preservation forest. Within this landscape-level plan a variety of perplexing and seemingly intractable issues emerges—how to plan small-scale agriculture, what is meant by ecologically managed large-scale agriculture, how to organize a rational timber extraction plan. Nevertheless one thing is clear from both the ecological literature on disturbance and the forestry literature on tropical forest extraction: the secondary damage done by unplanned agricultural activity is far worse than the initial damage done by the direct timber extraction, which brings me back to the original comparison between Costa Rica and Nicaragua.

Because of the political changes instituted in the early part of the 1980s, Nicaragua seemed to be moving toward a landscape-level system that resembles the second model. Conversely and evidently as an indirect consequence of the activities of some conservationists, Costa Rica seemed to be evolving toward a landscape-level system that resembles the first model. The previously cited data on rain forest deforestation rates leaves little doubt which of the two models was working best during the 1980s.

After the Elections

The Nicaraguan elections of 1990 changed that experiment significantly. The progressive agrarian reforms of the Sandinistas have been reversed, and once again landless peasants seek land, frequently after having been forcibly removed from land they had acquired under the Sandinista agrarian reform. So-called poles of development (land given to disarmed soldiers and Contras) form new bases of agricultural expansion as former Contra rebels search for parcels of land that will produce something.

A typical pattern is three years of cultivation after cutting a piece of forest, followed by a ten-year fallow period, followed by a single year of cultivation and a strong desire to move on to the hope of a better piece of real estate under the next piece of forest. The relative permanence provided by the Sandinista agrarian reform seems to have been completely destroyed. I have been traveling to this area since 1989, and the changes between then and now are spectacular. In 1989 (before the elections) Nicaraguans were having a major debate about how to use the natural resources of the area, the role of the government in regulating chain saws, how the government logging company (the only logging company operating) should be responsible to local communities, how agrarian reform must encompass technology to provide peasants with an alternative to burning for land preparation, and so on. Now a visitor hears the constant drone of chain saws almost everywhere, three lumber companies are actively engaged in cutting trees (openly flaunting a supposed ban on cutting), the autonomy of the region (supposedly now guaranteed by law) is blatantly violated by central government figures who accept bribes for logging concessions, fires are everywhere as landless peasants burn yet another piece of forest, and former Contra bands are engaged in robbing passing boats and fighting with one another.

Additionally, the economic situation has deteriorated significantly since the elections, and the peasantry that in the past at least held out hope that the government would help in times of crisis (e.g., if the crops failed they had a reasonable expectation that the government would supply basic grains) has become seemingly despondent and forced to do whatever it can to eke out a living, usually with little success.

Costa Rica's situation has also deteriorated considerably since the Nicaraguan elections of 1990. Feeling the effect of less U.S. aid (without the threat of the Sandinistas apparently the U.S. government felt less of a need to provide Costa Rica with the massive development aid that had characterized the 1980s), the government sought to expand traditional forms of export agriculture, especially the traditional export of bananas. The Sarapiqui Valley of Costa Rica is illustrative (although similar patterns are observable on the entire Atlantic coast). In 1990 a third of the valley was devoted to biological preserves (the La Selva Research Station, part of the Braulio Carrillo National Park, and the Barro del Colorado Reserve), whereas another third was in the legal hands of small peasant farming communities. The final third was a mosaic of small farms (most of which did not have title to their land), secondary and old growth forest, cattle pasture, and an occasional extensive fruit or ornamental plant plantation.

On the eastern periphery of the valley was an extensive banana plantation owned by the Standard Fruit Company, a major employer in the region. In the past several decades I have seen the ebb and flow of the banana business such that critical periods inevitably arise in which workers are laid off and forced to fend for themselves. Laid-off banana workers either migrate to the city, or they look for a small piece of land they can farm. Sometimes that small piece of land is a rain forest, thus contributing to deforestation, as described previously. Other times it is a small back corner of some large absentee landowner's cattle ranch, in which case either the homesteading family is forcibly evicted or the government agrarian reform institute (IDA) adjudicates a "fair" purchase for the peasant family, depending on complicated circumstances.

The past thirty or forty years have seen such an arrangement, with forest cover changing in the region from approximately 90% in 1950 to approximately 40% today, of which only a small amount is not part of the four large biological reserves. The Standard Fruit Company employs workers who migrate to the Sarapiqui region from other parts of Costa Rica. When sales slacken, workers are fired. The unemployed workers are still mainly rural people and basically have nothing to do but try to eke out a living farming, until Standard begins hiring again.

In 1991 the plot began to thicken. In response to an expected surge in demand for bananas resulting from the opening up of markets in Eastern Europe, five or six major banana companies began purchasing land in the Sarapiqui Valley as rapidly as possible and expanding banana production accordingly. Estimates ranged from 20,000 to 40,000 more hectares to be put into banana production within two years, and several visits to the area revealed intense activity in setting up new banana plantations. As in the past, workers were coming from all over the country. But this time apparently not enough Costa Rican workers were available to do the necessary work, and workers were being brought in from Nicaragua, Panama, and other Central American countries.

The expansion of bananas is largely viewed as a positive event by most actors in the Sarapiqui Valley and by most observers in the entire country. Local workers and peasants see jobs being created, local merchants see a potential surge in business, and local politicians see an increase in their power base. The Costa Rican government promotes the expansion, because it sees the increased tax revenues as helping to pay debt service on its international debt (one of the highest per capita debts in the world). Even the international conservation community has come on board and either remains blithely oblivious to what is going on around it or actively pursues a "neutral" position (see, for example, the statement by the Organization for Tropical Studies published in the national newspaper *La Nacion,* Janury 9, 1991). The only opposition is from a small and loosely organized local conservation movement, fighting against what would appear to be insurmountable odds.

Given this analysis, it is not hard to predict what will happen next. Banana prices will fluctuate on the world market as they always have. Furthermore the banana companies can be expected to reduce the cost of labor by laying off workers at points of economic downturn, just as they always have in the past. But this time where will those former workers go? In the past that mosaic of small farms, large cattle pastures, and second growth and primary forest stood between the organized agricultural settlements and the biological preserves. That area will now be taken up by banana plantations, and the only remaining area not devoted to some form of agriculture, and therefore available to displaced workers, will be the four biological preserves in the area. What is to prevent these landless peasants from cutting forest in the biological preserves?

Viewing the process with the general schema developed here makes the dynamics of the situation quite clear. Costa Rica really has no choice but to promote the expansion of bananas in the region. Its international debt requires that it seek tax revenues in whatever form it can. The banana companies (one of which is Costa Rican) continue to play their historical role as international accumulators of capital and temporary employers of peasants, thus continuing the dual and disarticulated economy. The peasants continue to arrive from other parts of the country and now even from Nicaragua, seeking the good life and willing to accept bare survival but without significant political representation. The loggers originally set the stage with their systems of logging roads, and the first wave of bananas ended with the periodic layoffs that force peasants into the forests. If the process continues with its basic overall structure, which I see no reason to doubt, there is little hope in the long run even for the rain forests that are under supposedly protected status (Vandermeer and Perfecto 1995).

In sum, the lessons of the 1980s seem to be clearly reflected in events subsequent to 1990. The pattern of the 1980s centered on Nicaragua's experiment in radical political reform that focused on social justice, with rain forest preservation an indirect consequence and an apparently slow evolution of a mosaic system of land use in its rain forest areas. Costa Rica's situation caused the evolution of a model of pristine islands in the midst of devastation. The Nicaraguan elections of 1990 changed the situation significantly. Broadly speaking, the patterns of rain forest destruction since that time follow what would be expected. The Nicaraguan mosaic model rapidly gave way to the more classical landless peasants moving continuously into rain forest areas, and deforestation appears to be proceeding at pre-1980 rates. Costa Rica's pristine model continues, but with increased pressure because of a shortfall of foreign assistance and an apparent approach to the limits of that model. The future of the rain forests of the Atlantic coast of Central America does not look good.

References Cited

Boucher, D. 1989. Growing back after hurricanes. *Bioscience* 40: 163–166.
Denslow, J. S. 1980. Gap partitioning among tropical rainforest trees. *Biotropica* 12 (Supplement): 47–55 .
Hartshorn, G. S. 1989. Application of gap theory to tropical forest management: Natural regeneration on strip clear-cuts in the Peruvian Amazon. *Ecology* 70: 567–569.
Hilje, L., L. E. Castillo, M., L. A. Thrupp, and I. Wesseling. 1987. *El uso de los plaguicidas en Costa Rica* (Use of pesticides in Costa Rica). San José, Costa Rica: Editorial Universidad Estatal a Distancia.

Hruska, Alan. 1986. The road from Switzerland to Honduras: The militarization of Costa Rica. In *Nicaragua: Unfinished Revolution*, P. Rosset and J. H. Vandermeer, eds., pp. 283–285. New York: Grove.

Johns, Andrew D. 1983. Ecological Effects of Selective Logging in a West Malaysian Rain Forest. Ph.D. diss., University of Cambridge, England.

Kamowitz, D. 1986. Nicaragua's agrarian reform: Six years later. In *Nicaragua: Unfinished Revolution*, P. Rosset and J. Vandermeer, eds., pp. 390–393. New York: Grove.

Lugo, Ariel E., M. Applefield, D. Pool, and R. B. McDonald. 1983. The impact of Hurricane David on the forests of Dominica. *Canadian Journal of Forest Research* 13: 201–211.

Perfecto, I. In press. Periodic disturbance and foraging behavior as determinants of interaction coefficients: *Solenopsis geminata* and *Pheidole radoszkowskii* in the coffee agroecosystem in Costa Rica. *Oecologia*.

Perfecto, I. and R. Snelling. In press. Ant diversity and the transformation of tropical agroecosystems. *Ecological Applications*.

Pickett, S. T. A. and P. S. White, eds. 1985. *Natural Disturbance and Patch Dynamics*. New York: Academic Press.

Poore, D. 1989. *No Timber Without Trees*. London: Earthscan.

Vandermeer, J. H. and I. Perfecto. 1995. *A Breakfast of Biodiversity: How Food Insecurity Destroys Tropical Rain Forests*. San Francisco: Institute for Food and Development Policy.

Vandermeer, J. H., N. Zamora, K. Yih, and D. Boucher. 1990. Regeneracion inicial en una selva tropical en la costa caribeña de Nicaragua después del huracan Juana (Initial regeneration of a tropical forest on the Caribbean coast of Nicaragua after Hurricane Joan). *Revista Biología Tropical* (Costa Rica) 38: 347–359.

Westoby, J. 1989. *Introduction to World Forestry: People and their Trees*. Oxford, England: Basil Blackwell.

Yih, K., D. Boucher, and J. H. Vandermeer. 1989. Efectos ecológicos del huracan joan en el bosque tropical humedo del sureste de Nicaragua a los cuatro meses: Posibilidades de regeneración del bosque y recomendationes (Ecological effects of Hurricane Joan on the tropical rain forest of the southeast of Nicaragua after four months: Possibilities for regeneration of the forest and recommendations). Informe final to CIDCA (Centro de Investigaciones y Documentacion de la Costa Atlantica). (Unpublished internal report.)

10

Logging in the Congo: Implications for Indigenous Foragers and Farmers

David S. Wilkie

Selective logging in the Congo has been promoted as a way to sustain economic development of rain forest communities, the inhabitants of which are isolated, barely assimilated into national culture, and the last to be provided with social services or included in market economies (Wilkie and Sidle 1990). Establishment of timber concessions in the Sangha region of northern Congo has thus been heralded as an effective way to provide much-needed income, education, and health care to local communities (figure 10.1). How sustainable are the local economic benefits, how equitably are these benefits distributed throughout the local communities, and what are the social and ecological costs of accruing these benefits?

This essay discusses some short- and long-term consequences of commercial selective logging by the Société Forestière Algéro-Congolaise (SFAC) on indigenous BaNgombe foragers and BaKouele farmers of the northern forests of the Republic of Congo.

All information described here was obtained from December 5, 1989, to January 12, 1990, during a social and ecological assessment (funded by the U.S. Department of Agriculture) of an African Development Bank loan to SFAC.

Congo Background

The Congo covers some 342,000 square kilometers, an area about the size of the state of New Mexico, and is inhabited by about 2 million people. The

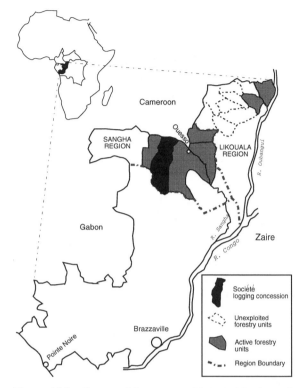

Figure 10.1   Forestry Management Units in Northern Congo

largest ethnic groups are BaKongo, BaTeke, M'Bochi, and Sangha. More than 60% of Congolese live in urban areas (48% in the capitol, Brazzaville, and the coastal port, Pointe Noire, and 12% along the railway that links these two principal cities). The remaining 40% are scattered in towns and small villages primarily in the central and southern portions of the country. Northern Congo is sparsely populated (fewer than three people per square kilometer), and most settlements are located along rivers.

About 60% of the Congo is covered by tropical moist forests. By the end of the 1960s timber was the country's principal export (from 1947 to 1980 the country produced 14.5 million cubic meters of timber from approximately 1.5 to 2 million trees), and much of the southern forests had been logged. Timber exports are less than 10% of the gross national product but remain the major focus of the Congo's five-year economic development plan.

With the exception of the vast inundated forest (largely in the southern Likouala region) and the northwestern corner, the northern Congo has

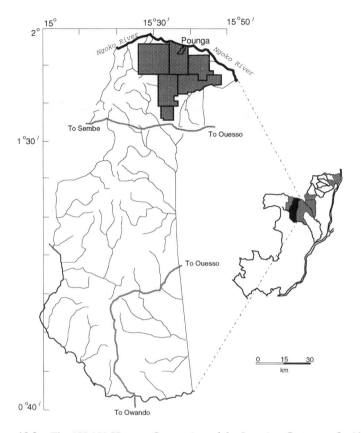

Figure 10.2   The 855,000-Hectare Concession of the Logging Company Société Forestière Algéro-Congolaise   The rectangular areas within the concession are the logging units exploited since 1985.

been divided into fifteen forestry management units (Unité Forestière d'Aménagement, UFA) (approximately one-sixth of the landmass) and ostensibly ceded to foreign logging concerns. This has negated the implicit land tenure rights of the local inhabitants and took from them any direct control over and investment in forest exploitation.

The Société Forestière Algéro-Congolaise, a semipublic company formed in 1983 within the framework of a twenty-year cooperative agreement between the the Democratic and Popular Republic of Algeria and the Republic of the Congo, received the UFA-Centre management unit (UFA-Centre)—855,000 hectares in the Sangha region (figure 10.2). Several companies are logging, including SFAC, Boissangha, Société Congolaise Arab-

Libyenne, Compagnie Industrielle des Bois, and Société Congolaise du Bois d'Ouesso, all in the Sangha region, and Société Nationale d'Exploitation des Bois and Forestiére du Nord Congo at Enneyele in the Likouala region. The remaining management units await foreign investment. Establishment of a new national park bordering the Sangha River in 1994 gave one or two management units a reprieve from commercial exploitation.

Since selective logging started in 1985, how has SFAC affected local populations living around its base camp at Pounga, and what does the future hold for them? How have settlement patterns and subsistence practices changed as a result of SFAC's presence? Has commercial logging in UFA-Centre improved social services and the economy in the area? Are these changes likely to continue in the future?

Indigenous Populations Within UFA-Centre

The SFAC concession is the traditional home of the BaKouele Bantu and the BaNgombe Pygmies (Bruel 1935; Dhellemmes 1985; Robineau 1971). A search of the National Library in Paris revealed a paucity of information on foragers and farmers within the Sangha region, west of Ouesso. Most information, with the exception of that contained in the research of L. Demesse (1978; 1980) and J. Pedersen and E. Waehle (1988), is anecdotal (Bruel 1935; Loung 1959) or describes primarily the BaBinga Pygmies north of the Ngoko River in Cameroon. I could find only two historical references on the BaKouele farmers and fishers of the Sangha region (Loung 1959; Robineau 1971). Information on the BaNgombe is also scarce (Dhellemmes 1985; Loung 1959). Consequently, this essay is based primarily on information gleaned from personal observations and from interviews conducted with men and women of the BaKouele and BaNgombe within the SFAC concession during December 1989 and January 1990.

Methods

I visited all the villages along the Ouesso-Sembe Road, along the Ngoko River, and within the SFAC compound area, took censuses of the population, and conducted semistructured interviews of people of different ages, sexes, and tribes. I walked to the villages along the road and surrounding the SFAC compound at Pounga and used a dugout canoe to reach those along the Ngoko (figure 10.3). I did not have enough time to visit the villages that border the road from Ouesso to Owando, which traverses the southeastern section of the SFAC concession. I used the Congo's 1984 cen-

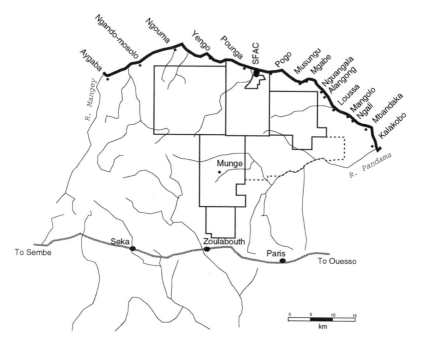

Figure 10.3   Settlements Censused Within the Concession Granted to the Société Forestière Algéro-Congolaise

sus to extrapolate all demographic data for these villages; the census is available from the offices of the Ministère du Plan et des Finances in Ouesso and Brazzaville.

I completed each village census by visiting each hut and asking how many men, women, girls, and boys slept in the hut that night. Restricting the number of hut occupants to those that slept there the previous night assures an accurate count and avoids multiple counting of family members who may sleep in several huts at different times. It does, however, risk not counting those family members who may be traveling outside the census region at the time. This is particularly relevant for those SFAC employees living temporarily in the forest as part of an inventory team.

I assessed their material culture by entering every house and noting the visible commodities. I specifically asked about certain items, such as traditional weapons, shotguns, musical instruments, clothing, and cooking equipment.

I interviewed Bantu and BaNgombe men and women about their subsistence practices, sources of revenue, education, health care, what they consider the benefits and drawbacks of living within the SFAC concession, and their views of their future.

I also examined SFAC employment records and queried employees about their tribal affinities (to assess the proportion of SFAC employees not from the Sangha region) and the opportunities available to them for training and advancement. At all villages that had a functioning school or infirmary I spoke with the school teachers and health care providers.

Historical BaKouele and BaNgombe Settlement and Subsistence Patterns

Before the colonial period (1900–1960) the BaNgombe lived a seminomadic existence similar to that of present-day Efe Pygmies of the Ituri Forest (Bailey 1985; Wilkie 1988). Individual BaNgombe were involved in a long-term exchange relationship with BaKouele farmers who were settled in patrilocal villages alongside perennial rivers and streams and practiced rotational slash-and-burn agriculture. The BaNgombe traded farm labor and forest products such as meat and honey for cultivated crops and commercial commodities such as salt, clothing, and tobacco. Although the BaNgombe spent most of the year camped close to BaKouele villages, they would travel into the forest for days or weeks at a time to hunt forest antelope and primates with crossbow, nets, and snares or to gather honey and the seeds of kana (*Panda oleosa*) or payo (*Irvingia* sp.).

During the French colonial period a road was built from the regional center of Ouesso west to Sembe and thence to Souanke (Zouanké). The BaKouele moved or were resettled along the road to help maintain the newly established cacao plantations that constituted a major export of the region. Because the BaKouele were involved in commercial cacao cultivation, they had access to money and a variety of exogenous commodities such as clothing, cookware, and soap. Villages at this time were still quite small and were often composed of only an extended family unit. Before independence twenty-five villages existed along the Ouesso-Sembe Road and twelve on the banks of the Ngoko within what is now the SFAC concession (Institut National 1980). Since the revolution in 1963 and more concertedly since President Ngouabi's ascension in 1968 the government has been collectivizing villages and integrating Pygmies into the more sedentary villager way of life. These policies were instituted to provide basic health and education services to the BaKouele and BaNgombe. As a result all BaNgombe live in permanent settlements and grow much of their own food. A decline in world cacao prices resulted in the collapse of the local cacao market in the early 1980s; although BaNgombe still work in BaKouele fields, they do so on an opportunistic basis and primarily to obtain food in exchange.

Settlements are clustered along the logging roads (within 3 kilometers of the lumber camp at Pounga), the two state roads, and along the Ngoko River. No permanent settlements are reported to exist within the forest interior. Several temporary hunting camps dot the forest interior but are used only as sleeping spots for one or two days when either the BaKouele or BaNgombe are checking or setting their snare or trap lines, or the latter are out searching for honey.

Only three large villages along the Ouesso-Sembe Road are within the SFAC concession; thirteen small villages border the Ngoko, and the 1984 Congo census shows five villages along the Ouesso-Owando Road between the Likoula and Mambili rivers. The SFAC compound consists of one large settlement (built by SFAC) surrounded by five smaller ones. No other settlements are reported within the SFAC concession. All villages have been at their current sites for at least four years, and most were established considerably before the appearance of SFAC at Pounga in 1985.

Effects of SFAC on Local Settlement Patterns and Economy

Approximately 2,745 people live within the SFAC concession in the Sangha region, a density of approximately three people per square kilometer, which is higher than that estimated by G. Bruel in 1935 but comparable to that of the 1984 national census, which predated the establishment of SFAC.

Bantu comprise 55% of the overall population, BaNgombe the other 45%. In villages outside SFAC's immediate influence, BaKouele make up 45% (93% of all Bantu), BaNgombe 51%, and other Bantu a mere 4% of the populations. This is in stark contrast with the ethnic composition within the SFAC compound, where BaKouele represent 26% (41% of all Bantu), BaNgombe 35%, and other Bantu 33% of the population; the remaining 6% include non-Bantu traders and expatriate SFAC employees. This indicates an ethnic bias in employment opportunities at SFAC, which primarily stems from the absence of technically trained BaKouele and BaNgombe. Of all the inhabitants of the SFAC concession, 39% live at or near the SFAC compound at Pounga, and of those at least one family member is employed by SFAC. One hundred and seventy eight (178) BaKouele and 212 BaNgombe moved from settlements within the UFA-Centre to the Pounga area as a result of the establishment of SFAC. Eighty-eight Bantu men and their families (257 people) from outside the UFA-Centre area have moved to Pounga as a result of the establishment of SFAC. Yet logging operations by SFAC have encouraged immigration that amounts to a mere 9% of the total population of UFA-Centre (2,745) and have resulted in only 14% of the area's population leaving its natal villages and moving to the SFAC compound. Unlike logging concessions in South America or Southeast Asia, SFAC has resulted neither in a substantial redistribution of local communities nor large-scale immigration.

Wage Labor and Alternate Sources of Income

Within the SFAC concession wage labor for the BaKouele consists of working for SFAC or for the state as a teacher or health care provider at the infirmary. The BaNgombe, the majority of whom are illiterate, can either work for SFAC or for the few BaKouele who still own and manage cacao plantations. Those BaKouele and BaNgombe not employed by SFAC earn their major source of revenue from the sale of game meat, fish, and agricultural products. The sale of meat and fish is without doubt the single most important source of income for non-SFAC employees.

The meat of wild game commands a high price, because it is a coveted dietary item and has a large pool of consumers at SFAC, Ouesso, and as far away as Brazzaville and Pointe Noire. Without a market for wild game no cash economy would exist along the Ouesso-Sembe Road or along the Ngoko River. All kitchen supplies, clothing, furniture, and building materials not made from local materials are purchased with the proceeds of wild game sales. The people also depend on the capture and sale of wild animals and fish so that they can buy such luxury items as soap, prestige dietary supplements such as canned sardines and corned beef, medicines, and notebooks and pencils for schoolchildren. If there were no market for wild game, the standard of living for non-SFAC employees would certainly be different.

The sale of agricultural products is a consistent but relatively minor source of income. Fresh plantains and prepared kwanga (manioc) are sold both by roadside and riverside BaKouele and BaNgombe. High perishability of fresh produce means that the market is largely local (SFAC and Ouesso) rather than regional or national. Because starches never command the same price as protein (wild game and fish), the profit level is much lower. Although Ngoko River farmers sell some dried manioc at the Ouesso market, the financial importance of this was difficult to assess.

Building pirogues, making palm thatch, and weaving sleeping mats provide a few artisans with some money. But again, in comparison to wild game sales, these are of little economic importance. Within the SFAC compound at Pounga, BaKouele, Cameroonian, and Zaïrois entrepreneurs have opened three bars, one general store, a carpentry shop, and a brothel. These enterprises provide SFAC workers' families with services and the few entrepreneurs with a credit-based income. During festivals and holidays many wives of SFAC employees make and sell distilled maize liquor (*Ngolongolo*) to augment their husbands' salaries.

Effects of SFAC as an Employer in UFA-Centre

The arrival of SFAC in 1985 had a profound effect on the local economy. Today logging concessions are the only employer along the Ngoko River and are the major economic and market force throughout the region. Although SFAC uses heavy machinery to build roads and to transport cut logs, the company, like all logging concessions, must still hire a relatively large number of workers (men) on a daily or monthly basis. Two hundred to 230 men were working in the concession each day during 1989–90.

Employment with SFAC is divided into several classes that express differences in tasks and skill level. Each employee within each category is paid a base monthly salary that is augmented according to his productivity. The supplementary component of an employee's wage can often double or triple his monthly earnings. Thus monthly salaries vary considerably, ranging from as little as 30,000 CFA francs to three or four times that amount.

The tribal compositions of SFAC work teams appear to have rather clear differences. Inventory and exploitation teams are primarily BaKouele and BaNgombe, and the drivers and mechanics are more often from regions other than the Sangha (Teke Plateau, Pool, Niari, Cuvette, and so on; see figure 10.4). It is not surprising that inventory teams are made up of Sangha residents, because they have the most intimate knowledge of the area and the trees within it. Similarly, until the arrival of SFAC BaKouele and BaNgombe within the area seldom had the opportunity to learn how to drive or how to repair vehicles. Thus it is not surprising that SFAC recruited other workers for these tasks in order to get into production as soon as possible. Several BaKouele are in positions of authority or those that require technical skills (chief of the inventory teams, driver, mechanic);however, the BaNgombe, except for chainsaw operators, remain in nontechnical positions.

The effect of SFAC on the local economy is exemplified by its personnel records—it employs 30% to 45% of all adult male BaNgombe and 20% to 25% of all BaKouele living within 30 kilometers of the base camp at Pounga. Most of these men are hired to clear inventory transects and to identify all commercially valuable timber trees within a given year's harvesting block. SFAC depends heavily on BaNgombe woodcraft, for without an accurate tree inventory the company would be unable to determine which trees to fell and could not design and build a cost-effective road system to extract the logs.

Because SFAC values forest skills, the average monthly wage for a BaNgombe man employed on an inventory team ranges from 50,000 to 90,000 CFA (U.S.$1 = 250 CFA); only technically skilled mechanics and drivers earn more. Many BaNgombe and BaKouele men therefore earn annual incomes more than two or three times that of the average Congolese.

Figure 10.4   A nonlocal Congolese employee of Société Forestière Algéro-Congolaise loading logs onto a logging truck.

How has this massive and relatively instantaneous influx of disposable income affected the lifestyles of the BaKouele and BaNgombe?

BaKouele and BaNgombe families with SFAC employees are more likely to have tin-roofed huts, new aluminum cooking pots, eating utensils, functioning flashlights, and new clothes and shoes than those families with no SFAC workers. Thus the material standard of living has changed for SFAC workers' families. Whether the ability to purchase commercially produced goods will result in the loss of traditional manufacturing skills is unclear. BaKouele farmers have used money from the sale of cocoa and other crops to buy these items for many years. The BaNgombe obtained the same items second-hand by trading with the BaKouele. However, the traditional material culture of both BaKouele and BaNgombe is depauperate, and interviewees mentioned that local artisans now rarely make hunting weapons (crossbows, nets, and spears), domestic utensils, and musical instruments. Roofing thatch (*Rafia* sp.), sleeping mats, and baskets are the most common items still produced by local craftspeople.

BaNgombe employees of SFAC are also less likely to work, or are less willing to work, in BaKouele fields. Interviewees commented that they no longer needed to work for the BaKouele, because they could now purchase their own commodities. Because many BaNgombe are unable to count, they depend on numerate BaKouele to intercede for them when purchasing items, a situation that often results in collusion between shopkeeper and BaKouele to defraud the BaNgombe.

As material possessions increase, so does the propensity to store or hide possessions. One SFAC worker said, "If I do not hide what I have bought at the market in Cameroon, my relatives will take/borrow them from me." Although this person commented that this attitude goes against traditional BaNgombe beliefs about the importance of sharing, he noted that the possessions that are hidden most frequently are such items as batteries, soap, tobacco, and palm oil— once borrowed, these cannot easily be returned in their original condition. This change of behavior has also been noted for the !Kung (Yellen 1990).

Access to Housing, Education, and Health Care Services Provided by SFAC

In contrast to the leaf or wattle-and-daub huts characteristic of the region, SFAC employee housing is made from wood planks, supported 1 meter above the ground with wood pylons. Houses have wooden doors and windows and corrugated metal roofs. Each SFAC house has a separate wooden kitchen and toilet. All SFAC wooden houses have electrical lighting that is available between sundown and sunup. Each house is divided into two sections (like a duplex); a male SFAC employee and his family occupy each half.

Although BaNgombe are valued as forest experts and are paid accordingly, there is an implicit prejudice against them when it comes to access to worker housing, and they rarely take advantage of education and health care services.

The state requires SFAC to provide housing, sanitation, fresh water, and health care to the workers and to provide primary education for workers' children. Although 35% of the work force is BaNgombe, none of them lives in SFAC-constructed housing at Pounga, a settlement of 64 two-family wooden dwellings. Tenancy is certainly not based on salary, because the monthly wage of many BaNgombe is equal to or greater than some Bantu's. BaNgombe workers and their families live in housing of their own construction in villages proximal to Pounga or, when working on tree inventory teams, in temporary canvas shelters in the forest.

No BaNgombe children from families with SFAC employees attend the primary school, whereas more than 60% of Bakouele children and 90% of nonlocal workers' children do attend. This is partly because the BaNgombe families who comprise the inventory teams do not live near the school at Pounga; however even the children of BaNgombe workers who live within 2 kilometers of the school do not attend.

Interviews with SFAC workers and the two health clinic nurses indicate that BaNgombe rarely visit the clinic to obtain medicines or to have traumatic injuries cared for. BaNgombe were much more likely to use traditional cures than were the BaKouele or other Bantu.

BaNgombe employees of SFAC are economically as well off as other SFAC workers but do not have access to SFAC housing and sanitation facilities and are not yet taking advantage of health and education services. Whether this will change is unclear.

Summary of SFAC Changes to Local Socioeconomy

Establishment of SFAC's logging operation has certainly improved health care services, primary education, and housing conditions for a small proportion of BaKouele who live at or within 2 to 3 kilometers of Pounga. These benefits are not realized by the BaNgombe, both as a result of implicit prejudice in the allocation of worker housing and their own failure to take advantage of health and education services.

Although SFAC has only an exceedingly localized effect in regard to providing social services, it has a more profound and far-reaching effect on the economy of the area. SFAC depends on the forest expertise of local BaNgombe and BaKouele and as a result employs a substantial proportion of local men living within 30 kilometers of Pounga. In doing so SFAC has raised the average worker's income from almost nothing to two to three times the national average. An increase in disposable income has allowed SFAC workers to improve their quality of life by purchasing better clothes, housing and cooking materials, medicines, and so on. SFAC employees have also used their wages to finance commercial wild game hunting by buying shotguns and cartridges. Although the proceeds of market hunting represent a substantial supplementary income for many workers, commercial hunting has already begun to have a profoundly deleterious effect on the fauna of the surveyed and logged sections of UFA-Centre (Wilkie, Sidle, and Boundzanga 1992).

From the perspective of rural development, SFAC has introduced quite substantial sums of money into the local economy, which has certainly enhanced the material quality of life for BaKouele and BaNgombe employees. Yet its effect on social services has been exceedingly local. Most important from the perspective of sustainable economic development and ecological conservation—of most concern—are the predictably transient nature of logging within UFA-Centre and the lack of indigenous peoples' control over how resources in their area are exploited and profits from resource exploitation are distributed.

Transient Nature of Selective Logging Operations

Although the SFAC concession agreement is for twenty years, the Pounga base camp is not likely to be active for that long. Legally and economically, SFAC cannot reexploit logged sections of the forest during the twenty years of the agreement. Thus, given the low density of commercially valuable trees (1 tree per 1–13 hectares), SFAC must inventory, exploit, and abandon large areas of forest within the concession each year (20,000–40,000 hectares). In two years' time SFAC will have exploited all forest within the concession that lies between the Ngoko River and the Sembe-Ouesso Road (figure 10.2).

SFAC uses the Ngoko, Sangha, and Congo Rivers to transport cut logs to Brazzaville (1,100 kilometers) and thence by rail to Pointe Noire (500 kilometers). However, as logging continues to move south through the concession area, the preferred timber transportation route is likely to change; it probably will go first to the Sembe-Ouesso Road and then to the Ouesso-Owando Road. SFAC believes that using road transportation will reduce the time from felling to sale to less than a month. This will cut three to five months off the period during which investment capital is frozen in unsold deteriorating timber.

By its very nature selective logging means that SFAC will have a transient presence within any given section of the forest during the twenty years of the concession agreement. Further, if the economics of timber exploitation do not change during that time, SFAC may not be interested in renewing the agreement in order to return to previously logged areas to extract those species considered to have no economic benefit or to cut those few trees that have reached harvesting size during the previous twenty years.[1] SFAC's intention to continue logging within UFA-Centre for the duration of the agreement is actually questionable, because SFAC is already lobbying to be allowed to swap the concession for the Nouabalé (or another) concession situated along the Sangha River. SFAC argues that the density of timber is not sufficient to provide a profit. Because SFAC does not intend to conduct the time-consuming and expensive surveys required to assess timber density in the concession to which SFAC hopes to move its operations, it is not clear why SFAC thinks that timber density will be higher than in UFA-Centre. Should the Ministère de l'Economie Forestière agree to SFAC's demands, discussing the long-term economic benefits of commercial logging within UFA-Centre is moot. Therefore it seems highly likely that SFAC will have only a short-term local presence within any given section of UFA-Centre or may abandon the concession altogether. How will this affect local economies?

Effect on Local Economy

Clearly, an increase in disposable income and material possessions is not in itself undesirable; in fact it is often a goal of development projects. However, SFAC's contribution to the local economy is likely to be quite short lived and cannot, at least at a local level, be considered sustained development. As SFAC moves to the more southerly reaches of the concession or to leave UFA-Centre altogether, it probably will take with it the workers' housing and abandon the schools and health clinics built on the banks of the Ngoko River in 1985. Equally important, the BaNgombe and BaKouele of the Ngoko River will no longer have access to a source of income: SFAC will be more interested in hiring workers living in the southern sections of the concession, because they will have a more intimate knowledge of that part of the forest.

How will five to ten years or more of high income affect the needs, aspirations, and social behavior of a given local population of BaNgombe and BaKouele? Will today's employees be able to return to a more basic lifestyle once SFAC moves out of their region? Will they be able to reattain traditional sharing patterns? Will young people have neglected to learn traditional techniques that once again become more important to daily subsistence? Will the state be able to assume the role of education and health care provider once SFAC moves out of an area? Will faunal populations be able to recover from such intensive market hunting and thus continue to provide the local population with a necessary source of protein?

Although I have no way to predict the effects of this boom-bust aspect of selective logging, it is clearly not a sustained form of rural development,[2] and this should be evaluated by countries intent on exploiting their tropical rain forests and by multilateral and unilateral banks that provide loans to logging companies.

The Future

What factors in the management of timber exploitation must be changed to ensure that the economic, educational, and health benefits for other SFAC employees will not be as short lived as those expected within Congo's UFA-Centre?

If the concern is for long-term development of tropical rain forest communities, its most visible economic resource—timber—must be exploited in a sustainable way (Lamb 1991).[3] Environmental economists assume that resources are more likely to used in an sustainable way when individuals or groups appropriate the full or true value of the resource. This is most often achieved when resource managers have ownership rights to the resource and can enforce those rights.

An Optimistic Scenario

If the Congolese government divides the Sangha and Likouala regions into national and private forest lands, the direct benefits of timber exploitation could accrue at both the local and national levels. Alternatively, the government could set aside national lands as preserves, with exploitation occurring only on private lands and the state obtaining revenue through taxes and fees. If individuals or groups of individuals own the timber on a specific area of forest, logging companies have to purchase exploitation rights from these landowners and would be responsible only for cutting and transporting the timber. With the technical assistance of forestry officers landowners would be responsible for managing the forest and for ensuring that logging practices are compatible with regeneration and sustainable use. Landowners could report to forestry officers any logging companies that are not complying with felling regulations, and the state could rescind a company's operating license if need be. Competition for a restricted number of exploitation licenses might even promote compliance among logging companies.

The government could obtain money for education and health care services by levying income taxes, and it could maintain the road system by charging logging vehicles fees based on tonnage hauled. This system would enable the government to promote development of previously disenfranchised populations, fund social services and maintain roads in remote areas, and encourage rational exploitation of a valuable resource.

Forest Conservation in the Real World

If we agree with the goals of the scenario presented, and we accept the contention that the most effective way to promote forest management is to convince the local population that forest conservation has an immediate and sustainable economic value to it, what steps need to be taken?

1. Establish a Congolese real estate market within which all citizens, including forest dwellers, can obtain secure and transferable titles to their land and property.

2. Develop markets for a diverse suite of forest commodities in Congo's urban centers and internationally.
3. Provide forest owners with the technical, legislative, and infrastructural tools to manage timber as a sustainable source of income. These tools must serve both as incentives for sustainable use and disincentives to overexploitation.

If these three steps can be implemented effectively, the benefits of resource exploitation are more likely to remain largely within the local community. It is hoped that this in turn will foster local individual and community stewardship of the forest's resources.

Successful implementation of these changes hinges on the willingness of the Congolese government to institute policy reforms that provide citizens with the legal right to own and sell land and property, formalize how customary lands are to be privatized and partitioned, and establish rules for conflict resolution associated with disputed landownership. Providing free and clear title to land requires that property boundaries be surveyed and demarcated and that an institutional infrastructure be established to record and maintain land and property titles. Whether the Congo has the political will or the economic or technical resources to establish a national real estate system is at best debatable.

As to diversifying forest resource exploitation, developing new markets for nontimber forest products is fraught with challenges. Given the isolation of forest-dwelling communities from the main urban markets in the Congo (where 60% of the consumers live), it may be exceedingly difficult to establish markets for perishable nontimber forest products that either do not command a high price or have readily available or inexpensive agricultural or industrial substitutes. Smoked and dried wild game meat commands a high price in urban markets (fresh wild meat is even flown to Paris), and as result the commercial infrastructure for obtaining, transporting, and selling wild game has developed spontaneously (Wilkie, Sidle, and Boundzanga 1992). Two forest tree nuts, kana (*Panda oleosa*), and payo (*Irvingia* sp.), are traded in local markets and may have potential for expanded sales in urban areas. Yet the fact that wild game traders have not incorporated kana and payo in their commercial transportation of forest products to urban markets suggests that it is not profitable to do so.

Last and most critical, it is a stretch of the imagination to believe that rural forest landowners are likely to have a planning horizon (personal discount rate) of four or five generations (100–125 years), the minimum time needed to bring old growth saplings into the large-diameter classes that are being felled right now. Personal discount rates may encourage landowners to clear-cut the forest now and invest the proceeds such that their rates of return exceed that expected from sustained growth and harvesting of timber. When livelihood insecurity issues shorten planning horizons to months or days, the prospect of sustainable forest management by poor forest farmers and fishers is bleak, even if the land tenure, forestry extension, and regulation enforcement issues are resolved.

Acknowledgments   This study was part of a timber production capacity extension project of the Société Forestière Algéro-Congolaise in the Republic of the Congo. The study was organized by the U.S. Department of Agriculture's Office of International Cooperation and Development (OICD), in conjunction with the African Development Bank (ADB), and was undertaken by the author, John G. Sidle, and George Claver Boundzanga.

I am grateful to many individuals who facilitated this study: the staff of OICD in Washington, D.C., of the U.S. Agency for International Development and ADB in Abidjan, of the American embassy and the Ministère de l'Economie Forestière in Brazzaville, and of SFAC in Pounga.

Notes

1. Because we have little data on the growth rates of commercially important species in African moist forests, and because tree inventories do not include trees smaller than 80 centimeters in diameter at breast height, no one knows how many trees are likely to reach harvesting size during the twenty-year period of the SFAC concession agreement. A recent study by the Wildlife Conservation Society showed that the average age of *Entandrophragma* sp. felled in a northern Congo logging concession was four hundred years and the oldest was more than nine hundred years (J. Michael Fay, personal communication 1990).

2. Since I collected the data for and wrote this chapter, the Société Forestière Algéro-Congolaise has declared bankruptcy, abandoned the UFA-Centre logging concession at Pounga, laid off or fired its employees, and left its business partner, the Congolese government, to pay off its loans and debts.

3. Although sustainable exploitation of tropical forest is favorable from a socioeconomic perspective, it is by no means clear that economically valuable tree species can be extracted on a sustainable basis. Similarly, any long-term extraction of timber changes plant species composition and the diversity and abundance of animals that the forest supports. Therefore it is important not to confuse the goals of sustained rural development with those of biodiversity conservation.

G. S. Hartshorn (1989) discusses the use of strip clear-cuts to extract timber in a way that mimics natural tree falls within the forest, making it likely that that technique would have the least adverse effect on forest regeneration. This method is an improvement over high grading. However, in the case of the Congo it assumes that a market for all felled timber exists and that the rate of clearing will not exceed the replacement rate—neither of which is assured.

References Cited

Bailey, R. C. 1985. The Socioecology of Efe Pygmy Men in the Ituri Forest, Zaire. Ph.D. diss., Harvard University.

Bruel, G. 1935. *La France équatoriale Africaine.* Paris: Larose.

Demesse, L. 1978. *Changements techno économique et sociaux chez les Pygmées Babinga Nord-Congo et Sud-Centrafrique.* 2 vol. Paris: Societé d'Études Linguistiques et Anthropologiques de France.

———. 1980. *Techniques et économie des Pygmées Babinga.* Paris: Institut d'entholo-gie du Musée de l'Homme.

Dhellemmes, I. 1985. *Le père des Pygmées.* Paris: Flammarion.

Hartshorn, G. S. 1989. Application of gap theory to tropical forest management natural regeneration on strip clear cuts in the Peruvian Amazon. *Ecology* 70: 567–576.

Institut National de Recherché et D'Action Pedagogiques. 1980. *Géographique de la République Populaire du Congo.* Brazzaville: Institute National de Recherché et D'Action Pedagogiques.

Lamb, D. 1991. Combining traditional and commercial uses of rain forests. *Nature and Resources* 27 (2): 3–11.

Loung, J. F. 1959. Les Pygmées de la forêt de Mill: Un groupe de Pygmées Camerounais en voie de sédentarisation. *Les Cahiers d'Outre-Mer* (12): 1–20.

Pedersen, J. and Waehle, E. 1988. The complexities of residential organization among the Efe Mbuti and the BaMgombi Baka: A critical view of the notion of flux in hunter-gatherer societies. In *Hunters and Gatherers: History, Evolution, and Social Change,* T. Ingold, D. Riches, and J. Woodburn, eds., pp. 75–90. Oxford, England: Berg.

Robineau, C. 1971. *Évolution économique et sociale en Afrique centrale: L'exemple de souanke.* Paris: ORSTROM.

Wilkie, D. S. 1988. Hunters and farmers of the African forest. In *People of the Tropical Rain Forest,* J. S. Denslow and C. Padoch, eds., pp. 111–126. Berkeley: University of California Press.

Wilkie, D. S. and J. G. Sidle. 1990. Social and Environmental Assessment of the Timber Production Capacity Extension Project of the Société Forestière Algéro Congolaise in the People's Republic of Congo. Unpublished report. Washington, D.C.: U.S. Department of Agriculture, Office of International Cooperation and Development.

Wilkie, D. S., J. G. Sidle, and G. C. Boundzanga. 1992. Mechanized logging, market hunting, and a bank loan in Congo. *Conservation Biology* 6 (4): 1–11.

Yellen, J. E. 1990. The transformation of the Kalahari !Kung. *Scientific American,* April, pp. 96–105.

V

Contributing to Solutions

In part 5, Contributing to Solutions, the authors discuss various approaches to assessing and combating deforestation. These range from the use of satellite imagery technology to ground surveys and the empowerment of local communities in the forests.

In the Philippines most remaining forest is in the uplands, where about one-third of Filipinos live. In chapter 11 James Eder examines Palawan, often considered the last frontier in the Philippines. He shows that in the Philippines the main causes of deforestation are commercial logging and lowland migrant farmers. The latter migrate to upland areas to escape poverty, land scarcity, and population pressure. They bring knowledge of sustainable farming in permanent plots and agroforestry, but they engage in shifting cultivation, which contributes to deforestation, mainly because of the insecurity of land tenure. Thus neither ignorance nor greed underlies the role of these migrant farmers in deforestation; it is simple need. Eder concludes that these farmers must be considered part of the solution as well as part of the problem and that the most important factor in resolving the dilemma is for government policy to effectively provide fair agrarian reform and land rights.

Papua New Guinea and Irian Jaya are important regions of tropical forests in which both biological and cultural diversity are extremely high. In chapter 12 Carolyn Cook provides an insightful comparison of

Papua New Guinea and Irian Jaya; the striking contrasts in the two governments lead to major differences in the rates of land and resource use as well as deforestation. Deforestation rates once were similar, because the people and their interactions with natural resources have the same cultural roots; because they are on the same island, they have common environmental elements in their ecosystems. Using as a basis her extensive fieldwork with the Amung-me, Cook reviews several possibilities for economic development and concludes that intensification of one traditional seasonal staple crop, pandanus, offers the best prospects. In this case the usual development procedure would be reversed—it would not be outsiders involving the local people in development but the local people involving some outsiders in their development. This procedure could work on both sides of the island if the officials of the respective governments can see this approach as a win-win solution.

In chapter 13 Robert Sussman, Glen Green, and Linda Sussman consider Madagascar, a megadiversity country because its high percentage of endemic species results from its long geographic isolation. In this nation the main causes of deforestation are unsustainable farming and use of wood for fuel and charcoal production, which in turn are related to population pressure and poverty. The limited success of using protected areas as a conservation strategy stems from the neglect of the local people, their needs, and culture. The authors use satellite imagery to determine rates of deforestation since 1960 and to locate hot spots of deforestation and monitor the process today. This high-tech approach is integrated with the exploration of the situation on the ground through anthropology's time-tested methods of ethnographic fieldwork. The authors also identify important regional differences in the deforestation process in the dry and wet forests of Madagascar.

In chapter 14 Robert Bailey concludes the book with his long-term field observations in central Africa. He argues that conservationists and developers seldom give adequate attention to involving local people in the design, implementation, monitoring, and assessment of their projects. Partly this results from the artificial dichotomy between the biological and social components of environmental conservation. He also presents field data to demonstrate that population pressure does not come from the people within the forests but instead from immigrants and other outsiders. Because to some degree many forests are anthropogenic, and because local individuals and groups identify with specific areas of forests, the usual conservation strategy of protected areas—especially when based on the belief that forests are natural—is not effective. Bailey concludes that the local people are not simply part of the problem but must also be realistically considered part of the solution. This must involve the empowerment of the local communities through meaningful participation in conservation and development. Conservation and development that come exclusively from urban bureaucrats and technocrats, to the disregard of local community knowledge, needs, interests, and rights, are unlikely to succeed.

11

After Deforestation: Migrant Lowland Farmers in the Philippine Uplands

James F. Eder

Large-scale immigration by land-seeking lowland farmers threatens the remaining forest cover of mountainous Palawan Island, the Philippines' last agricultural frontier. The agricultural practices of these migrant farmers are commonly viewed as an ecologically destructive form of swidden (slash-and-burn) agriculture. In one upland Palawan community settled during the 1940s and 1950s, however, as in other more recently settled upland communities, many migrant farmers aim to plant tree crops and to pursue other ecologically sound farm-management practices. This essay explores why these aims are not often realized—and why as a result lowland migrants to the uplands are so readily portrayed as rapacious shifting cultivators.

Palawan has a series of migrant communities whose upland farming systems are in principle ecologically sound (or approach ecological soundness). Just as poverty and unsolved agrarian problems in densely populated agricultural lowlands drive frontier colonization to begin with, such factors as poverty, land tenure insecurity, and lack of access to credit—rather than greed, ignorance, or ecological insensitivity—explain the environmental damage inflicted by many colonists upon their arrival in Palawan.

My purpose here is not to propose ways to slow the alarming rate of deforestation. Rather the indigenous farming systems that at least some lowland migrants bring to the uplands exhibit some unexploited potential for ameliorating some of the more destructive ecological and social consequences of deforestation.

Background

The Philippines has one of the highest rates of deforestation in the world today (Porter and Ganapin 1988:23). The country's total land area is 30 million hectares, inhabited by a population of approximately 64.6 million people growing at a rate of 2.5% per year (Population Reference Bureau 1993). In 1900 (when the population numbered 7.6 million) forest covered most of the archipelago, and two-thirds of it may have remained forested at the end of World War II (Myers 1988:205). However, these forests have disappeared at a steady rate since World War II. Recent estimates of the remaining area of "adequately stocked forests" range from 6.6 million hectares (Myers 1988:205) to 6.8–7.3 million hectares (Porter and Ganapin 1988:24). These figures are on the order of 22% to 24% of national territory. Based on satellite imagery, the Swedish Space Corporation (1988) estimates that forested land of all types accounts for 25% of national territory. This remaining forest is also disappearing, at a current rate on the order of 100,000 to 300,000 hectares per year (Anderson 1987:250; Porter and Ganapin 1988:24). It is generally assumed that the most valuable forests, both commercially and biotically, will be gone by the end of the century (see, for example, Bee 1987).

This destruction has two major causes. One is the relentless practice of destructive logging. The major forest type in the Philippines is composed primarily of species of the family *Dipterocarpaceae*; among these, red lauan (*Shorea negrosensis*) provides the greatest timber volume (Garrity, Kummer, and Guiang 1993:561). Dipterocarp timbers in general are prized for construction purposes and primarily bound for the export market (Bee 1987:11). Commercial logging in the Philippines dates to the late Spanish and American colonial periods. After independence the new government viewed continued exploitation of the country's forest reserves as a good way to raise money (Repetto 1988:59; Porter and Ganapin 1988). This practice, and the export of "primary products" in general, has continued as the government struggles to earn the foreign exchange necessary to repay staggering foreign debts (Broad and Cavanagh 1989:18).

The second major cause of forest destruction has been lowland migrants, in recent decades driven into uplands throughout the Philippines by landlessness and lack of employment opportunities in their regions of origin (Porter and Ganapin 1988:28–29). The highly visible slash-and-burn (Tagalog: *kaingin*) agriculture characteristically used by these migrants is immediately to blame. But the ultimate cause of this sort of forest destruction is unsolved agrarian problems in the lowlands and the filling in, earlier in this century, of the lowland forest frontiers that historically served as a safety valve for the pressure these problems put on land resources (Anderson 1987:253). Note, then, that in the Philippines today the particular issue is upland forest destruction. What to do about the uplands is in fact one of the country's most pressing public policy concerns, and numerous government agencies and nongovernmental organizations are attempting to make upland agricultural systems—still evolving and poorly understood to begin with—more economically productive and ecologically sustainable (Garrity, Kummer, and Guiang 1993).

Thus large numbers of people inhabit the Philippine uplands—and precisely how many is a matter of considerable disagreement. Past government estimates are generally considered far too low (Porter and Ganapin 1988:6). As many as 18 million people, almost one-third of all Filipinos, may reside in the uplands broadly defined, and as many as 8 million may in turn live within the public forest reserve (Cruz, Zosa-Feranil, and Goce 1986). Public forest land in the Philippines includes—by definition and regardless of the presence of tree cover—all land with slopes greater than 18 percent. Such land may not be legally sold or transferred.

Regardless of numbers, a basic distinction needs to be drawn among the throngs of relatively recent immigrants from the lowlands, mostly Hispanicized peasant peoples, and the various indigenous peoples, often called Tribal Filipinos, who have lived in the uplands on ancestral lands for generations. (The total Tribal Filipino population is on the order of 5 million people; most today live in upland areas.) This distinction is not absolute: some peoples of lowland origin have resided in upland areas for years or even generations, and some displaced tribal peoples are recent arrivals in upland areas to which they have been forced to immigrate. The Philippine government in any case often ignores such distinctions, indiscriminately labeling all forest land as public land and in turn labeling all occupants of these lands as squatters, regardless of length of occupancy (Lynch and Talbott 1988:682).

Filipinos are widely aware of depletion of forest resources and sometimes engage in acrimonious debate about who is to blame. In particular, they debate whether loggers or shifting cultivators are more culpable. Ooi Jin Bee (1987:44–46), for example, argues that logging only "degrades" the forest, whereas the encroachment of shifting cultivators on newly logged areas causes actual deforestation, or total forest resource depletion. Gareth Porter and Delfin Ganapin (1988:28) in contrast cite a local estimate that perhaps 25% of deforestation is the result of shifting cultivation, with logging accounting for the remainder. Vigorous defenses of logging, usually including criticism of shifting cultivators, appear regularly in the media and range from self-serving advertisements taken out by logging concessionaires to supportive statements by their political allies. One prominent member of Congress from Palawan, for example, recently claimed that the benefits of logging in his province extended to the "guarding" of the forest against *kaingineros* (swidden farmers) by the private forest guards of a concessionaire crony (Clad and Vitug 1988:50).

However, complicating this debate is an important but elusive distinction between "good" and "bad" shifting cultivation (e.g., Lopez 1987:230; Porter and Ganapin 1988:28–29). The first, termed *harmonic* by Harold C. Olofson (1981:3) and *integral* by Harold C. Conklin (1957), is the long-fallow and ecologically sound shifting cultivation practiced by indigenous upland peoples with long histories of swiddening on their ancestral lands. The second, the *disharmonic* (Olofson 1981:3) or *partial* (Conklin 1957) sort, is the environmentally inappropriate and destructive slash-and-burn cultivation allegedly pursued by recent lowland migrants. This distinction helps serve efforts to defend the land and lifeways of Tribal Filipinos, efforts that are praiseworthy and should be vigorously pursued. But as I hope to make clear, the distinction ill serves the public (and academic) image of lowland migrants.

Further, the distinction is breaking down. Many indigenous uplanders have been displaced by government forest-management policies from the forest environments in which they traditionally practiced their environmentally sound agricultural technologies. Locked in unequal competition for scarce land resources with their more aggressive lowland migrant counterparts, these people have turned to destructive, short-fallow shifting cultivation on degraded or otherwise marginal lands. Furthermore, many indigenous uplanders have in fact married lowland migrants, thereby blurring the cultural basis on which the distinction between harmonic and disharmonic shifting cultivation is said to rest. However, the basic point here is that population pressure and land scarcity, not ethnicity or major cultural differences in knowledge of or concern about the environment, explain the prevalence today of the more destructive forms of shifting cultivation. Indeed at least in the Philippine context, shifting cultivators who appear to be living in harmony with their environments have approximately the same status as hunter-gatherers said to be living in leisured affluence and in harmony with their environments: rare species at best and of a type constituted as much by 1960s-era neofunctionalism as by ethnographic reality.[1]

The Research Setting

Mainland Palawan, located in the southwest of the Philippines, is the largest of the 1,768 islands forming Palawan Province. The island is long and narrow, extending 425 kilometers from northeast to southwest and varying in width from about 40 kilometers to only 5 kilometers (Finney and Western 1986:45; see figure 11.1). Total area is approximately 1.2 million hectares, most of it mountainous or hilly upland. The island is richly endowed with natural resources: good agricultural land, tropical hardwood forests, well-stocked fishing grounds, and a variety of valuable min-

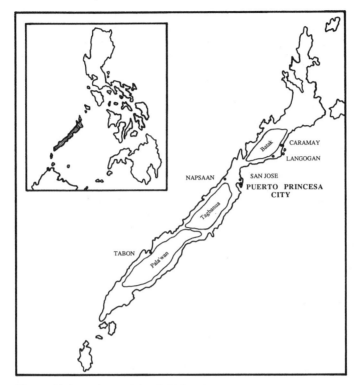

Figure 11.1   Palawan Island, Philippines

eral deposits (Finney and Western 1986:46). Palawan is also well known for its natural beauty, seen in outstanding beaches and coral reefs, a remarkable underground river, lakes and waterfalls, and a variety of wildlife, some of it unique (Finney and Western 1986:46). Palawan Island is home to three indigenous tribal peoples: the Pala'wan, or Palawano, a shifting cultivation folk inhabiting the mountains in the south; the Tagbanua, another group of shifting cultivators inhabiting the riverbanks and valleys of the central mountains; and the Batak, a hunting-gathering people inhabiting the north central part of the island. Figure 11.1 shows the approximate ancestral territories of these three indigenous peoples and the locations of several of the recently established settler communities.

Only after 1872, when the town of Puerto Princesa was founded on a small east coast bay in the middle part of the island, did Palawan in fact become a destination for significant numbers of migrant lowlanders. They came first from Cuyo Island (located to the east in the Sulu Sea, near Panay Island) and later from throughout the Philippines. Even then Palawan (which little interested Spain during its 350-year rule) remained a little-known area on the periphery of Philippine economic and political life until well into the twentieth century (Conelly 1983:39). In 1903 the total population of Palawan Island was estimated to be only 10,900 people, most of whom were probably Tagbanua or Pala'wan. By the standards of the rest of the Philippines Palawan remains sparsely populated and heavily forested, and it is commonly regarded as the country's last frontier. Like all Philippine frontiers before it, it is rapidly becoming settled; in 1980 the island's population was approximately 270,000, and today it exceeds 400,000 (National Census 1980, 1990).

Just as settlement of Palawan has occurred relatively late by the standards of the rest of the Philippines, so too has forest clearance. Most lower hills and plains have been cleared, but most higher uplands (which account for much of the island's area) are still forested (Finney and Western 1986:47). Precisely how much land is still forested is uncertain. The government's *Strategic Environmental Plan for Mainland Palawan* estimates that 68% of the island is still forested (National Council [NCIAD] 1985:12), but this figure (apparently the basis for Finney and Western's estimate [1986:47] that "about two-thirds" of the land area is still under forest) is based on 1979 Landsat imagery. A more recent estimate of the island's forest cover is 54%, based on the Philippine-German Forest Resources Inventory Project (Forest Management Bureau 1988).

What is clear is that the forest that does remain is disappearing at a shocking rate, somewhere on the order of 19,000 hectares per year (Clad and Vitug 1988:48) to 47,000 hectares per year (Broad and Cavanagh 1989:20). As elsewhere in the Philippines, the immediate causes are commercial logging and settlement by immigrant lowlanders and displaced tribal peoples. Dipterocarps are not as dominant in Palawan's forests as elsewhere in the Philippines; *apitong* (*Dipterocarpus grandiflorus*) is the principal commercially important species. But Palawan is also home to relatively dry monsoonal forests of a type known as molave forest (Kummer 1992:43). Here are found such beautiful and durable hardwoods as *narra* (*Pterocarpus indicus*) and *ipil* (*Instsia bijuga*), highly prized on both the local and export markets for furniture manufacture (Bee 1987:10–11). The principal logging operation in Palawan is Pagdanan Timber Products; its politically influential owner holds timber license agreements covering a remarkable 168,000 hectares in northern Palawan, more than 25% of the island's remaining forest area. The future of this particular operation is in doubt; reflecting both national and local environmentalist pressures, a moratorium on commercial logging has been in effect since 1992. Although some commercial logging continues illegally, the role of logging in the deforestation of Palawan has apparently attenuated in the last few years.

In contrast, the effect of immigration on forest resources remains acute. Palawan's annual growth rate was about 4.6% during the 1980s (Finney and Western 1986:47), and the current growth rate is generally assumed by local government officials to be on the order of 5%. Such estimates of course combine the contributions of natural increase and in-migration to population growth, but the latter has clearly played a substantial role; about 20% of Palawan's population growth between 1980 and 1990, for example, is attributable to migration from elsewhere in the Philippines (Michael Costello, personal communication 1994). Population growth and development have severely affected other Palawan resources, particularly fishing grounds, coral reefs, and mangrove swamps. Attempts by the Manila-based Haribon Foundation, a leader in the country's nascent conservation movement, to protect these resources have led to a fierce national debate over Palawan's resources and the alliances of politicians, government officials, and businesspeople involved in their exploitation (Clad and Vitug 1988).

Beyond gross numbers, relatively little is known about the nature of migration to Palawan.[2] The government-sponsored resettlement project at Narra, on the eastern coastal plain near the central part of the island, has received the most explicit attention to migration (e.g., Fernandez 1972; James 1979, 1983), but migration to Palawan has also been a major theme in several studies of the development of settler communities (J. F. Eder 1982; Chaiken 1983; J. F. Eder and Fernandez 1990). The influx of migrants and attendant forest clearance have also been major themes in virtually all recent anthropological accounts of the indigenous peoples of Palawan (e.g., Brown 1990; Conelly 1983; J. F. Eder 1987; Lopez 1985, 1987).

Lowland Migrants in Palawan's Uplands: Aims and Realities

I turn now to the agricultural activities characteristically pursued by these migrant lowlanders upon their arrival in Palawan. My interest in disputing the public image of migrant farmers as destructive shifting cultivators entails more than a desire to tone down the rhetoric characterized by such terms as *destructive,* for I believe it is also questionable whether many of these migrants are in fact appropriately characterized as "shifting cultivators" at all.

Olofson (1980) has argued that the Tagalog term *kaingin,* widely used by social scientists and public officials in the Philippines to refer only (and often pejoratively) to slash-and-burn or shifting cultivation—in other words, to swiddens, as customarily understood by anthropologists—in fact embraces both swiddens and fixed dryland hillside plots, the latter relatively permanent and ecologically conservative. As examples of sustainable dryland hillside farming, Olofson cites cases of self-described kaingineros in the southern Tagalog provinces of Laguna and

Quezon where hillside "kaingins" had not in fact been shifted or enlarged for decades but had nevertheless sustained their fertility and yielded a steady stream of harvestable (and marketable) produce (1980:174).

Olofson's important distinction, between shifting and fixed modes of upland farming, is relevant here, because in my view it is the second sort of farming—in fixed, dryland hillside plots—that is often the aim (albeit less often the practice) of migrant settlers to Palawan (see figure 11.1).

The first community, San Jose, was established on nonirrigated flatland near the outskirts of Puerto Princesa City during the 1930s and 1940s by migrant homesteaders from the Cuyo Island. These migrant Cuyonons used slash-and-burn to clear the forest cover from their homesteads and to produce an annual crop of upland rice. But over the longer term they were less interested in using the extensive land-management techniques of shifting cultivators than in establishing permanent farms similar to those they had left behind in Cuyo. By the early 1970s when I first studied San Jose, production for the regional market had become more important than production for subsistence. Average farm size was on the order of 3 hectares per household, and although short-fallow shifting cultivation of upland rice was still common, market-oriented production of vegetables, tree crops, field crops, and livestock dominated the agricultural landscape (J. F. Eder 1982).

In more recent years further population growth and stronger integration with regional markets has brought increasingly intensive, and increasingly productive, use of land (see, e.g., J . F. Eder 1991). Despite the extensive and reported use of swidden agriculture in earlier years, there is little grassland today; cattle control what grass there is. The remaining fallow land is gradually being brought under some form of permanent cultivation, particularly arboriculture or annual cropping of corn, root crops, or legumes. First appearing in commercial vegetable gardening several decades ago, chemical fertilizers have played an increasingly important role in these various patterns of cultivation.

The second community, Napsaan, was settled on a hilly area of the west coast of Palawan beginning in the late 1950s, mainly by settlers from the Visayan region of the Philippines (Conelly 1983). Unlike the smallholding and upland-oriented Cuyonon migrants in the first case, the migrant Visayans have come mostly from backgrounds of tenancy and irrigated rice cultivation; they have been joined in more recent years by settlers of similar background from the island of Masbate, near the southern tip of Luzon. As on other Philippine frontiers, migrants to Napsaan learned the proper techniques of slash-and-burn land clearance and cultivation of upland rice from local tribal peoples—in the case of Napsaan, from the indigenous Tagbanua (Conelly 1983:58; see figure 11.2).

However, after an initial stage of forest clearance, most Napsaan settlers

Figure 11.2   A migrant farmer from the Ilocos region of the northern Philippines clears the forest cover from his homestead on the Napsaan coastal plain in the early 1970s.

aim to establish a fixed and productive upland farm. For example, in 1981 a stand of coconuts occupied one corner along the coast at a 3-hectare "upland homestead" belonging to a Visayan family that arrived during the early 1970s. The remainder of the homestead consisted of a house lot, the family's swidden, low- and high-fallow second growth, and some uncleared forest. Broadly dispersed across this landscape were banana and a variety of fruit trees of varying maturities (Conelly 1983:312). At the time of Conelly's study, cashew (*Anacardium occidentale*) was the most common tree crop in this part of Palawan. Although few Napsaan farmers still grew it on a commercially successful scale, cashew (and tree crops generally) were recognized as an ecologically desirable and economically remunerative form of land use (1983:309–310; see figure 11.3).

Lowland settlers in upland communities up and down Palawan Island have likewise attempted to establish stable and productive upland farms. The particular farming systems vary with settlers' backgrounds and local

Figure 11.3   Cashew trees are an important part of the Palawan people's evolving upland agricultural landscape. Here a farmer's cattle graze on the regrowth in his cashew orchard.

environmental conditions. In the Langogan and Caramay River valleys northeast of Puerto Princesa City, coffee grows well, and small coffee farms are scattered along river banks and on surrounding hillsides once cleared by swiddening and otherwise vulnerable to erosion. In the community of Tabon, near the town of Quezon on the west coast of the island, migrant farmers from the Batanes Islands in the extreme northern Philippines cultivate permanent hillside farms. Here, as throughout the Quezon area, bananas are a major cash crop, but Tabon farmers also plant such root crops as manioc (*Manihot esculenta*), sweet potato (*Ipomoea batatas*), taro (*Colocasia esculenta*), and yam (*Dioscorea* sp.) and such legumes as peanuts (*Arachis hypogaea*) and mung beans (*Phaseolus aureus*) for sale in the Quezon public marketplace. Many of Tabon's hillside farms have been in the same place for years.

All the foregoing agricultural landscapes are clearly still evolving. Also, all have clearly been associated with some significant environmental degradation: the loss of forest cover, or reduction in native populations of fauna, and (probably) a decreased capacity to store and cycle water and other nutrients. But these cases also indicate that other significant environmental deteriorations—"green deserts" of cogon grass (*Imperata cylindrica*), erosion, flooding—are not a necessary consequence of forest clearance by "migrant swidden farmers." Rather, and at least where these farmers have been in part oriented toward tree crops, relatively stable and productive forms of land use have emerged. In short, then, the basic dynamic associated with human activity in these pioneer settings on Palawan is the conversion of forest to farmland—not to grassland, abandoned land, or some other form of "waste" land.

Discussion

However, the point of the foregoing case studies is not to argue that ecologically stable and economically productive upland farms are common in Palawan but that the capabilities to establish such farms do in fact exist, embodied in the knowledge and aspirations that many settlers bring to the frontier. A growing body of comparative evidence suggests that migrants to Palawan's uplands are little different in this regard from migrants to upland regions throughout the Philippines. David M. Kummer (1992) argues that most agriculturalists in the upland Philippines in fact are, or aim to be, sedentary farmers. Some are doing rather well. Sam Fujisaka (1986), for example, describes a community of pioneer shifting cultivators, Calminoe, located near Mount Banahaw in Laguna Province on Luzon and settled by Tagalogs and other lowland migrants after a period of commercial logging. Less than half the migrant households Fujisaka studied actually had previous occupational experience with shifting cultivation (1986:146). But he found that over time farming knowledge and practices in Calminoe "evolved" locally and in an ecologically appropriate fashion to produce a productive and potentially sustainable agricultural landscape of mixed annual and perennial, cash, and subsistence crops (1986:138).

Although Fujisaka detected some worrisome trends, particularly increasingly intense resource use and a tendency for farmers to seek private gain at the expense of the public good, he also observed a variety of resource-conserving measures: intercropping with fruit trees, composting of household wastes, green manuring, and so forth. Such observations led Fujisaka to conclude that Calminoe farmers are "actively trying to ensure the sustainability of the system" on which they are dependent for survival (1986:160–161).

The Calminoe case is not unique. Olofson (1985) describes a similar situation in a pioneer region in Rizal Province, elsewhere on Luzon, where migrant Tagalog farmers from Lake Taal in Batangas Province are pursuing a variety of different agroforestry systems to manage their homesteads in a resource-conserving manner. Sophisticated indigenous agroforestry knowledge in fact appears to be widespread in the Philippines (Olofson 1983).

Nevertheless stable and productive upland farmers, like stable and productive swiddens, are in actuality uncommon in the Philippines, and they are uncommon in Palawan. Fruit trees are not planted, despite knowledge of their importance; erosion-control measures are not undertaken, despite awareness of their necessity. Why? At least in the Palawan case the interactions of poverty, market conditions, and land tenure insecurity, the last rooted in disadvantageous government policies, appear most to blame. In San Jose and Napsaan, for example, many farmers must focus almost exclusively on their immediate subsistence needs. They find prohibitively expensive the opportunity costs of regularly clearing the undergrowth from around immature tree crops (a task essential to successful arboriculture). Furthermore, many Napsaan farmers are reluctant to plant tree crops in the first place, lest they further diminish their limited stock of swiddenable land. In some cases, Conelly reports, low-fallow second growth already planted in tree crops is nevertheless slashed and burned again because of a lack of alternative swidden sites (Conelly 1983:311–317). Writing of pioneer coffee-growing peasants in the Peruvian Andes, Jane Collins similarly observes that household labor scarcity reflecting relative poverty, rather than farmer ignorance or the inherent fragility of the environment, explains why colonists fail to maintain the soil and thereby set environmental degradation in motion (Collins 1989:257).

In the Palawan case, however, land tenure insecurity may be the single most important factor underlying settlers' failure to develop more productive and sustainable farms. There are several related issues. First, the government's reluctance to release forest land for homesteading makes many settlers de facto squatters on public land. Political and moral issues aside, the scale of such encroachment on public land by would-be homesteaders is in any case far larger than the ability of the government to control it. Second, where public land is eventually made available for homesteading (usually after forest clearance and settlement), the land titling process is lengthy or, if a private surveyor is hired to expedite the process, prohibitively expensive. Yet until land titles are awarded, settlers cannot obtain commercial credit for farm development purposes, and they may fear land grabbing in the interim by the economically powerful or politically influential.

In such circumstances it is not surprising that many settlers appear reluctant to plant tree crops or to otherwise invest for the longer term in their farms. Indeed relating government policies to the Philippine deforestation crisis as a whole, Owen J. Lynch and Kirk Talbott (1988) argue that the single most effective step that the government could take to ensure sustainable development of the remaining Philippine forests would be to recognize the currently undocumented private property rights of the millions of people (both migrant and indigenous) who already reside in upland forest areas. Christopher Gibbs, Edwin Payuan, and Romulo del Castillo (1990:260–261) similarly report that lack of tenure security is the obstacle that upland people most commonly perceive as blocking their own development initiatives.

As the Peruvian case illustrates, of course, merely investing in tree crops or other permanent crops rather than annual crops does not ensure that upland farms will be protective of soil or environment. However, even here economic conditions and government policies, rather than farmer mentality, may play the decisive role. Collins in fact argues that the Peruvian farmers she studied did not invest in sustained-yield coffee production (through such ecologically desirable practices as multicropping, maintenance of ground cover, and terracing) because monopsonistic market conditions, limits on landholding, and concerns about the security of existing land claims discouraged farmers from making long-term investments (Collins 1989:256). Similarly, Andrew P. Vayda and Ahmad Sahur report that the pepper, clove, and coconut farms that migrant Bugis farmers establish in east Kalimantan after an initial phase of swiddening are not sustainable and cause considerable ecological damage (1985:104–107). But Vayda and Sahur argue that certain changes in Indonesian government policies would make the Bugis more oriented toward sustainable development (1985:110).

That indigenous and migrant farmers in the Philippine uplands must be seen as part of the solution, rather than as part of the problem, is not surprising; it is precisely this philosophy that today informs the social forestry programs now enthusiastically pursued throughout the Philippines (see, e.g., Aguilar 1986). For years public programs for Philippine forest occupants were essentially punitive; the focus of forest-management policy was on the prohibition of settlement. This policy proved ineffective, and with assistance from universities and private and government agencies the Forest Management Bureau (then known as the Bureau of Forest Development) during the 1970s initiated a number of programs designed to settle upland occupants on their landholdings and simultaneously enlist their cooperation in adopting agroforestry and other soil and water conservation measures. Since 1972 these various programs have been reorganized into the Integrated Social Forestry Program (ISFP), today administered by the Department of Environment and Natural Resources (DENR) (Gibbs, Payuan, and del Castillo 1990).

The ISFP is still evolving, but the underlying philosophy that remains is that supportive rather than punitive efforts to deal with forest settlers will lead to less environmentally damaging farming systems. Although the tenure issue remains only partially resolved, it is at least recognized as central. A standard feature of ISFP today is the awarding of twenty-five-year renewable individual and communal stewardship contracts (Gibbs, Payuan, and del Castillo 1990:260–261). Filomeno V. Aguilar (1986) and Rosemary M. Aquino, Romulo A. del Castillo, and Edwin Payuan (1987) have reviewed the accomplishments and problems of social forestry programs in the Philippines.

What have such programs accomplished to date in Palawan? By national standards the ecological and socioeconomic problems attending the settlement of Palawan are only recent, and government attention to them has been late in coming. In the case of forest settlement social forestry has been vigorously pursued only since 1982, as part of DENR's participation in the Palawan Integrated Area Development Project (PIADP). Fifth in a series of integrated area development projects in the Philippines, PIADP is the first to have an explicit environmental protection component. Environmental concerns have been central to DENR's own participation in PIADP, which since 1982 has consisted of the Upland Stabilization Program (USP), aimed at controlling shifting cultivation in Palawan's uplands by helping immigrant and indigenous residents to farm in a more ecologically sound manner.

The centerpiece of USP has been a series of three pilot projects on public forest land in central and southern Palawan. These projects cover a total area of 6,873 hectares and involve as at least nominal participants a total of 489 households, including both migrant lowlanders and indigenous Pala'wan and Tagbanua, living within project boundaries (DENR 1990). Each project site includes a nursery, intended as a seedling distribution center for project participants, and several crop deforestation fields. The basic USP agroforestry package is a form of vegetable and bench terracing that involves simultaneous cultivation of trees and agricultural crops. "Community organizers" and other staff try to encourage the farmers to adopt these farming practices through a variety of means (training programs, formation of farmer-beneficiary cooperatives). The ultimate goal is that these practices, once spread to project participants with the assistance of project staff, will spread beyond project boundaries as other upland farmers adopt them in turn and on their own.

However, at least by 1988 the evidence that these goals were being realized was scant. Project participants, much less their neighbors, were not adopting terracing and other agroforestry practices.Although many project participants had found periodic wage employment at the project centers, working in the nurseries or on the demonstration farms, the latter appeared to be little more than expensive showcases, of little utility or relevance to the lives of everyday upland residents, indigenous or migrant (F. F. Eder 1988). Two years later DENR's own final project report identified the low participation rate of the communities as one of USP's principal problems (DENR 1990:29).[3]

On the other hand, the 489 USP households did ultimately receive individual stewardship contracts for land parcels of 2 to 7 hectares each (DENR 1990:26).[4] Project staff are somewhat discouraged that this important accomplishment, given the effort and expense of the pilot projects, has not resulted in more sustainable agricultural practices by the contract recipients. At least two important issues are involved here. First, tenure security is not the only obstacle to more sustainable upland farming. The capital and labor costs of the project's state-of-the-art vegetative and bench terracing systems are quite high, running to thousands of U.S. dollars per hectare. The issue of adoption costs would therefore have been important in any case, but it was made particularly acute by PIADP's decision to locate the three USP project sites in areas inhabited by indigenous (albeit somewhat displaced) tribal peoples as well as by lowland migrants. This decision reflected well on PIADP's broader goal of including all segments of Palawan society, and it recognized, correctly, that the indigenous and migrant inhabitants of Palawan's uplands today are in basically the same circumstances. But the particular task of USP was also made more difficult; as a rule Palawan's tribal peoples are still more impoverished and marginalized than its lowland migrants. In any event USP did ultimately accomplish the bench and vegetable terracing of 884 hectares, about 13% of the total project area (DENR 1990:27). But even this modest achievement, it was said, came only because project personnel eventually paid participants to terrace their farms.[5]

Second, by 1991 even the awarding of stewardship contracts was not proving adequate to secure tenure. Several PIADP staff members reported that some indigenous people participating in the project moved on, apparently intimidated into "selling" their theoretically non-negotiable leases to recently arrived lowland migrants. This observation raises serious doubts about the adequacy of the stewardship concept as a land-management strategy, given the cultural, political, and economic realities of the Philippine uplands.

Regardless of the ultimate outcome of these and other social forestry programs in the Philippines, they can reach only a fraction of upland residents. Thus the agroforestry knowledge that at least some settlers bring to the frontier becomes even more important. This consideration, and the case material discussed earlier, suggest that fundamental equity-oriented changes in state policy governing new lands settlement, rather than showcase projects, would be the more effective way to make the lowland migrants to Palawan's uplands more ecologically accountable.

Meanwhile anthropologists can contribute to this effort by carefully attending to the decision-making environments that such migrants confront and by helping to make their circumstances and motivations more understandable to the makers of public policy.

Acknowledgments   A National Institute of Mental Health Pre-Doctoral Research Fellowship supported my fieldwork in Palawan during 1970–1972. A sabbatical leave from Arizona State University and a grant from the Social Science Research Council supported my fieldwork there during January to July 1988. I would like to thank Elaine Brown, Thomas Conelly, Janet Fernandez, David Kummer, and Harold Olofson for their helpful comments on an earlier draft of this article.

Notes

1. Neofunctionalism is the theory, popular among ecological anthropologists in the 1960s and 1970s, that tribal customs, even those that seem bizarre to Westerners, may function ecologically to keep local populations adapted to their environments.

2. Despite its last frontier image, until about 1970 and from the standpoint of the Philippines as a whole, Palawan was not the demographically important settler destination that such provinces as Davao del Sur, Zamboanga del Sur, Cotabato, Bukidnon, or even Rizal were (Pryor 1979). This meant that data on Palawan were historically lumped in with data on other provinces or regions. Thus Peter C. Smith's 1977 analysis of 1970 census data concerns a variety of interregional migration patterns but discusses Palawan only in the context of aggregated data on in-migration to Region IV, the southern Tagalog region, which includes Batangas, Laguna, and Rizal, much of which is suburban or even urban.

3. Adoption rates for terracing and other soil-conserving technologies are often quite low in agroforestry projects; see Sam Fujisaka (1989).

4. DENR has awarded 1,300 such stewardship contracts to date in Palawan. PIADP's Strategic Environmental Plan estimates that more than 20,000 households inhabit Palawan's forested uplands, with more than 6,000 in the "critical mountain zone" (NCIAD 1985).

5. Again, the issue of adoption costs is not unique to USP. Agroforestry research in the Philippines has been quite limited, given the enormous variety of ecological and economic circumstances that upland farmers confront, and few inexpensive and locally appropriate agroforestry techniques are available for adoption (Gibbs, Payuan, and del Castillo 1990:261).

References Cited

Aguilar, Filomeno V. 1986. Findings from eight case studies of social forestry projects in the Philippines. In *Man, Agriculture, and the Tropical Forest: Change and Development in the Philippine Uplands,* Sam Fujisaka, Percy Sajise, and Romulo del Castillo, eds., pp. 189–222. Bangkok: Winrock International Institute for Agricultural Development.

Anderson, James N. 1987. Land at risk, people at risk: Perspectives on tropical forest transformation in the Philippines. In *Lands at Risk in the Third World: Local Level Perspectives*, Peter D. Little and Michael Horowitz, eds., pp. 249–267. Boulder, Colo.: Westview.

Aquino, Rosemary M., Romulo A. del Castillo, and Edwin Payuan. 1987. *Mounting a National Social Forestry Program: Lessons Learned from the Philippine Experience*. Honolulu: Environment and Policy Institute, East-West Center.

Bee, Ooi Jin. 1987. *Depletion of the Forest Resources in the Philippines*. Singapore: Institute of Southeast Asian Studies, Field Report Series No. 18.

Broad, Robin and John Cavanagh. 1989. Marcos's ghost. *Amicus Journal* (Fall): 18–29.

Brown, Elaine. 1990. Tribal Peoples and Land Settlement: The Effects of Philippine Capitalist Development on the Palawan. Ph.D. diss., State University of New York, Anthropology Department, Binghamton.

Chaiken, Miriam S. 1983. The Social, Economic, and Health Consequences of Spontaneous Frontier Resettlement in the Philippines. Ph.D. diss., University of California, Anthropology Department, Santa Barbara.

Clad, James and Marites D. Vitug. 1988. The politics of plunder. *Far Eastern Economic Review* (November 24): 48–52.

Collins, Jane. 1989. Small farmer responses to environmental change: Coffee production in the Peruvian high selva. In *The Human Ecology of Tropical Land Settlement in Latin America*, Debra A. Schumann and William L. Partridge, eds., pp. 238–263. Boulder, Colo.: Westview.

Conelly, W. Thomas. 1983. Upland Development in the Tropics and Alternative Strategies in a Philippine Frontier Community. Ph.D. diss., University of California, Anthropology Department, Santa Barbara.

Conklin, Harold C. 1957. *Hanunoo Agriculture*. Rome: United Nations.

Cruz, Maria Concepcion, Imelda Zosa-Feranil, and Cristela L. Goce. 1986. *Population Pressure and Migration: Implications for Development in the Upland Philippines*. Los Banos, Philippines: Center for Policy and Development Studies, Working Paper No. 86–06.

Department of Environment and Natural Resources (DENR). 1990. *Project Completion Report. Component: Upland Stabilization Program*. Puerto Princesa City, Philippines: Department of Environment and Natural Resources.

Eder, Florenia F. 1988. *Communication Research on Tribal Farmers: A Preliminary Report and Some Recommendations*. Puerto Princesa City, Philippines: Palawan Integrated Area Development Project Office.

Eder, James F. 1982. *Who Shall Succeed? Agricultural Development and Social Inequality on a Philippine Frontier*. New York: Cambridge University Press.

——. 1987. *On the Road to Tribal Extinction: Depopulation, Deculturation, and Adaptive Well-Being Among the Batak of the Philippines*. Berkeley: University of California Press.

——. 1991. Agricultural intensification and labor productivity in a Philippine vegetable gardening community: A longitudinal study. *Human Organization* 50 (3): 245–255.

Eder, James F. and Janet O. Fernandez. 1990. Immigrants and emigrants in a Philippine frontier farming community, 1971–1988. In *Patterns of Migration in Southeast Asia,* Robert R. Reed, ed., pp. 93–121. Berkeley, Calif.: Center for Southeast Asian Studies.

Fernandez, Carlos A. 1972. Blueprints, realities, and success in a frontier resettlement community. In *View from the Paddy,* Frank Lynch, ed., pp.176–185. Quezon City, Philippines: Institute of Philippine Culture.

Finney, Christopher E. and Stanley Western. 1986. The economic analysis of environmental protection and management: An example from the Philippines. *Environmentalist* 6 (1): 45–61.

Forest Management Bureau. 1988. *Natural Forest Resources of the Philippines.* Manila: Philippine-German Forest Resources Inventory Project.

Fujisaka, Sam. 1986. Pioneer shifting cultivation, farmer knowledge, and an upland ecosystem: Co-evolution and systems sustainability in Calminoe, Philippines. *Philippine Quarterly of Culture and Society* 14 (2): 137–164.

———. 1989. The need to build upon farmer practice and knowledge: Reminders from selected upland conservation projects and policies. *Agroforestry Systems* 9: 141–153.

Garrity, Dennis P., David M. Kummer, and Ernesto S. Guiang. 1993. The Philippines. In *Sustainable Agriculture and the Environment in the Humid Tropics,* Committee on Sustainable Agriculture and the Environment in the Humid Tropics, ed., pp. 549–624. Washington, D.C.: National Academy Press.

Gibbs, Christopher, Edwin Payuan, and Romulo del Castillo. 1990. The growth of the Philippine social forestry program. In *Keepers of the Forest: Land Management Alternatives in Southeast Asia,* Mark Poffenberger, ed., pp. 253–265. West Hartford, Conn.: Kumarian Press.

James, William E. 1979. An Analysis of Public Land Settlement Alternatives in the Philippines. Ph.D. diss., University of Hawaii, Economics Department, Honolulu.

———. 1983. Settler selection and land settlement alternatives: New evidence from the Philippines. *Economic Development and Cultural Change* 31 (3): 526–586.

Kummer, David M. 1992. *Deforestation in the Postwar Philippines.* Geography Research Paper No. 234. Chicago: University of Chicago Press.

Lopez, Maria Elena. 1985. The Pala'wan: Land, Ethnic Relations, and Political Process in a Philippine Frontier System. Ph.D. diss., Harvard University, Anthropology Department.

———. 1987. The politics of lands at risk in a Philippine frontier. In *Lands at Risk in the Third World: Local-Level Perspectives,* Peter D. Little and Michael Horowitz, eds., pp. 230–248. Boulder, Colo.: Westview.

Lynch, Owen J. and Kirk Talbott. 1988. Legal responses to the Philippine deforestation crises. *New York University Journal of International Law and Politics* 20 (Spring): 679–713.

Myers, Norman. 1988. Environmental degradation and some economic consequences in the Philippines. *Environmental Conservation* 15 (3): 205–214.

National Census and Statistics Office. 1980. *Census of Population and Housing (Palawan)*. Manila: National Census and Statistics Office.

——. 1990. *Census of Population and Housing*. Report No. 2–73D (Palawan). Manila: National Census and Statistics Office.

National Council for Integrated Area Development. (NCIAD). 1985. *Strategic Environmental Plan for Mainland Palawan*. Manila: National Council for Integrated Area Development, Palawan Integrated Area Development Project Office.

Olofson, Harold C. 1980. Swidden and *kaingin* among the southern Tagalog: A problem in Philippine upland ethno-agriculture. *Philippine Quarterly of Culture and Society* 8 (2/3): 168–180.

——. 1981. Introduction. In *Adaptive Strategies and Change in Philippine Swidden-Based Societies*, Harold C. Olofson, ed., pp. 1–12. Los Banos, Philippines: Forestry Research Institute.

——. 1983. Indigenous agroforestry systems. *Philippine Quarterly of Culture and Society* 11 (2/3): 149–174.

——. 1985. Traditional agroforestry, parcel management, and social forestry development in a pioneer agricultural community: The case of Jala-jala, Rizal, Philippines. *Agroforestry Systems* 3: 317–337.

Population Reference Bureau. 1993. *World Population Data Sheet*. Washington, D.C.: Population Reference Bureau.

Porter, Gareth, with Delfin Ganapin Jr. 1988. *Resources, Population, and the Philippine Future: A Case Study*. Washington, D.C.: World Resources Institute, Paper 4.

Pryor, Robin J. 1979. The Philippines patterns of population movement to 1970. In *Migration and Development in South-East Asia: A Demographic Perspective*, Robin J. Pryor, ed., pp. 225–243. Kuala Lumpur: Oxford University Press.

Repetto, Robert. 1988. *The Forest for the Trees? Government Policies and the Misuse of Forest Resources*. Washington, D.C.: World Resources Institute.

Smith, Peter C. 1977. The evolving pattern of interregional migration in the Philippines. *Philippine Economic Journal* 16 (1–2): 121–159.

Swedish Space Corporation. 1988. *Mapping of the Natural Conditions of the Philippines*. Solna, Sweden: Swedish Space Corporation.

Vayda, Andrew P. and Ahmad Sahur. 1985. Forest clearing and pepper farming by Bugis migrants in East Kalimantan: Antecedents and impact. *Indonesia* 39: 93–110.

12

The Divided Island of New Guinea: People, Development and Deforestation

Carolyn D. Cook

Papua New Guinea in the east and Irian Jaya in the west comprise the two halves of the world's second-largest island, New Guinea, and a significant amount of the world's remaining forest (see figure 12.1). More than one-fourth of Indonesia's remaining forest is in Irian Jaya, which has a forest cover of 354,360 square kilometers. Papua New Guinea (PNG) contains 363,530 square kilometers of forest (Collins, Sayer, and Whitmore 1991). The rugged terrain of the mountainous areas of the island makes travel and trade from one area to another difficult and is no doubt a factor contributing to the development of the diverse customs, technologies, and languages (more than one thousand) of the New Guinea people.[1] In the coastal areas fishing and sago gathering are the means of subsistence, whereas in the highland areas cultivation of sweet potatoes and taro and pig husbandry are practiced.

The east and west parts of New Guinea have similar climates, soils, forests, and indigenous peoples, but they have very different forms of government. Whereas Irian Jaya has been likened to a Fourth World country under Third World imperialism (Nietschmann 1985–86:86), PNG boasts a new government in the hands of the indigenous people. We would expect that different governments with different policies regarding natural resources would result in two distinctly different rates of deforestation. New Guinea makes an excellent test area for this hypothesis. The estimated annual decrease of forest cover in Irian Jaya is 1,637 square kilometers per year (Sutter 1989:136), whereas in PNG the decrease has been only 120

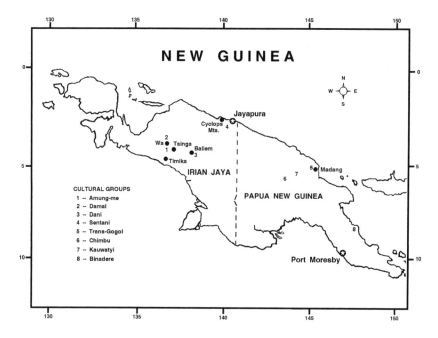

Figure 12.1   The Divided Island of New Guinea

square kilometers per year since 1985 (Myers 1989). (Hurst's figure of 210 square kilometers per year for PNG [1990:129] is for the period from the 1970s to 1990.) These estimates indicate that the current rate of deforestation in Irian Jaya is at least seven times higher than in PNG. This discussion of the effects of different political systems is intended to provoke thinking and stimulate further research into the subject. The information regarding indigenous peoples' forest management and complications of outside influence is provided to emphasize the role these people have played in the past and their potential for the future of global forests.

Irian Jaya

Nearly 98% of the land in Irian Jaya is classified as forest land and consequently falls under the jurisdiction of the Indonesian Forestry Department. Seventy percent of this has been allotted to logging concessionaires.

Advocates for indigenous people—such as anthropologists and representatives of such Indonesian nongovernment organizations as SKEPHI (*Sekretariat Kerjasama Pelestarian Hutan,* Joint Secretariats for Indonesian

Forest Conservation), WALHI (*Wahana Lingkungan Hidup Indonesia,* Indonesian Environmental Forum), and Irian Jaya Development Foundation—often cite poorly planned development projects as reasons for deforestation and disastrous depletion of soil fertility.[2] The government claims that swidden farmers (i.e., slash-and-burn or shifting horticulturalists) bear the major responsibility for forest loss in Indonesia (Hurst 1990:27). Also, some researchers blame deforestation partly on the indigenous peoples' swidden practices.[3] See, for example, Anung Kusnowo and Amru Hydari Nazif (1992:28), David McGrath (1987:240), Bryant J. Allen and Robert Crittenden (1987), and Sandra Moniaga's discussion (1991:120) of Indonesian government attitudes toward indigenous agriculture. No one would dispute that densely populated areas, as in Chimbu or in the North Baliem, have little remaining forest, but traditional swidden agricultural systems appear to be sustainable over long periods of time when population densities are not high (Conklin 1957; Carneiro 1960) and even increase biodiversity (Balée 1994).

Nongovernment organizations (Barnett 1990; Aditjondro 1984; Korwa and Blowfield 1984; Korwa 1985), researchers (Brookfield 1988), and local people cite multinational corporations—which export resources and leave the locals with little in exchange—as greater perpetrators of deforestation than indigenous cultivators. Those directly affected blame deforestation on land alienation and separation from the benefits of their natural resources.

Here we address why New Guinea is being deforested by briefly reviewing theories about why indigenous people use land the way they do, some case studies illustrating indigenous forest management and modern problems, and differences and similarities in PNG and Irian Jaya policies and development and their resulting effects on deforestation. Table 12.1 summarizes the forests, deforestation, and populations of Irian Jaya and New Guinea.

Table 12.1
Forests and People of New Guinea

	Irian Jaya	PNG
Total area (km²)	410,660	451,710
Forested area (km²)	354,360	363,530
Population	1,649,000	3,874,000
Rate of deforestation	1,637	120

Sources: Collins, Sayer, and Whitmore (1991); Government of Indonesia (1991); United Nations (1990); Sutter (1989); Myers (1989).

Extensive sources in ecological anthropology and human geography reveal that many if not most traditional cultivation systems were viable before the introduction of modern technology. After his thorough study of new and old swiddens of the Kauwatyi people in montane PNG, H. I. Manner (1981:359) concluded that their system of subsistence, which emphasized diversity and conservation practices such as the planting and protection of certain forest regrowth species, ensured the regeneration of the forest. William Clarke (1976:108) found that swidden cultivators in New Guinea anticipated benefits in the forest-fallow process and used certain management techniques that intentionally favored forest regrowth. They did few and selective weedings, used short cropping periods, planted trees, and prevented fires in regrowth areas. Also, they actively preserved old forest areas as species reservoirs. These people were aware that the more intensively they worked the land, the smaller the economic return they would realize for their labor. Michael Robinson (1988:358) also observed that the indigenous knowledge and use of many foods and materials of the tropical forest could reveal clues to maintaining sustainable yields without environmental destruction. For an extensive discussion of research results that confirm that most soil depletion in tropical forests is not caused by cultivation, refer to F. Bruce Lamb (1987). Also relevant is Roy A. Rappaport's discussion (1984:90) of "absolute degradation" versus "anthropocentric degradation," which is a human-caused change in terrain that lowers the productivity of biomass useful to humans per unit of area or labor or both.

Each case of forest use and cultivation must be examined individually before drawing conclusions about its sustainability. But many indigenous communities in the outer islands of Indonesia clearly have been managing forest resources sustainably for generations (see, for example, Dove 1983, 1988; Colfer 1983; Kartawinata 1984; Alcorn 1990). My own studies in Irian Jaya, Indonesia, have led me to agree with these researchers and to conclude that the Amung-me, a group of highland Papuan people (see figure 12.1), regulate the degree to which they allow the land to become depleted of nutrients before they begin a rejuvenating fallow process for it. The Amung-me people do not allow the land to pass the point of no return to fertility. They actively protect fallow land until it regains fertility and is again ready to be cleared for gardening. Rotational reuse of the land for cultivation prevents the opening of additional forest land. Yet the government favors logging, mining, and national parks over traditional land use.

Case Studies of Deforestation

The Amung-me people number about six thousand. They call their language *Amung-kal*. Older literature refers to these people as a part of the

Uhunduni tribe (Heider 1979). Amung-kal is a separate dialect from that spoken by the other half of the tribe, the Damal, who live on the northern side of the mountain range.[4] The land and natural resources of the Amung-me people cover an area of seventeen valleys on the southern slopes of the Surdirman Range (Nemang Kawi). This area is today affected by the following outside factors: (1) Freeport Indonesia's copper and gold mining project, (2) the Indonesian government's World Bank–funded relocation and transmigration project, (3) missions outside the area, (4) the new 1.56 million-hectare Lorentz National Park, and (5) logging concessions. The logging concessions are mostly in the lowland rain forest and the lower montane zone where access by road is feasible (albeit far from easy).[5]

The Amung-me who live in the areas undergoing the most rapid deforestation were relocated there by the Dutch government before 1962, by the Indonesian government after 1963, and by Freeport Indonesia Incorporated (FII) after 1979. The latter was an effort to rid mining camps in the mountains of local people. The Amung-me of Wa Valley were "encouraged" to relocate at first, and when many refused to move, they finally were forced to migrate "for their own good." The options were to move to Timika or Banti in the lowlands or relocate north of the mountains. The benefits of moving to Timika were alleged to be education, health care, job opportunities, and access to a market system. In Timika young Javanese men just out of college were enlisted to "teach" farming to the newly relocated Amung-me and others in the community. The community (or "dream city," as the government officials called it) consisted of Amung-me from several areas, Moni people, some Dani and Ekari people recruited to help the Amung-me learn intensive farming, relocated coastal people of the Kamoro tribe, and spontaneous migrants from Sulawesi, Kei, and other parts of Indonesia. This situation was a stewpot of conflict for many reasons.

The Kamoro people had been the first people to use the lowland rain forest; after 1960 they shared it with a diverse range of peoples. The Kamoro had been primarily a fishing and sago-gathering group. Their greatest demands on trees of the forest, other than the sago for food, had been trees for canoes. Even their houses did not require the felling of large trees. In 1980 the government, suspecting Papuan Freedom Movement (OPM) activities in the Kamoro villages, forced them to move to Timika, which is inland from their sago-gathering area.[6] Also in Timika were coastal people and several groups of mountain people. The Amung-me, Dani, Ekari, and Moni were all from areas of higher elevation, and they had to learn through experiment about their new surroundings (figure 12.1).

Beginning in 1980 the Indonesian government, in its proclaimed goal to better the lives of the people, sent people from other islands of Indonesia to teach the Irianese to live as nuclear families in little square box houses with dirt floors, to plant gardens in their yards, do intensive agriculture using commercial fertilizer, make fish ponds, raise goats, rabbits, and chickens, and wash their clothes at neighborhood wells. The most fatal problem was malaria. Many mountain people died of this tropical lowland disease. By 1985 the Amung-me were trying to retreat into the forest to a land they considered theirs by traditional law. They succeeded temporarily but by 1992 had been forced to yield to P. T. Komanden Raya, an Indonesian logging company holding a government concession to log the area.

While the forest and mining land in the lowlands was changing in 1992, FII and the Indonesian government were in the final stages of negotiating a new border between the mining concession land and the Lorentz National Park from the coast to the highlands. The outcome, not yet made public, will determine how much land will be susceptible to deforestation by logging companies, how much will be opened for mining, and how much may be protected by park regulations. Initial recommendations by the International Union for the Conservation of Nature (IUCN) were that Lorentz become a world heritage site, but the Indonesian government did not accept this recommendation because of the strict regulations it would place on flexibility of boundaries and limitations on its use.[7] Ronald G. Petocz (1984) lists Lorentz reserve as the largest, most diverse, and richest protected area in Indonesia. The word from the World Wide Fund for Nature (WWF) is that the area will be classified as national park, rather than a world or national heritage site, giving Indonesia more control and leeway to develop the area (Smith 1992 personal communication).

The Tsinga Valley Amung-me: Isolated but Affected

To find the Amung-me follow the rivers from the Arafura Sea, past Timika, and up into the highlands of the isolated mountain valley of Tsinga, where I spent 1992 (figure 12.1). These people were perplexed and angered by activities occurring in their corner of the world. Surveys for ore bodies and the Lorentz park boundary were being conducted without adequate consultation with the local people.

On August 7, 1992, the day of celebration for the opening of Tsinga Valley's first school, rumors were that the army was on its way into the area to recapture some OPM men who had surrendered and been given permission to visit their families in Tsinga. The men had allegedly not returned to Timika as requested by the army. Soldiers had been sighted on the trail, and tension built as people tried to decide how to protect themselves. As the food distribution for the feast was completed, a cry of warning sounded on the mountainside near the bridge to the village. A bedraggled "army" of three Javanese and twenty Irianese came into site. The villagers froze. Tension broke when the Javanese men introduced themselves as a team of Indonesian Forestry Department cartographers. The fear of arrest allayed, the Amung-me began to puzzle over why cartographers had come to their valley.

The cartographers explained to the people that they would be putting the boundaries of the Lorentz reserve on paper and marking boundaries with large wooden posts. The work went smoothly so long as the cartographers stayed with their paperwork, but when they planted the posts, the Amung-me became upset and began to destroy the markers. The Amung-me had immediately sensed that they were losing control over their land. They refused to allow the mapping team to enter the sacred ground on the east side of Dolailningok-in. This area is taboo to all but the Kum and Uamang clans, and even they must observe traditional ritual when entering the area. Thus the team was not completely successful in its endeavor, for areas within the park boundary were inaccessible—either because Freeport Indonesia would not allow or help the cartographers to enter or because the Amung-me forbade it.

The park divides Amung-me land in the Tsinga Valley. Thus what until 1993 had been governed traditionally will be controlled by two new sets of regulations: one will be susceptible to mining and logging concessions, and one will limit resources taken from the forest. The Amung-me have ambushed mining surveyors and staged protests against them to gain attention and force communication with FII and government officials. The Amung-me still have no access to information regarding their options and the advantages and potential dangers involved in the two new directions in which they have been forced to go. What will happen to their traditional ways, which have served them so well over the past centuries? In this case the park may protect the land within its boundary against deforestation, but will this force indigenous people to clear forest to try to stake out an irrefutable claim to their remaining land? Or will the logging companies take out more forest than is preserved by park regulations—leaving roads for encroachers who will clear land that the Amung-me would have left for nontimber-product harvest?

How was it that the Amung-me were able to live and prosper in Tsinga Valley without destroying the forest environment? It may be that their social values, spatial arrangement of villages, and taboos on sex at various times and for various reasons helped to maintain a relatively low population in the valley. They have pre- and postpartum sexual taboos, as well as underlying beliefs that intercourse before important activities like hunting and ritual will bring bad luck. Sexual intercourse is restricted to a woman's home, where there are often children and other female relatives. The men usually sleep in men's houses, and in general those men who wish to be viewed as strong and worthy do not spend a lot of time with women.

My estimate of current Tsinga population density is two people per square kilometer, which I believe they have maintained both purposely and coincidentally. Medically trained personnel who work with Amung-me have found them to be relatively healthy people (personal communications and Tembagapura Hospital Records up to 1992). Studies of fertility and life expectancy have not been undertaken in Tsinga, but these may be factors contributing to the low population density.

Apart from low population density, Amung-me traditional land use plays an important role in long-range availability of both forest and farming land. The state of the Tsinga Amung-me land and forest in 1992 was as follows: approximately 17% of the land within 10 kilometers from the villages was in active garden, 66% in one- to ten-year fallow stages, and 17% in ten- to twenty-year fallow, which was dense secondary forest. A few patches of old forest were maintained within this 10-kilometer radius, but most of the old forest was farther away. Roughly two-thirds of the total Tsinga area is in forest and rocky mountain slopes.

The Amung-me live in eight small villages throughout Tsinga Valley. When it becomes difficult for the villagers to walk far enough to cut and carry home the supply of wood they use for cooking, heating, and building houses, they move the village. In the past this has occurred every twenty to thirty years. The old village sites are not totally vacated but are often occupied by a few remaining families (or their descendants) until former villagers decide the wood supply has been replenished enough for them to return. Because of this practice each Amung-me has at least two sites available for living, foraging, hunting, and farming. It is a rotation system rather than one of progressive forest cutting. Of course to establish the gardens and fallows they had to "deforest" the area first, but they seem to have developed a balance in their technology, population, and amount of land they require, to the point that they do not cut old forest.

The highest ground, which includes glacier, tundra, and craggy mountain walls, has been the most sacred ground for the Amung-me. They believe the spirits of their ancestors dwell there, watching over the land for mortals. Resources they use from this area are minimal and consist mostly of small tundra herbs that they use for secret medicines. The rivers and streams that run through the valleys originate in the high mountains, and legends regarding their origin are a part of Amung-me oral tradition. For example, the Beanaikogom River, the largest river near the village of the same name, came into being when Keneming-ki lost her enormous pig to an earth spirit. Although most fluid from this pig flowed underground, water also sprang from the body openings of the pig and joined to form the great river. Sacred trees were formed from the feces of Keneming-ki's pig. When the Amung-me enter this area, they are obligated to observe taboos on food, clothing, and behavior. For example, they may not cut certain trees, wear fragrant beads, spit, defecate, or have sexual intercourse.

The Amung-me value the variety of foods, fuel, and other materials for their daily lives that the secondary and old forests provide. They actively perpetuate forest land by their cultivation limitations and their taboos on

Figure 12.2   Pandanus fruit being prepared for a snack on a trail through high mountain forest.

certain areas and resources. Traditional land tenure practices also limit which groups of people have access to any given area.

The main activities performed in the forest are hunting animals, collecting various nontimber products, gathering firewood, and trimming and harvesting pandanus. They collect the nuts of three varieties of pandanus (*aliu* [*Pandanus* sp.]) and at least five more types are important for their useful leaves (figure 12.2).[8] Forests also contain the sacred "dream ground" where people go to seek visions that will help to guide their decision making and their daily lives. Areas such as this are preserved by taboos on cutting trees and harvesting plants.

According to Amung-me tradition, the Tsinga forests have been divided among the three major clans: Magal, Beanal, and Kum. Each group has several areas for pandanus harvest, but activities such as hunting and salt gathering are open to all three clans. Beyond this, if clans from elsewhere wish to hunt, they first ask permission. Salt from Pelnok Manogom pool is used for trade with other clans or tribes. All Amung-me may use the sacred dream grounds, but they first must report to the clan leader of the land on which it is located, and usually a clan member will accompany them.

Figure 12.3   Nut pandanus orchard above
Emtawaroki village in Tsinga Valley.

Foraging and farming are practiced within the context of social, political, and spiritual aspects and according to the provisions of various terrains.

In addition to growing taro (*Colocasia esculenta*) at high elevations, the Amung-me establish orchards of pandanus nut trees (*Pandanus julianettii*, *P. brosimos*, and possibly others). From the forest the Amung-me have taken seedlings or pieces of favored pandanus trees with which to plant their own groves. They use both sexual and asexual propagation methods and have developed twenty-four varieties or provenances of the pandanus nut, a favorite and nutritious food (see figures 12.3 and 12.4). The Amung-me refer to domesticated nut pandanus as *kweng*. They replant volunteer seedlings from forest and grove, use vegetative propagation by cutting

Figure 12.4   Girl from Emtawaroki selling pandanus
nuts at a public meeting in Tsinga's central village,
Beanaikogom.

tops from old trees and replanting them, make seedling nurseries on roofs
of homes, and protect a special marsupial (*amat*) that "plants" and hides
seed from other rodents. These elevation-dependent groves require minimum pruning and clearing of grass beneath the stands to prevent rodents
from harvesting the proceeds. Some pandanus grow in sacred areas where
old trees may never be cut.

The Amung-me believe that cutting protected or taboo pandanus trees
can lead to severe punishment from the earth spirit protecting the land.
An example of this occurred while I was in the field. The mining company was clearing a new place for storing machinery and its Amung-me

day-labor man, Umarki, was ordered to cut some pandanus trees that were in the area of Hanya jum ki, the male earth spirit. He wanted to refuse but feared being fired. After cutting the trees, he lost his sight and went into shock. He was taken to the company hospital where he was treated but remained sightless and disoriented. After hearing the circumstances of Umarki's illness, his wife prepared a ritual appeasement offering. He partially recovered but was dismissed from his job and had to leave the area.

Amung-me of Tsinga Valley have a traditional system of resource management that has served them and their environment well through the last several centuries. However, they soon will be the victims of progress as a result of loggers, miners, and government—people who do not understand or respect their knowledge and lifestyle. The Amung-me hope to benefit from the resources that are being taken by outsiders and to partake in the newly introduced economic system. At the same time they want their sacred grounds to remain sacred and what is left of their land to remain theirs.

Development in Irian Jaya

The Amung-me are not alone in their struggle; similar cases occur throughout Irian Jaya. Before it became an Indonesian province in 1963, Dutch New Guinea, as it was then called, had lost timber to the Dutch, who in many cases did not compensate the local people for their resources and their labor (Aditjondro 1984:62). The story is the same today—appropriation of Irianese tribal land continues. George A. Aditjondro (1984:63), former Indonesian director of the Irian Jaya Rural Development Foundation, cites three main reasons for threats to forest biomes in Irian Jaya: land clearing for Javanese transmigrants, timber-cutting concessions, and mining and infrastructure projects.

If it proceeds according to plan, transmigration in Irian Jaya will soon turn the population ratio between the lowlands (which now has the lowest concentration) and the highlands upsidedown. With no lowland agricultural soil to exploit the highland horticulturalists will be forced to deforest steeper slopes, because the valley bottoms are already overcrowded and overexploited. Aditjondro (1984:63) notes that at least 400,000 hectares (4,000 square kilometers) of forest land will have to be cleared for the transmigrants' rice fields or oil palm plantations; he raises the question of how many more trees will have to be cut to supply the lumber needed to build the transmigrants' homes. The total number of transmigrants was expected to reach 1.5 million by the end of the century, but because of high rates of failure of land to support the new forms of agriculture, the venture has been scaled back. In addition to needing trees for homes, the transmigrants illegally pirate logs from the indigenous neighbors' forest as a source of additional income. I witnessed this on both the northern and southern coasts of Irian Jaya. However, this is insignificant when compared to the number of trees cut by logging concessions.

A Potential Solution

What if the government and the people were to cooperate to prevent deforestation? Management programs for Irian Jaya's state forest land or forest reserves have been underway since the early 1980s. The Cyclops Mountains Nature Reserve was established to protect the flora and fauna of its tropical rain forests and to provide a water catchment for Jayapura, the villages surrounding the reserve, and Lake Sentani to the south. Although the Indonesians have discussed the buffer zone concept since before 1980, the plan for the Cyclops Mountains represents one of the first attempts to establish community-managed zones within and around the periphery of a nature reserve (Mitchel, de Fretes, and Poffenberger 1990:238).

Several hundred Dani and Ekari people have been resettled in an area near the Cyclops reserve by the Social Affairs Department and by mission groups. These highland groups have cleared most of the new gardens that are within the reserve and along its boundary. This creates conflict and causes serious problems for forest protection, because the non-Sentani Irianese settlers are not bound to Sentani traditional law and do not use the same resources normally regulated by it. On the other hand, their own law has not had time to adjust to the new social and biological environment that now invalidates traditional means of regulating the people-nature balance. However, Sentani, Dani, and Ekari have begun to cooperate with the National Social Forestry Working Group to formulate an agroforestry system in which the immigrants will assist the Sentani in maintaining new tree crops on the same plots on which outsiders have their annual crop gardens. The working group is Jakarta based and responsible for promoting local decision making regarding indigenous peoples' use of the forest. Profits from the long-term tree crops will go to the Sentani landowners, and the Dani and Ekari will have continued use of the land, expansion of cultivation plots will be restricted, and boundaries will be clearly identified. This sounds like a positive endeavor and indeed has helped to bring tenure problems to the fore, but it is marred by a top-heavy bureaucracy and lack of follow-through, which have stifled individual initiative (Mitchel, de Fretes, and Poffenberger 1990:247). It may also be flawed because it does not provide for the Dani and Ekari to earn a profit.

Contrasting the Development Policies of PNG and Irian Jaya

Indonesia's Government Policy Regarding Irian Jaya's Forests

Most of Irian Jaya's forest land is claimed under indigenous people's law. Conflict has arisen between government and the original forest resource managers, because the Indonesian Forestry Department also claims all the forested areas. It has issued permits to logging concessions covering more than 70% of the island and has assigned most remaining land to national parks.

Many groups in Irian Jaya have managed to maintain significant amounts of their forest land, and they still have time to reinforce traditional values and promote sustainable forest management. People like the Amung-me, who have maintained traditional conservation measures and low levels of population, have the potential to strengthen their environmental conservation policies and their economic stability. This will require encouragement and support from the surrounding communities, nongovernment organizations, and the government but not a great outlay of capital. The question is, will the government allow a program that promotes self-sufficiency, does not provide material for taxation or royalties, and helps to establish indigenous land tenure?

Indonesia's Forestry Department determines what is best for the nation as a whole (not Irian Jaya alone) in making decisions regarding how forest land is to be used. Traditional land claims are recognized but only if they do not conflict with the interest of the government. The basic Forestry Law of 1967 (Government of Indonesia 1967) classifies all forests as production, protection, and wildlife and other reserves. It is organized to keep timber profit within Indonesia. Concessionaires' contracts include obligations to replant and establish wood-processing plants. Philip Hurst (1990:12) asserts that the Forestry Department is mainly concerned with timber production and has little influence over more general forest development. For example, government policy on agriculture in forest areas is implemented through three departments: agriculture, social affairs, and interior; control of national parks falls to the Department of the Environment. On the positive side, the Forestry Department does have some social forestry programs such as the Cyclops reserve project (Mitchel, de Fretes, and Poffenberger 1990) that could be effective in conservation if they can survive the bureaucracy. Indonesia suffers from having many agencies with unclear responsibilities. Heavy bureaucratic involvement in projects, such as delineating boundaries, protecting forest, and and running community participation projects, makes project management cumbersome and unnecessarily complicated.

People and Deforestation Case Studies in PNG

Trans Gogol Valley in the northern coastal province of Madang regularly floods and experiences drought. I cannot be specific about the effects of the extensive commercial tree cutting that is underway, because we do not know enough about the role of the forest in stabilizing the valley and its influence on the microclimate. Because human presence in the area may go back fifty thousand years (De'Ath and Michalenko 1988:168), it seems logical to assume that the scattered cuttings of subsistence farmers have relatively few effects compared to those of the large commercial operations.

JANT (Japan and New Guinea Timbers) entered New Guinea in the late 1960s when the Australian government was a United Nations trustee for the territories of Papua and New Guinea. Its commitment to reforestation was not clearly spelled out; neither were rigid obligations for agricultural or industrial diversification. The agreement did not include strict monitoring of company operations or require social, technological, and ecological impact studies before operations began. Colin De'Ath and Gregory Michalenko (1988:170) have described the complex social and ecological problems that have arisen from this enterprise.

Ecologically, the area of Madang Province (PNG) was affected in a negative way. Reforestation did not begin until 1978—and then on a very limited scale. The economic system, which was relatively self-sufficient in terms of food and material culture, is being transformed into one of dependency, with the Japanese reaping most of the benefits (De'Athand Michalenko 1988:172). Outsiders have not recognized or respected local people's value systems. Upholding their values of sharing and reciprocity, the indigenous people shunned the exploitative behavior of the outsiders, even before the foreign timber firm arrived. They managed to keep most of their most valued land and forest intact until 1973.

Elsewhere in PNG, the Binandere people of the Northern Province have been struggling successfully against becoming victims of resource exploitation by foreign companies (Waiko 1977:407). This struggle goes as far back as 1895, when they resisted gold prospectors in their land south of the Wuwu River. Powerful foreign investors first tried to get a toehold in 1971 when a timber company's officials chose as a town site on land belonging to the local tribe. These Papuans, who in the past had used traditional political and religious organization to defend their land, added a new weapon: education. A local "schoolboy," John Waiko, helped to communicate his people's hopes and fears to foreign timber investors and the government.

He also provided information for the Binandere people to use in analyzing the situation and making wise decisions. When his work was published in 1977, the people had managed to ward off the logging companies and to plan other, more sustainable development for their community.

PNG Government Policy

The PNG Department of Forestry was established under Australian administration during World War II when timber demand in Australia was high. So long as PNG was under Australian administration, logging concessions were rarely granted to companies with no sawmill. This system, although intended to provide local employment, prevent rapid export of logs, and obtain higher profits, also prevented local people from forming groups and controlling their own resources, because they did not have the capital to establish sawmills.

A report of the International Bank for Reconstruction and Development (IBRD) in 1965 was a stong influence on PNG development. The report proclaimed that PNG must earn foreign capital and recommended the exportation of unprocessed logs. Large foreign companies were urged to operate in PNG and thus it came to pass. Australia has continued its commercial interest in PNG's timber, and Japan moved in during the 1970s. Even in 1989 local companies had rights to only one-fifth of PNG's total area of timber concessions (Myers 1989).

In PNG many foreign companies rushed to enter agreements with Michael Somare's coalition government when it entered the state of self-determination in 1973. The foreign companies were eager to exploit the rich mineral, forest, and human resources of New Guinea, because other parts of the world were closing their doors to export of primary resources.

PNG's constitution, similar to that of Indonesia's, called for natural resources to be conserved and used for the collective benefit of all. But unlike the situation in Indonesia, where the government owned the forest land, PNG's forests fell to the hands of community-based ownership with various degrees of private tenure rights. Strong legislation protects tribal land rights. It not only maintains full rights but contains mechanisms that attempt to ensure that tribal groups receive a fair price for their resources. Hurst (1990:138) finds these mechanisms relatively effective in preventing the forestry minister from imposing logging in areas where local people oppose it. He believes this has made a major contribution to forest conservation in PNG. However, Godfried Yassafar (1994:2) points out that these procedures are not immune to such problems as forged signatures of landowners, as illustrated by the Madang case.[9] As of 1973, the government could purchase timber rights and would sell them to timber companies. At first the government returned 25% of the royalties to the indigenous people, a figure now increased to 75%.

Still, the forest and land policy in PNG has its drawbacks. For example, it allows some clear-cutting, which is carried out by JANT (Japan and New Guinea Timber) and other timber concessionaires, and permits some export of logs. Forest losses are extremely intense in areas where logging and other land development projects take place. The hope of employment often dwindles as floods of outsiders come to compete for jobs. Royalties for the land are almost always paid to the people, but in no way do they compensate for the loss of land and access to forest resources.

The largest log dealers in PNG usually sell their logs to major buyers from Japan, Korea, and Taiwan. The price agreed is usually the current world market price for the particular wood type. Hurst (1990:134) believes that there is definite evidence of a price-fixing cartel of five companies, accounting for more than 62% of log exports from PNG. The *Barnett Report* (1990) notes that transfer pricing has been a serious problem for PNG. Many logs shipped to other countries are transshipped to Japan, following manipulation of prices en route (Thompson 1994). However, the government is often forced to cover up malpractices it finds in order to continue to receive financial aid from foreign countries extracting PNG resources.

Roughly 80% of PNG's population still live as subsistence farmers. They compete with timber concessions and plantations of oil palm, coffee, rubber, and other export crops for their land (Gradwhol and Greenberg 1988:135). Plantation use, combined with logging and intense slash-and-burn practices by novice farmers, will soon eliminate traditional swidden agriculture as an option. Growing ranks of urban poor have been partly the result of foreign investment that squeezes people out of their land and/or attracts them to urban centers. The government recently began a campaign to encourage these people to return to their villages to farm. Judith Gradwhol and Russell Greenberg (1988:136) urge that city dwellers who are relocated to the villages be retrained in the traditional agroforestry ways in order to make the land last longer.

Policies regarding forest management and indigenous peoples will play a large part in the future of the forest cover and diversity. Do the human dimensions of politics and government make a difference in indigenous people's ability or desire to maintain forests? Drastically different governments have resulted in similar problems for indigenous people and resource conservation, but the differences also are significant.

Haberle, Hope, and de Fretes (1991:38) suggest that in precolonial times tribal warfare promoted preservation of buffer zone forests. The Dutch were not as adamant about cessation of warfare as the Australians were, but in both cases warfare was outlawed. In remote areas where the colonial governments had no control, wars continued through the 1970s.

The World Bank has played major roles in developing both PNG and

Irian Jaya. It formulates plans and contributes expertise and money to carry out projects. The *Barnett Report* (1990:4) criticizes the World Bank's Tropical Forestry Action Plan Review for PNG by noting that its general status and position commit the World Bank to capitalist economics and profit motive. Thomas Barnett finds a fundamental contradiction between profit and conservation as they relate to rain forests and argues that the choice has to be made for conservation. Beyond this, he argues that the World Bank plan is statist in orientation and neglects the community.

The World Bank, a division of IBRD, has influenced both halves of the island to secure foreign capital. This eagerness to seek profit "for the national good" frequently means that government officials are willing to rescind or waive rules and restrictions. Corruption of the systems and overlooking violations continue to plague both sides of the island, while individuals and groups vie for the favor of international trading partners. Japan is the largest trading partner of both PNG and Indonesia and together with other foreign investors affects much of New Guinea's remaining forests. Plantations of oil palm, coffee, rubber, and other export crops tie up the best agricultural land on both sides of the island. Then, as the people move out to poorer soils to farm, they run into conflict with timber concessions.

Potential Solutions to New Guinea Deforestation

Each geographical area is unique in its combination of peoples, resources, and government. A solution for one area may not work in another, but a key factor in making beneficial changes involves understanding the roles each of these three elements play. For example, Tsinga Valley's people will have to deal with culture, technology, and environment, as well as mining and logging company extractions, the Indonesian government, and the United Nations. But there is a way to simplify forest management and stabilize the situation for the Irianese people. It would involve modifying and legitimizing the *adat* (indigenous law) settlement, subsistence, and tenure patterns in the forest to reflect a sustainable multi-use forest-management system that benefits both the local people and the government. A second way might be to map out areas of adat settlement and subsistence rights and exclude them from the forest domain, registering the land and bringing it into the legal regime of the Basic Agrarian Law (which applies to non-forestry lands). However, this would leave the Forestry Department with little to work with. Both possibilities have been in an experimental stage for some time, but nothing has been finalized.

The third alternative (totally unacceptable to the Indonesian government) is for the Irianese people to free themselves from Indonesian rule and develop their own policy for forest management. This movement has been in the experimental stage since 1963 with little progress. Even if they did gain freedom, if they follow the precedent set by Papua New Guinea, there is no assurance that they can avoid deforestation. Although preliminary estimations indicate a much lower deforestation rate in New Guinea today, both systems are affected by outside investors. The main difference may lie in the fact that the indigenous people of PNG receive more reimbursement for their losses. An issue that remains to be sorted out is how to deal with the distribution of "profits" within the communities, for New Guinea's diversity includes is a wide variety of land tenure systems. Charles V. Barber, Nels Johnson, and Emmy Hafild (1994:99–107) suggest that Indonesia change its policies for forests; they strongly urge more local participation and greater public access to information related to forest policy decision making.

The situation in PNG is far from ideal.[10] Egalitarian backgrounds of the tribal peoples make it difficult to form a cohesive central government. Yet, despite problems with corruption and profit orientation, Hurst (1990:149) finds hope in PNG as a basic model for political control of forests. He concludes there is a lot of room for improvement but that minimizing government ownership of forests will contribute much to creating a sustainable future.

Indigenous people often do have something valuable to offer the world: their knowledge of land use and forest protection. As such it provides a rich resource for sustainable development. For example, Australia is now calling for involvement of Aboriginal information in nature conservation and land management (Kean, Richardson, and Trueman 1988), and traditional medicines of Indonesia are providing a portion of the nation's health care as well as a potential for new bioactive compounds and medicines (Sedik 1994).

Putting the human dimension of deforestation into perspective means that governments, organizations, and groups of people who profit from indigenous peoples' land would be wise to take more care to have clear and respectful communication with them and to provide opportunities for them to give environmental advice to outsiders. The indigenous peoples of New Guinea should be the first to reap the bounty of their land, and it is unethical for others to alienate them from this right.

Acknowledgments   I carried out my formal field study of the Amung-me in Timika in the summers of 1985 and 1987, under the sponsorship and financial support of Minister B. J. Habibie and the Indonesian Technology Department (BPPT) and with permission from the Indonesian Institute of Sciences (LIPI), both of which are based in Jakarta. BPPT and LIPI again sponsored my field study in 1992 in Jakarta; my sponsor in Irian Jaya was Cenderawasih University, and BBPT and LIPI sponsored my field study in Irian Jaya. I am grateful to these institutions for being open to pinpointing problems and finding solutions. Funding for the study came from the East-West Center in Honolulu. I obtained background knowledge from 1977 to 1983 when I lived in the Freeport Indonesia mining camp of Tembagapura and assisted BPPT with the Timika Relocation and Transmigration Project. This manuscript benefited from the editing and critiques of Carol Carpenter, Leslie Sponsel, Thomas Headland, and Anton Ploeg.

Notes

1. New Guinea today has 1,109 living languages: 862 in Papua New Guinea and 247 in Irian Jaya (Grimes 1992:585, 877).

2. The Foundation of the United Nations for the Development of West Irian was established with a donation from the Dutch government and contributions from Canada and Great Britain. It attempts to assist grassroots development projects but is also subject to the activities, policy, and bureaucracy of the Indonesian government. One way it assists grassroots development projects is by networking developers from nongovernment as well as government agencies.

3. Swidden cultivation may be to blame for deforestation where population pressures (often caused by outsiders or relocated people) have interfered with the balance of people and resources. Some Papuans—for example, the Dani and Ekari, who have large populations and who practice swidden farming systems that border on intensive—will need to use more intensive cultivation to correct deforestation. S. G. Haberle, G. S. Hope, and Y. de Fretes (1991:38) suggest that one reason for the spread of people into previously preserved forests may have been the cessation of warfare, as decreed by missionaries and Dutch and Indonesian governments. This left what once was buffer zone vulnerable to settlement and further deforestation.

4. The Amung-me people are included in Summer Institute of Linguistics's classification as Uhunduni Family-Level Isolate language speakers and are considered a subgroup of Damal (Silzer and Clouse 1991:45). However, they classify themselves as a separate group having the same origin as the Damal people. The Summer Institute is a leading authority on languages of New Guinea.

5. I gathered the data for the Amung-me case study over a period of fifteen years, eight of which I spent living in their homeland.

6. OPM, or Operasi Papua Merdeka (Papuan Freedom Movement) is the liberation movement of West Irian that has been in existence since 1962, when the Indonesians began to take over the Dutch half of the island. Irian Jaya formally became an Indonesian province after the vote in 1969 that was called the Act of Free Choice by the Indonesians and the Act of No Choice by the West Papuans. In Bahasa Indonesia the term for this vote is *Penetuan Pendapatan Rayat*, which translates literally as 'Declaration of People's Opinion.' See Kees Lagerberg (1979) for a detailed description of how the United Nations, the United States, and the government of Indonesia carried out the annexing of Irian Jaya.

7. IUCN is an international conservation group that works closely with the

World Wide Fund for Nature (WWF–also known as the World Wildlife Fund) and the World Bank. The goal of IUCN is to ensure sustainable use of natural resources.

8. *Aliu* is the Amung-kal term for screw pine, one of the forest pandanus trees that produces nuts. Benjamin C. Stone has described *Pandanus brosimos* as the forest tree and *P. julianettii* as the semidomesticated species but warns that we do not know enough about pandanus to determine whether these are separate species or varieties of a single species (Stone 1982:412–413).

9. Yassafar (1994:2) discusses the process of designating specific areas as timber supply areas (TSAs), which become the supply for foreign multinational contractors. Landowners' signatures are required for land to become TSAS.

10. Bas Louman, writing in *Tropical Forest Management Update* (1993) finds PNG's "customarily owned" lands a problem in regard to development of forest plantations. However, I would argue that distribution of benefits—for the indigenous people as well as the global environment—needs to be examined more closely in establishing such plantations.

References Cited

Aditjondro, George A. 1984. An environmental sketch of Indonesia's biggest timber empire. *Kabar Dari Kampung* 1 (2): 12–13.

Alcorn, Janis 1990. Indigenous agroforestry strategies meeting farmers' needs. In *Alternatives to Deforestation: Steps Toward Sustainable Use of Amazon Rainforest*, Anthony B. Anderson, ed., pp. 141–151. New York: Columbia University Press.

Allen, Bryant J. and Robert Crittenden. 1987. Degradation and a pre-capitalist political economy: The case of the New Guinea highlands. In *Land Degradation and Society*, Piers M. Blakie and Harold C. Brookfield, eds., pp. 145–156. New York: Methuen.

Balée, William. 1994. *Footprints of the Forest*. New York: Columbia University Press.

Barber, Charles V., Nels Johnson, and Emmy Hafild. 1994. *Breaking the Logjam: Obstacles to Forest Policy Reform in Indonesia and the United States*. Washington, D.C.: World Resources Institute.

Barnett, Thomas. 1990. *The Barnett Report: A Summary of the Report of the Commission on Inquiry into Aspects of the Timber Industry in Papua New Guinea*. Hobart, Tasmania: Asia Pacific Action Group.

Brookfield, Harold. 1988. The new great age of clearance and beyond. In *People of the Tropical Rainforest*, Sloan Denslow and Christine Padoch, eds., pp. 209–224. Los Angeles: University of California Press.

Carneiro, Robert L. 1960. Slash-and-burn agriculture. In *Selected papers of the Fifth International Congress of Anthropological and Ethnological Sciences*, Anthony Wallace, ed., pp. 229–234. Philadelphia: University of Pennsylvania Press.

Clarke, William. 1976. The maintenance of agriculture and human habitats within the tropical forest ecosystem. In *Human Ecology* 4 (3): 247–259.

Colfer, C. P. 1983. Change and indigenous agroforestry in East Kalimantan. *Borneo Research Bulletin* 15 (1): 3–20.

Collins, N. Mark, Jeffrey A. Sayer, and Timothy C. Whitmore. 1991. *Conservation Atlas of Tropical Forests: Asia and the Pacific.* New York: International Union for the Conservation of Nature and Simon & Schuster.

Conklin, Harold C. 1957. *Hanunoo Agriculture.* Rome: United Nations.

De'Ath, Colin and Gregory Michalenko. 1988. High technology and original peoples: The case of deforestation in Papua New Guinea and Canada. In *Tribal Peoples and Development Issues,* John H. Bodley, ed., pp. 166–180. Mountain View, Calif.: Mayfield.

Dove, Michael. 1983. Theories of swidden agriculture and the political economy of ignorance. *Agroforestry Systems* 1: 85–99.

———. 1988. Introduction: Traditional culture and development in contemporary Indonesia. In *The Real and Imagined Role of Culture in Development,* Michael Dove, ed., pp. 1–40. Honolulu: University of Hawaii Press.

Government of Indonesia. 1967. Undang-undang No. 5, 1967 Tentang ketentuan-ketentuan pokok kehutanan (About basic forest regulations). In *Himmpunan peraturan perundangan dibidang kehutanan Indonesia I* (Compilation of regulations of the Indonesian forestry sector 1 [revision II]), pp. 1–21. Jakarta: Yayasan Bina Raharja Departmen Kehutanan (Forestry Department Foundation for Prosperity).

Government of Indonesia. 1991. *Irian Jaya: Statistik Indonesia 1991* (1991 Indonesian statistics). Jakarta: Pusat Biro Statistik (Central Bureau of Statistics).

Gradwhol, Judith and Russell Greenberg. 1988. *Saving the Tropical Forests.* Washington, D.C.: Island Press.

Grimes, Barbara F., ed. 1992. *Ethnologue: Languages of the World, Twelfth Edition.* Dallas: Summer Institute of Linguistics.

Haberle, S. G., G. S. Hope, and Y. de Fretes. 1991. Environmental change in the Baliem Valley, montane Irian Jaya, Republic of Indonesia. *Journal of Biogeography* 18: 25–40.

Heider, Karl G. 1979. *Grand Valley Dani: Peaceful Warriors.* New York: Holt, Rinehart, and Winston.

Hurst, Philip. 1990. *Rainforest Politics: Ecological Destruction in South-East Asia.* Atlantic Highlands, N.J.: Zed Books.

Kartawinata, K. 1984. The impact of development on interactions between people and forest in East Kalimantan: A comparison of two areas of Kenyah Dayak settlement. *Environmentalist* 4 (Supplement 7): 87–98.

Kean, J. S., G. Richardson, and N. Trueman. 1988. *Aboriginal Role in Nature Conservation.* Emu, South Australia: Emu Conference.

Korwa, Abner. 1985. *Dampak operasi Perusahaan Kayu PT. You Lim Sari* (Impacts of the operations of the You Lim Sari Logging Company). Report 4 (2). Jayapura, Indonesia: Irian Jaya Rural Community Development Foundation.

Korwa, Abner and Mick Blowfield. 1984. Irian Jaya helps Korean development. *Kabar Dari Kampung* 2 (1): 20–23.

Kusnowo, Anung and Amru Hydari Nazif. 1992. *Yogotak hubulik motok hanorogo* (Tomorrow will be better than today). Jakarta, Indonesia: Lembaga Ilmu Pengetahuan Indonesia.

Lagerberg, Kees. 1979. *West Irian and Jakarta Imperialism.* New York: St. Martin's Press.

Lamb, F. Bruce. 1987. The role of anthropology in tropical forest ecosystem resource management and development. *Journal of Developing Areas* 21: 429–458.

Louman, Bas. 1993. The possible contribution of social forestry to forest development in Papua New Guinea. In *Tropical Forest Management Update* 3(1): 7–8. A publication of the U.N. Food and Agriculture Organization.

Manner, H. I. 1981. Ecological succession in new and old swiddens of montane Papua New Guinea. *Human Ecology* 9 (3): 359–377.

McGrath, David. 1987. The role of biomass in shifting cultivation. *Human Ecology* 15 (2): 221–242.

Mitchel, Arthur, Yance de Fretes, and Mark Poffenberger. 1990. Community participation for conservation area management in the Cyclops Mountains, Irian Jaya, Indonesia. In *Keepers of the Forest,* Mark Poffenberger, ed., pp. 237–252. West Hartford, Conn.: Kumarian Press.

Moniaga, Sandra. 1991. Towards community-based forestry and recognition of *adat* property rights in the outer islands of Indonesia: A legal and policy analysis. In *Voices from the Field: Fourth Annual Social Forestry Writing Workshop,* Jefferson Fox, Owen Lynch, Mark Zimsky, and Ed Moore, eds., pp. 113–133. Honolulu, Hawaii: Environment and Policy Institute, East-West Center, Land Air and Water Program, Forest and Farms Project.

Myers, Norman. 1989. *Deforestation Rates in Tropical Forests and their Climatic Implications.* London: Friends of the Earth .

Nietschmann, Bernard. 1985–86. Indonesia, Bangladesh: Disguised invasion of indigenous nations. *Fourth World Journal* 1 (2): 89–126.

Petocz, Ronald G. 1984 . *Conservation and Development in Irian Jaya.* WWF/IUCN Project No. 1528, Irian Jaya Conservation Programme. Bogor, Indonesia: World Wide Fund (WWF) for Nature and International Union for the Conservation of Nature (IUCN).

Rappaport, Roy A. 1984. *Pigs for the Ancestors: Ritual in the Ecology of a New Guinea People.* New Haven, Conn.: Yale University Press.

Robinson, Michael H. 1988. Are there alternatives to destruction? In *Biodiversity,* Edward O. Wilson, ed., pp. 355–360. Washington, D.C.: National Academy Press.

Sedik. 1994. The current status of Jamu and suggestions for further research and development. *Indigenous Knowledge and Development Monitor* 2: 13–15.

Silzer, Peter J. and Helja. H. Clouse. 1991. *Index of Irian Jaya Languages: A Special Publication of Irian Bulletin of Irian Jaya.* Jayapura: Program Kerjasama Universitas Cenderawasih and Summer Institute of Linguistics.

Smith, Andrew. 1992. Interview on May 5 with Andrew Smith of World Wide Fund for Nature, Jayapura, Irian Jaya, Indonesia.

Stone, Benjamin C. 1982. New Guinea *Pandanaceae*: First approach to ecology and biogeography. In *Monographiae Biologicae* 42: 401–435.

Sutter, Harald. 1989. *Forest Resources and Land Use in Indonesia.* UTF/INS/065: Forestry Studies Field Document No. I-1. Jakarta: Government of Indonesia, Ministry of Forestry, Directorate General of Forest Utilization, and Food and Agriculture Organization of the United Nations.

Thompson, Herb. 1994. Herb Thompson ponders the question of rainforests in Papua New Guinea. *FrontLine,* September 1994, pp. 1–2. A monthly newspaper published in Collingwood, Australia.

United Nations. 1990. *World Population Chart.* Revised. New York: United Nations Publication.

Waiko, John D. 1977. The people of Papua New Guinea, their forests, and their aspirations. In *The Melanesian Environment,* John Winslow, ed., pp. 407–427. Canberra: Australian National University Press.

Yassafar, Godfried. 1994. Forged signatures compiled in genealogy survey of Madang area. *Times* (Papua New Guinea), May 12, p. 2.

13

The Use of Satellite Imagery and Anthropology to Assess the Causes of Deforestation in Madagascar

Robert W. Sussman, Glen M. Green,
and Linda K. Sussman

Madagascar is located approximately 400 kilometers off the southeast coast of Africa. It is an island continent, 1,600 kilometers long and 580 kilometers wide at its broadest point, with an area of 590,000 square kilometers. Madagascar separated from Africa about 175 million years ago and has been in approximately the same position for about 120 million years (Rabinowitz, Coffin, and Falvey 1982, 1983). Many plants and animals are endemic since the initial separation. Many others apparently reached Madagascar across a water barrier after the mid-Cretaceous, and these forms have been relatively isolated for the past 50 to 55 million years (Darlington 1957; Raven and Axelrod 1974; Leroy 1978; Tattersall 1982).

This isolation corresponds with the early stages of evolution and dispersal of certain taxa that are widespread today. Angiosperms and prosimians, for example, were just beginning to diversify when Madagascar became isolated (Leroy 1978; Tattersall 1982; Gentry 1988; R. W. Sussman 1991). Until the arrival of humans fifteen hundred to two thousand years ago (Dewar 1984), these forms had little competition from mainland species, and for millions of years the plants and animals of Madagascar had an independent evolutionary history.

Topographically, Madagascar falls into three major zones (Battistini 1972), a narrow eastern plain, plus the steep escarpment that demarcates it to the west; a rugged high central plateau; and a vast sedimentary plain in the west and northwest. The variety of relief and the great size of the island give rise

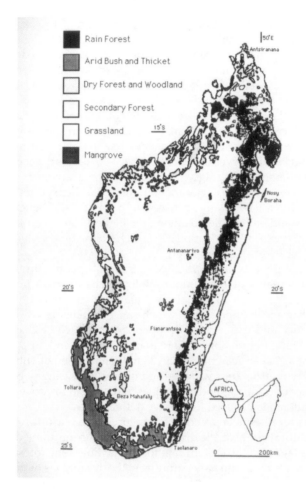

Figure 13.1   Vegetation Zones of Madagascar

to wide regional and local variations in climate, and there are a number of vegetation types (figure 13.1). The isolation of Madagascar and its variety of climates have led to an exceptional richness of plant species. Recent estimates indicate that a total of 8,000 to 12,000 species of plants occur, as compared to a total of only 30,000 species for the much larger area of mainland tropical Africa (Koechlin, Guillaumet, and Morat 1974; White 1983; Guillaumet 1984; Raven 1985; Jenkins 1987). Furthermore the flora of Madagascar exhibit a remarkable level of endemism. At least five plant families are restricted to the island, and well over half of its species are endemic (Leroy 1978), a total representing nearly 2% of all species in the world. Unfortunately, much of the primary vegetation of Madagascar rapidly is being degraded.

In addition to the extremely high endemism of the flora Madagascar has unique and highly endemic fauna, much of which also are in danger of extinction. For example, the island is home for the world's twenty-nine species of lemurs, as well as five of the fourteen primate families (Tattersall 1982; Richard and Dewar 1991). Forty percent of the world's chameleons are found in Madagascar, and more than 95% of its reptiles are endemic (Blanc 1984). Since the arrival of humans at least sixteen species of lemur have become extinct (Tattersall 1982; Albrecht, Jenkins, and Godfrey 1990). The largest land bird that ever roamed the earth also has disappeared since the arrival of humans, as has a pygmy hippopotamus, an endemic aard-vark, a giant tortoise, a large eagle, and innumerable other species (Dewar 1984; Goodman 1994). Many conservation biologists believe that Madagascar should be considered among the highest conservation priori-ties (Raven and Axelrod 1974; National Research Council 1980; Myers 1988; Jolly 1989; McNeely et al. 1990; Mittermeier et al. 1992).

However, the poverty that afflicts Madagascar's people threatens to destroy what remains of this unique biology. The average income is roughly $200 per year (Population Reference Bureau 1992). Yet Madagascar has a $2.5 billion debt, which nearly equals its yearly gross national product, and real income dropped 25% between 1985 and 1989 (Jolly 1989). With a population of more than 13 million and a population growth rate of 3.3% per year, more than 33 million people will inhabit the island by the year 2020. Thus widespread poverty, increasing population, and the absence of resources and techniques to reuse and improve agricul-ture and pasturelands have led to massive deforestation.

This article represents an attempt to explore some reasons for defor-estation in cultural context. We draw our examples from three different regions of Madagascar: the eastern rain forest, the limestone forests of the west, and the xerophytic and gallery forests in the dry regions of the south. These examples range from broad regional trends to quite localized areas of deforestation.

Rain Forest of Eastern Madagascar

In Madagascar the need for land and an expanding population, not large-scale timbering, have been the major cause of rain forest clearing (Rauh 1979; Jolly and Jolly 1984; R. W. Sussman, Richard, and Ravelojaona 1985; Jenkins 1987). Figure 13.2 is a photograph of the east coast taken by space shuttle astronauts. The fluffy white areas are clouds, but the white smoke plume coming from the upper part of this small island is derived from swid-den (i.e., small-scale shifting) agriculture of the sort that has cleared much of the eastern rain forest (Humbert 1927; Jolly and Jolly 1984; Jolly 1986).

Figure 13.2   Photograph taken by space shuttle astronauts of a smoke plume from a burning swidden.

In forest areas such as these the nutrients are quickly leeched from the soil, and the resulting fields provide only a few years of subsistence before they are depleted of nutrients (Betsch 1972; Berry and Johnson 1986). Unfortunately, because of the expanding population, fields are not abandoned for periods long enough to allow forest regeneration (Jolly 1980; Food and Agriculture Organization and United Nations Environment Programme [FAO and UNEP] 1981; Jenkins 1987). Soil no longer protected by forest is subject to rapid erosion, and annual watershed erosion rates as high as 250 tons per hectare have been reported in Madagascar (Helfert and Wood 1986). This leads to siltation and flooding on many rivers.

Landsat can be used to map the history, progress, and processes of defor-

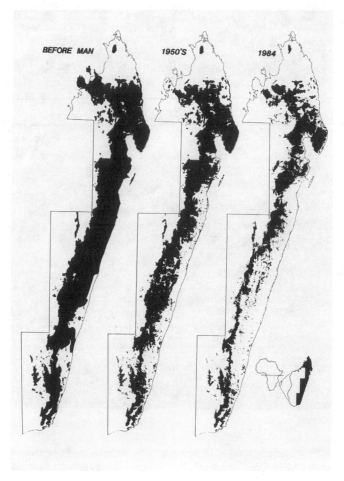

Figure 13.3   Deforestation history of eastern Madagascar
derived from aerial photographs and satellite images.

estation in Madagascar (Green and Sussman 1990). Figure 13.3 shows three
maps of the progressive deforestation that has taken place during the past
two thousand years or so. The map in the middle is derived from vegetation
maps prepared by French botanists H. Humbert and G. Cours Darne (1965),
using aerial photographs taken in 1949 and 1950. Humbert and Darne also
estimated the original extent of rain forest, which appears in the map at left.
We used Landsat data from 1985 to generate the map on the right.

    We estimate that rain forest covered 11.2 million hectares of the east coast at colonization, of which 7.6 million hectares remained by 1950. By 1985 only 3.8 million hectares remained. Thus in 1985 only 50% of the rain forest existing in 1950, and only 34% of the original extent, was still standing. This yields an average rate of clearance of 111,000 hectares (1.5%) per year between 1950 and 1985 (Green and Sussman 1990).

What factors influence deforestation in eastern Madagascar? We found that rates of deforestation are directly related to population density (table 13.1) and to the slope of the land. In the southeast the only forest remaining is that located on very steep slopes. The northern region still has large areas of low-lying tropical forests, but these forests remain because the population density in the north is generally much lower than in the south. However, deforestation is proceeding in all areas, and the remaining forests in the north are now also being cleared. In fact, the population in the north is now as high as it was in the south thirty-five years ago (National Institute of Geodesy and Cartography 1969, 1984).

Table 13.1

Area of the Eastern Rain Forest of Madcagascar, for Before Human Arrival, 1950 & 1985, and for High, Medium, and Low Population Densities

Year	Aerial extent (ha. x 10^6)	Forest remaining (%)	Deforestation rates from 1950 to 1985 (ha. x 10^3/year)
High density (> 10 people per aquare kilometer)			
Before humans	4.7	100	
1950	2.4	50	43
1985	0.89	19	
Medium Density (5 to 10 people per square kilometer)			
Before humans	3.4	100	
1950	2.5	76	37
1985	1.6	51	
Low Density (< 5 people per square kilometer)			
Before humans	3.1	100	
1950	2.7	86	31
1985	1.6	51	
Totals			
Before humans	11.2	100	
1950	7.6	67	111
1985	3.8	34	

Source: Adapted from Green and Sussman (1990)

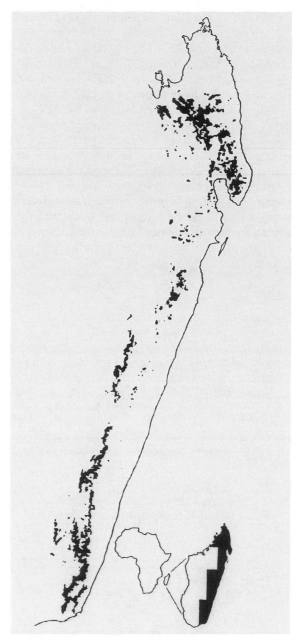

Figure 13.4   Computer-generated map of extent of rain
forest in eastern Madagascar in the year 2030.

Using rates of deforestation since 1960 and current rates of population growth, we have attempted to predict what the eastern rain forests will look like thirty-five years from now if deforestation continues at its current pace (figure 13.4). Because forest is cleared predominantly for subsistence agriculture, we assume that future needs for land and, thus deforestation rates, will not change so long as subsistence techniques and the social, political, and economic contexts remain stable. By the year 2020 only about 38% of the rain forest remaining in 1985, and 12.5% of the original extent, will still exist, leaving an area of only 1.4 million hectares (Green and Sussman n.d.). Not only will the total area of forest be reduced but, as can be seen in figure 13.4, the forests will have been fragmented into many small parcels. Both changes will have a profound effect on rates of extinction in Madagascar (Frankel and Soulé 1981; Pollock 1986). Also, particular forest types are preferentially destroyed, particularly the lowland flat-lying forests, none of which will remain in thirty-five years.

A number of reserves were established in Madagascar in the 1920s and 1930s. Most are largely intact. In fact, many will still be relatively untouched in thirty-five years because they are in areas of steep slope and high elevation. These reserves have been protected not by conservation efforts, or by sustainable agriculture surrounding them, but solely by the natural topography. The locations of natural reserves were actually chosen in the first place because they were remote and difficult to access—"in little populated or mountainous areas which would be shielded from the pressure of a population in constant search of new crop land" (Andriamampianina 1984:219).

This points out a fundamental problem with current conservation thinking. Conservation efforts on the east coast of Madagascar have focused mainly on protected reserves (MacKinnon and MacKinnon 1987; Nicoll and Langrand 1989; McNeely et al. 1990; Mittermeier et al. 1992). Reserves are viewed as fortresses established to keep people out and thereby preserve biodiversity. These areas are protected only so long as they are remote, or presumably with armed guards once the population increases (see, for example, McNeely et al. 1990). Finally, when the situation becomes extreme or commercial needs become attractive, reserve boundaries are ignored. For example, in 1964 one of the twelve reserves in eastern Madagascar was declassified in favor of commercial exploitation (Andriamampianina 1984).

In reality, to slow deforestation and maintain an integral forest in the east, conservation efforts must be focused at the fronts of deforestation and ultimately involve a cooperative effort by conservationists and local people to develop ways to establish sustainable use of land that has already been cleared. Satellite imagery can be used to locate these fronts and to monitor the success or failure of efforts to slow their advance.

Figure 13.5   Satellite image of limestone forests of southwestern
Madagascar.

Dry Forests of Southwestern Madagascar

The southwestern dry forests of Madagascar have no steep slopes to protect the forest, and the area has only two large reserves. Figure 13.5, an
image provided by satellite, shows the dry limestone forests of southwestern Madagascar. Many of these forests were pristine until the early 1970s.
The major city on the west coast is Toliara. The dark areas in the photograph are forest cover. Notice the whitish patch stretching to the east of the
city. This is a recently deforested area along the road that connects Toliara
to the capital, Antananarivo.

This extensive deforestation has occurred since 1970 and is related to an
economic downturn in Madagascar (Vérin 1990) and to global increases in
fuel prices. Since 1970 huge numbers of people have moved from the countryside into the cities (Salomon 1977; Hoerner 1981, 1986). This also correlates with a collapse of the agricultural infrastructure in much of southwestern Madagascar and an increase in the incidence of cattle rustling
(Hoerner 1982; Vérin 1990).

Most cooking in Madagascar is done with charcoal from various hardwoods or with dried deadwood (Rafidison 1987). With the population
increase in the large cities like Toliara, surrounding regions could not produce enough deadwood to supply fuel needs. Thus a large-scale charcoal
industry was begun in the early 1970s.

A bag of charcoal costs almost U.S.$3 and will last a family approximately two months. Deforestation for charcoal has led to bleak-looking landscapes on which only bare limestone rocks remain; after clearing, the land is basically unused. We estimate that since 1972 more than 100,000 hectares of limestone forest bordering Toliara have been cleared (Green, Feinan, and Nelson n.d.). There are alternatives to cooking with charcoal, such as fast-growing gourds that have replaced the use of charcoal in a number of regions throughout the world (Bragg, Duke, and Shultz 1987). However, we know of no major effort to reduce the use of charcoal and the destruction of limestone forest in southwestern Madagascar (Rafidison 1987).

Beza Mahafaly

A small reserve, Beza Mahafaly, was established in the south in 1978 as part of a cooperative effort by the University of Madagascar, Washington University, and Yale University to encourage research, conservation, development, and education. It was inaugurated as a Special Government Reserve in 1986 and has been funded largely by World Wildlife Fund and the U.S. Agency for International Development (USAID) (Richard, Rakotomanga, and Sussman 1987). In 1990 the World Wildlife Fund (although originally only one of many funding agents for the activities at the reserve) took it upon itself "to transfer the Conservation and Development in Southern Madagascar Project to the Malagasy people" (Wyckoff-Baird, personal communication 1990), although the World Wildlife Fund (WWF) and USAID would still control, manage, and monitor the funding of the project. This eliminated any official scientific participation by researchers unaffiliated with the funding agents.

Figure 13.6 is an aerial photograph taken in 1987 of the region in which the reserve is located. The dark area in this picture is dense gallery forest. The lighter area to the east of the forest is the dry bed of the Sakamena River. Note the great contrast between the west and east sides of the river. Figure 13.7 is the same area in 1968. The rich gallery forest on both sides of the river at that time is obvious. Thus this forest was cut recently, within the last twenty years or so. In fact, we have learned from satellite images that fewer than 4,500 hectares of gallery forest remain in all of southwestern Madagascar. During an ethnographic study in 1987–1988, L. K. Sussman (1994) questioned the people about their use of forest resources and the recent necessity of converting the gallery forest into crop land and pasture; the reasons are counterintuitive.

Unlike in the highly populated areas like Toliara, residents in the region of Beza Mahafaly generally do not cut wood for firewood. Women normally collect brush and deadwood in forests adjacent to the villages. The

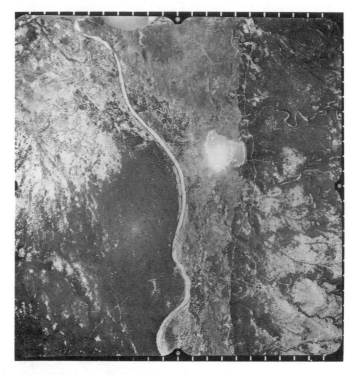

Figure 13.6   Aerial photograph of the region surrounding the
Beza Mahafaly Reserve in 1987.

major reasons for cutting trees are building houses and corrals and clearing fields for subsistence agriculture. Beza Mahafaly is located in a semiarid area (R. W. Sussman and Rakotozafy 1994), and the general assumption is that any forest clearing in this region is related to periodic droughts
and the need for more dry season crop lands (e.g., Hoerner 1977; Barbour
1988). However, we estimate from residents' reports in this region that one
dry season field has in fact been productive for at least eighty years, and it
supplies much of the sustenance for the residents of the seven hamlets east
of the Sakamena River, closest to the reserve. This field is reportedly what
drew people to the region sixty to eighty years ago and is the backbone of
the local subsistence economy. It is used for the cultivation of corn, manioc,
onions, tomatoes, various leaves, and beans. In the wet season much of this
field floods, and the people fish in it (figure 13.8). They also attempt to

Figure 13.7   Aerial photograph of the region surrounding the
Beza Mahafaly Reserve in 1968.

grow rice in some portions of this field during the wet season, although
usually unsuccessfully. Fields in other areas are used for corn, manioc, and
sweet potato production (L. K. Sussman 1990).

The recent clearing of forest in the immediate area surrounding the Beza
Mahafaly Reserve had little to do with the need for more dry season crops.
The major agricultural problems occur in the rainy season, especially when
the corn crop fails. Ultimately, the lack of fertility of some wet season fields,
which were reported by residents to have been used for thirty to forty
years, necessitated the clearing of new fields along the Sakamena before
1987 (L. K. Sussman 1994). In a follow-up study in 1993 L. K. Sussman
found that although some tracts in the main dry season field are not cultivated, during the previous five years people had cleared additional tracts
of forest for dry season use. This expansion of agricultural land is related
to a considerable increase (50%) in the population in the area, resulting
from both natural increase and older village residents' returning to their
natal homes after long absences (L. K. Sussman 1994).

Figure 13.8   People fishing in the dry season field at Beza Mahafaly during the wet season.

Further, with the economic downturn in the early 1970s cattle rustling became common and widespread in southwestern Madagascar. This forced the people of the Sakamena Valley to concentrate their cattle, so they could better guard them close to the villages. When they lose cattle, which aid in labor and serve as the banking system for the rural Malagasy, the people need more crops to create surplus for generating cash. Cattle graze in fallow fields, secondary forest, primary gallery forests, and dry woodland forests adjacent to the villages. Because of overgrazing regeneration is poor in unprotected gallery and dry woodland forests.

Only the combination of reconstructing the history of deforestation in the region using remote sensing and detailed ethnographic study led us to an understanding of the patterns, processes, and motivation behind the clearing of this forest. However, the response to this problem thus far has been to send agronomists to study the situation during the dry season and to plan the construction of a canal in an attempt to provide irrigation for wet rice agriculture. The people of the Sakamena Valley, adjacent to the reserve, have had little or no experience with wet rice agriculture or with large irrigation systems (L. K. Sussman 1990). Furthermore no one knows whether the canal will actually provide water to villages near the Beza Mahafaly Reserve or in fact flood the productive dry season field and destroy the local economy. Although cattle rustling and finding a way to protect cattle and provide sustainable pasture are major local concerns, they have been essentially ignored. There also has been no attempt to find ways to ensure the continued fertility of the wet and dry season fields currently in use.

Recommendations

In reference to these examples of the complexity of factors involved in deforestation and development, we make the following broad recommendations:

1. Given a paucity of ethnographic research, detailed ethnographic studies must be done before any meaningful development or conservation projects can begin. These must focus specifically on use of resources and resource and conservation needs. To date most communication between Western conservation and development agencies and the people of Madagascar has taken place in the capital. For example, at the Beza Mahafaly Reserve the wwf has turned over everyday management to the University of Antananarivo and the Department of Waters and Forests. The local Malagasy have little or no say about the goals or direction of the project. In most cases the Malagasy agencies in the capital are as remote and uninformed about rural Madagascar as are people in the West.

In a study for the World Bank of sixty-eight rural development projects, C. P. Kottak (1990) found that the success rate in economic terms of those projects that are socioculturally compatible was twice those that disregard or give inadequate attention to local culture (see also Anderson and Huber 1988). In this context inappropriate actions by conservation and development agencies at Beza Mahafaly include introducing new dry season crops (such as okra, squash, and beets) that are not used by the local people (who normally have ample food during the dry season, in any case), paying little or no attention to the major problems of cattle rustling and the rapid decrease of suitable pastureland, and attempting to introduce wet rice agriculture to replace corn, manioc, and sweet potatoes as major crops.

2. We believe that the culture of the conservation and development agencies themselves must change. Between 1978, when we began this project at Beza Mahafaly, and 1990, the World Wildlife Fund underwent four major administrative changes. During the same period usaid changed its personnel three times. In both institutions one administration gives the next very little information about specific projects, and each change calls for a complete reeducation of new personnel. It is the policy of usaid and the World Bank to transfer personnel from one country to another every three years; thus personnel switch from one unrelated project and country to another. The reason given is that these people must stay "objective" (Mahar 1990); the result is that for the most part they remain ignorant.

Between 1985, when Beza Mahafaly Reserve was inaugurated as a special government reserve, and 1990 WWF and USAID sent at least ten separate development or conservation evaluation missions to the reserve. These included two to nine individuals, few of whom spoke Malagasy. They visited the site for no more than two weeks, usually for less than forty-eight hours. Few members of these missions communicated with members of previous missions, and none knew about or bothered to contact L. K. Sussman, the only ethnographer to conduct research in the region and whose project was partially funded by WWF and USAID.

3. We believe that the concept of development itself must change in these institutions. They do not support small projects, yet we feel it is necessary to steer away from large, overinnovative projects that are based on a Western model of development. Kottak (1990) has suggested using Romer's Rule—The goal of stability is the main impetus for change—as a principle to guide development projects (Romer 1959:93–94, 327). In many regions, including the Sakamena Valley, people want to maintain the lifestyle that they have had for generations. However, they have to change in order to do this.

In 1987 with a USAID grant the Beza Mahafaly Project began to reconstruct a portion of a road to villages bordering the reserve. People used this road to take agricultural crops to the local markets, but it had become nearly impassable. This was one of two priority projects requested by local residents so that they could maintain established commerce networks. The completion of this road was canceled because of infighting between two Malagasy political agencies and because new WWF and USAID administrators believed the sole purpose of the road was to allow easier access to the reserve.

The conservators and developers have extremely little understanding of or interest in grassroots development. Most people involved in conservation and development in Madagascar have little interest in staying in rural areas long enough to really understand the problems and the needs. Projects in these areas ultimately will involve a great deal of time rather than a great deal of money, and these large agencies have a lot of money and little time.

4. We believe there is a need to establish organizations to monitor and empirically and systematically evaluate projects (see, for example, Seidman and Anang 1992). Most conservation and development agencies (both government and nongovernment organizations) have only in-house reviews. Anthropologists and other social scientists and environmental and conservation biologists should continue to conduct independent research in regions in which conservation and development projects are being implemented. This research probably would require funding by scientific granting agencies or private foundations independent of the major conservation and development agencies.

Ideally, the results would convince these large agencies (or the people or governments that fund them) of the necessity for basic research in conservation and development before starting large-scale projects. Further, integrated conservation and development plans need to be developed. Beza Mahafaly has some conservation projects (such as fencing and guards) and some development projects (such as schools and a road). However, no conservation-oriented plan exists to guide development of sustainable resources, and no development-oriented plan exists to guide conservation. These two problems must be integrated rather than parallel and unrelated.

Objective criteria must be developed to evaluate projects. For instance, it is possible to generate predictive maps such as those we have produced and then use satellite images to monitor progress in areas surrounding planned projects. After fronts of deforestation are identified, social scientists could research the social, economic, and political contexts related to this clearing. Using basic ethnographic and social research, environmental scientists, agronomists, and local people could try to determine whether alternative sustainable land-use practices are feasible. Furthermore, by continuously monitoring satellite images of the region the effectiveness of these projects in slowing rates of deforestation can be measured. Although conservation and development agencies have spoken repeatedly in the last ten years of using satellite images to map deforestation in Madagascar and organizing international committees to oversee this research, they have done nothing. It took independent funding and a small-scale collaborative effort for us to develop our maps of eastern Madagascar. The total cost was approximately $5,000.

5. Landsat data must become more accessible. The Landsat system was instituted in 1972 but was privatized in 1986. Since privatization the prices of Landsat products have increased. We believe that the high cost has reduced the use of satellite images in research on deforestation. Furthermore data acquired in the early 1970s are stored on digital tape in Sioux Falls; we estimate that nearly 40% of that digital data has deteriorated and is no longer usable.

In Madagascar we have less than thirty-five years to save most of the rain forest areas of the east coast. If we leave this task solely in the hands of major conservation and development agencies without outside participation and monitoring, we fear that our predictions may come true.

Acknowledgments   This project would not have been possible without the generous aid of the Ministries of Higher Education and Scientific Research, the School of Agronomy of the University of Antananarivo, and the Muséum d'Art et d'Archeologie. We especially thank B. Rakotosamimanana, B. Andriamihaja, B. and V. Randrianasolo, J. Andriamampianina, P. Rakotomanga, J.-A. Rakotoarisoa, and A. Rakotozafy. We also gratefully acknowledge the hospitality and assistance of the people of Analafaly and the reserve guards. We thank Michael Fienen and Andrew Nelson for field and laboratory assistance. The research was funded in part by the World Wildlife Fund, Fulbright Senior Research Grants, the National Science Foundation, the National Geographic Society, Pew Midstates Science and Mathematics Consortium, BRSG Grant #S07 RR077054 awarded by the Biomedical Research Support Grant Program, Division of Research Resources, National Institutes of Health, and Washington University.

References Cited

Albrecht, G. H., P. D. Jenkens, and L. R. Godfrey. 1990. Ecogeographic size variation among the living and subfossil prosimians of Madagascar. *American Journal of Primatology* 22: 1–50.

Anderson, R. S. and W. Huber. 1988. *The Hour of the Fox: Tropical Forests, the World Bank, and Indigenous People in Central India.* Seattle: University of Washington Press.

Andriamampianina, J. 1984. Nature reserves and nature conservation in Madagascar. In *Key Environments: Madagascar,* A. Jolly, P. Oberlé, and R. Roland, eds., pp. 219–227. Oxford, England: Pergamon Press.

Barbour, R. 1988. Plan de travail pour les activités de developpement à Beza Mahafaly. Unpublished report to World Wildlife Fund.

Battistini, R. 1972. Madagascar relief and main types of landscape. In *Biogeography and Ecology of Madagascar,* R. Battistini and G. Richard-Vindard, eds., pp. 1–25. The Hague: W. Junk.

Berry, L. and D. L. Johnson. 1986. Geographical approaches to environmental change. In *Natural Resources and People,* K. A. Dahlberg and J. W. Bennett, eds., pp. 67–105. Boulder, Colo.: Westview.

Betsch, J. M. 1972. La microfaune du sol à Madagascar, témoin de la santé des sols. Comptes rendus de la Conférence Internationale sur la Conservation de la Nature et de ses Ressources à Madagascar, Tananarive, Madagascar, 1970. Morge, International Union for the Conservation of Nature.

Blanc, C. P. 1984. The vegetation: An extraordinary diversity. In *Key Environments: Madagascar,* A. Jolly, P. Oberlé, and R. Roland, pp. 105–114. Oxford, England: Pergamon Press.

Bragg, W. G., D. L. Duke, and E. B. Shultz Jr. 1987. Rootfuel: Annual roots as cookstove fuel in the arid Third World. Paper presented at the 28th Annual Meeting of the Society for Economic Botany, June 22–25, Chicago.

Darlington, P. J. Jr. 1957. *Zoogeographgy: The Geographical Distribution of Animals.* New York: Wiley.

Dewar, R. E. 1984. Recent extinctions in Madagascar: The loss of the subfossil fauna. In *Quaternary Extinctions: A Prehistoric Revolution,* P. S. Martin and R. G. Klein, eds., pp. 574–593. Phoenix: University of Arizona Press.

Food and Agriculture Organization and United Nations Environment Programme (FAO and UNEP). 1981. *Tropical Forest Resources Assessment Project. Forest Resources of Tropical Africa, Part 2: Country Briefs.* Rome: Food and Agriculture Organization.

Frankel, O. H. and M. E. Soulé. 1981. *Conservation and Evolution.* Cambridge, England: Cambridge University Press.

Gentry, A. H. 1988. Distribution and evolution of the Madagascar *Bignoniaceae*. In *Modern Systematic Studies in African Botany*, P. Golblatt and P. P. Lowry II, eds., pp. 175–185. Monographs in Systematic Botany from the Missouri Botanical Garden, Vol. 25. St. Louis: Missouri Botanical Garden.

Goodman, S. M. 1994. The enigma of antipredator behavior in lemurs: Evidence of a large extinct eagle on Madagascar. *International Journal of Primatology* 15: 129–134.

Green, G. M. and R. W. Sussman. 1990. Deforestation history of the eastern rain forests of Madagascar from satellite images. *Science* 248: 212–215.

——. n.d. Predicted future deforestation of the eastern rain forests of Madagascar: An assessment of conservation priorities. (To be submitted to *Science*)

Green, G. M., M. Feinan, and A. Nelson. n.d. Deforestation history of the forests of Southern Madagascar. (In preparation)

Guillaumet, J.-L. 1984. The vegetation: An extraordinary diversity. In *Key Environments: Madagascar*, A. Jolly, P. Oberlé, and R. Roland, eds., pp. 27–54. Oxford, England: Pergamon Press.

Helfert, M. R. and C. A. Wood. 1986. Shuttle photos show Madagascar erosion. *Geotimes* 31: 4–5.

Hoerner, J. M. 1977. L'eau et l'agriculture dans le sud-ouest de Madagascar. *Madagascar Révue de Géographie* (30): 63–104.

——. 1981. Tulear et le sud-ouest de Madagascar: Appoche demographique. *Madagascar Révue de Géographie* (39): 9–49.

——. 1982. Les vols de boeufs dans le sud Malgache. *Madagascar Révue de Géographie* (41): 45–105.

——. 1986. *Géographie régionale du sud-ouest de Madagascar.* Antananarivo, Madagascar: Association des Géographes de Madagascar.

Humbert, H. 1927. *La destruction d'une flore insulaire par le feu.* Mémoires de L'Academie Malgache, Fascicule V. Tananarive, Madagascar: L'Academie Malgache.

Humbert, H. and G. Cours Darne. 1965. *Carte internationale du tapis végétal: Madagascar, 1:1,000,000.* Toulouse: French Institute of Pondichéry.

Jenkins, M. D., ed. 1987. *Madagascar: An Environmental Profile.* Gland, Switzerland: International Union for the Conservation of Nature.

Jolly, A. 1980. *A World Like Our Own.* New Haven, Conn.: Yale University Press.

——. 1986. Lemur survival. In *Primates: The Road to Sustainable Populations*, K. Benirschke, ed., pp. 71–98. New York: Springer Verlag.

——. 1989. Banking on Madagascar. *Earthwatch*, April, pp. 11–13.

Jolly, A. and R. Jolly. 1984. Malagasy economics and conservation: A tragedy without villains. In *Key Environments: Madagascar*, A. Jolly, P. Oberlé, and R. Roland, eds., pp. 211–217. Oxford, England: Pergamon Press.

Koechlin, J., J.-L. Guillaumet, and P. Morat. 1974. *Flore et végétation de Madagascar*. Vaduz, Liechtenstein: Cramer.

Kottak, C. P. 1990. Culture and "economic development." *American Anthropologist* 92: 723–731.

Leroy, J.-F. 1978. Composition, origin, and affinities of the Madagascan vascular flora. *Annals of Missouri Botanical Garden* 65: 535–589.

MacKinnon, J. and K. MacKinnon. 1987. Étude du réseau des aires protégées du domaine Afro-tropical: L'exemple Malgache. In *Priorités en Matière de Conservation des Espèces à Madagascar*, R. A. Mittermeier, L. H. Rakotovao, V. Randrianasolo, E. J. Sterling, and D. Devitre, eds., pp. 21–26. Gland, Switzerland: International Union for the Conservation of Nature.

Mahar, D. 1990. Economic Development or Destruction in the Amazon? Paper presented to the United Nations Association Mini-Conference, May 17, St. Louis.

McNeely, J. A., K. R. Miller, W. V. Reid, R. A. Mittermeier, and T. B. Werner. 1990. *Conserving the World's Biological Diversity*. Gland, Switzerland, and Washington, D.C.: International Union for the Conservation of Nature, World Resources Institute, Conservation International, World Wildlife Fund—United States, and World Bank.

Mittermeier, R. A., W. R. Konstant, M. E. Nicoll, and O. Langrand. 1992. *Lemurs of Madagascar: An Action Plan for Their Conservation 1993–1999*. Gland, Switzerland: International Union for the Conservation of Nature.

Myers, N. 1988. Threatened biotas: "Hotspots" in tropical forests. *Environmentalist* 8: 1–20.

National Institute of Geodesy and Cartography. 1969. *Atlas de Madagascar*. Antananarivo, Madagascar: National Institute of Geodesy and Cartography.

——. 1984. *Population Density Map, 1984, 1:6,000,000*. Antananarivo, Madagascar: National Institute of Geodesy and Cartography.

National Research Council. 1980. *Research Priorities in Tropical Biology*. Washington, D.C.: National Academy of Sciences.

Nicoll, M. E. and O. Langrand. 1989. *Madagascar: Révue de la conservation et des aires protégées*. Gland, Switzerland: World Wildlife Fund.

Pollock, J. 1986. Towards a conservation policy for Madagascar's eastern rain forests. *Primate Conservation* 7: 82–86.

Population Reference Bureau. 1992. *World Population Data Sheet*. Washington, D.C.: Population Reference Bureau.

Rabinowitz, P. D., M. F. Coffin, and D. Falvey. 1982. Salt diapirs bordering the continent margin of Northern Kenya and Southern Somalia. *Science* 215: 663–665.

——. 1983. The separation of Madagascar and Africa. *Science* 220: 67–69.

Rafidison, M. 1987. *Contribution à l'étude des production et consumation de combustibles forestiers dans un centre urbain et sa périphérie (cas de Betioky sud 1987)*:

Mémoire de fin d'etudes. Antananarivo, Madagascar: Université d' Antananarivo.

Rauh, W. 1979. Problems of biological conservation in Madagascar. In *Plants and Islands*, D. Bramwell, ed., pp. 405–421. London: Academic Press.

Raven, P. H. 1985. Overview of Plant Conservation in Africa. Paper presented at the 11th Plenary Meeting of the Association for the Taxonomic Study of the Flora of Tropical Africa, June 10–14, St. Louis.

Raven, P. H. and D. I. Axelrod. 1974. Angiosperm biogeography and past continental movements. *Annals of Missouri Botanical Garden* 61: 539–673.

Richard, A. F. and R. E. Dewar. 1991. Lemur ecology. *Annual Review of Ecology and Systematics* 22: 145–175.

Richard, A. F., P. Rakotomanga, and R. W. Sussman. 1987. Beza-Mahafaly: Recherches fondamentales et appliquées. In *Priorités en Matière de Conservation des Espèces à Madagascar*, R. A. Mittermeier, L. H. Rakotovao, V. Randrianasolo, E. J. Sterling, and D. Devitre, eds., pp. 45–49. Gland, Switzerland: International Union for the Conservation of Nature.

Romer, A. S. 1959. *The Vertebrate Story*. Chicago: University of Chicago Press.

Salomon, J. N. 1977. Tulear: Un exemple de croissance et de structure urbaine en milieu tropical. *Madagascar Révue de Géographie* (30): 33–61.

Seidman, A. and F. Anang. 1992. *Twenty-First-Century Africa: Towards a New Vision of Self-sustainable Development*. Trenton, N.J: African World Press.

Sussman, L. K. 1990. Unpublished report to World Wildlife Fund.

——. 1994. Effects of a conservation/development project on the residents in southwestern Madagascar: Whose goals are being met? Paper delivered at the 93rd Annual Meeting of the American Anthropological Association, November 30–December 4, Atlanta.

Sussman, R. W. 1991. Primate origins and the evolution of angiosperms. *American Journal of Primatology* 23: 209–223.

Sussman, R. W. and A. Rakotozafy. 1994. Plant diversity and structural analysis of a tropical dry forest in southwestern Madagascar. *Biotropica* 26: 241–254.

Sussman, R. W., A. F. Richard, and G. Ravelojaona. 1985. Madagascar: Current projects and problems in conservation. *Primate Conservation* 5: 53–59.

Tattersall, I. 1982. *The Primates of Madagascar*. New York: Columbia University Press.

Vérin, P. 1990. *Madagascar*. Paris: Karthala.

White, F. 1983. *The Vegetation of Africa. A Descriptive Memoir to Accompany the UNESCO/AETFAT/UNSO Vegetation Map of Africa*. Natural Resources Research 20. Paris: United Nations Educational, Scientific, and Cultural Organization.

14

Promoting Biodiversity and Empowering Local People in Central African Forests

Robert C. Bailey

More concerted efforts than ever before are being made to protect substantial areas of tropical rain forest in central Africa. In recent years various African governments, in coordination with nongovernmental conservation organizations and international funding agencies, have established or proposed to establish new parks and reserves specifically designed to arrest deforestation and to protect the biodiversity of central African forests.[1] Part of the process of creating protected areas of forest in Africa has been recognizing that forest-dwelling peoples depend on the forests for their physical survival and learning that such peoples can serve critical functions in the sustainable management and conservation of forest biodiversity (Halpin 1990).

Indeed over the last several years every policy statement, every directive, every proposal, and every management plan promoting development and conservation in central Africa has contained clear and forceful statements promoting the practical and ethical virtues of integrating local peoples' opinions in the planning and management of protected areas.[2] Some level of input from local peoples is now nearly universally espoused as a sine qua non of effective conservation management. In fact, the creation and management of protected areas is no longer seen by many as motivated purely by the desire to halt deforestation and to protect biodiversity but rather as part of a broader strategy for engineering effective resource management and rural development (Batisse 1990; Wells, Brandon, and Hannah 1990). Conservationists and wildlife managers increasingly recognize that what happens outside protected areas is as important in preventing deforestation and maintaining biodiversity as what goes on within them. They realize that successful protection means being responsive to activities and concerns of people beyond the fence line (Hackel n.d.; Peterson 1992), and active support and participation by local peoples can be crucial to the success of any project (McNeely et al. 1990; Oldfield and Alcorn 1991; West and Brechin 1991).

Despite such strong and pervasive agreement among conservationists, as well as all the major international funding agencies, concerning the desirability of bringing local peoples into the process of protected area planning and management, it has proved much easier to press for such concerns on paper than it has been to put them into practice. Reasons for this are myriad.

The Difficulties of Integrating Local People and Conservation

Expecting Too Much of Conservationists

Most issues surrounding development and conservation are not simple. There are countless examples of rural development programs in Third World countries that have attempted to tackle problems that have proved intractable to gargantuan efforts by a host of governments, international agencies, and local organizations. We may be expecting too much of any one group of experts, even of those who specialize in development, not to mention a community composed mostly of people trained as biologists, zoologists, foresters, or wildlife managers.[3] The concepts involved in making conservation and rural development compatible require extraordinary breadth of expertise and adaptability beyond the skills of most conservationists, given the great ecological, economic, and social complexity of our world.

Conservation organizations, especially internationally active nongovernmental agencies such as the Wildlife Conservation Society, Conservation International, World Wildlife Fund, and others, have developed many skills necessary to create and manage protected areas. They have become highly skilled at public relations, increasingly able to take advantage of the full variety of media to promote conservation generally and their latest project in particular. They have been able to attract substantial sums of money from a great variety of sources, and they have developed creative financing schemes for ever increasing numbers and sizes of projects. They are skilled at heightening the awareness of host governments to the value of conservation and are now well practiced at cooperating with government agencies and host country officials to designate new areas for protection.

They are using new technologies (e.g., global information systems) to become even better at monitoring ecological change and identifying entire ecosystems, as well as specific geographical areas, that require most urgent protection. Primarily conservation biologists, they are especially skilled at conducting ecological research aimed at assessing and projecting floral and faunal populations within various ecosystems. In addition, conservationists are becoming increasingly effective at developing educational programs to teach people of other cultures about the value of biodiversity and to instill a conservation ethic. Indeed to be effective conservation agencies have had to develop a great variety of skills, and their ability to do so, often under very difficult circumstances, has been truly impressive.

The Top-Down Approach

What conservationists and wildlife managers have not yet done effectively is to bring local people into the conservation process. In part this is undoubtedly the result of their necessary preoccupation with the many other challenges to conservation. However, it is also likely the result of the process they have used to identify, create, and manage protected areas. Typically, the process conservationists have pursued has been to activate change from the top down.

Conservation projects in central Africa are invariably conceived and initiated by a person or group conducting ecological research in a particular geographic area. I know of no instances in which an area in central Africa has been targeted for preservation on the a priori basis that it is a "hot spot" or crucial location for maximizing overall African forest biodiversity. This means that a conservation effort is not based on an Africa-wide or global strategic analysis but rather is the result of the personal stake that an individual researcher or small group of Western scientists has in promoting the preservation and management of a given area. Arguments advanced to promote a reserve and models used in project design are thus often based on ecological data collected selectively from research stations. Information gathered from or about local people is usually cursory, impressionistic, or lacking.

An individual or group of researchers seeking to preserve a given area invariably turns first to conservation agencies for assistance and then to international funding agencies in Europe and the United States. Representatives of these organizations tend to then discuss plans with host government officials residing in the capital city, far from the area of interest. Sometimes surveys are conducted to further assess the suitability of particular areas for protection. Such surveys, almost invariably conducted by people trained in biology, zoology, botany, forestry, or ecology, focus nearly exclusively on the flora and fauna of the area. Local populations are generally mentioned with reference to the degree to which they might complicate the establishment of any reserve. Reports are funneled back to the capital city and Europe and the United States, and the desirability of the project is evaluated with passing reference to and little knowledge of the residents of the targeted area.

When discussions finally reach Africans living in and around a projected reserve, plans for the reserve and its management are already well developed. The first people living near the projected reserve to be consulted are usually authorities who have been appointed from other areas of the country and the chiefs or other elite members of the local populations. The people who are going to be most affected by the creation of the reserve—those who actually depend on the resources of the area for their subsistence and consider it their land—are seldom solicited for suggestions in any meaningful way until all the deals have been cut and creation of the protected area is a fait accompli. Project plans may be altered at some late stage, but usually only slightly and only to conform to funding agency policy directives designed to protect the rights of indigenous peoples.

Priorities for developing the institutions to promote and support conservation in African countries also tend to come from the top-down perspective. International conservation agencies and their funding sources (e.g., the World Bank, U.S. Agency for International Development, and the European Economic Community) are rightfully concerned about building lasting infrastructures to support conservation efforts in African countries. Thus they have emphasized creating the institutional structure as well as the training necessary for continuous management of protected areas. There has been a strong bias toward supporting or even creating new government-sponsored institutions and training high-level bureaucrats and national university students in hopes that they will direct and manage biodiversity within their own country in the future. Building institutions and training good people to manage them are legitimate and important goals that will surely prove salutary to conservation efforts in central Africa in the long term. At the same time the international agencies promoting these efforts need to be aware that building such infrastructures cannot be a substitute for promoting local peoples' participation in the conservation and development process.

There has been a tendency to underdifferentiate various distinct segments within the same African nation. In most cases local people are very distinct from other levels and groups within the larger society. Supporting the government structures that can promote conservation in the country as a whole is not the same as supporting local peoples' participation in specific projects. Indeed in some cases supporting national-level structures may actually undermine the chances that local people will participate in conservation and development decisions, because authorities often see local groups and local organizations as impediments to centralization and consolidation of their power. Thus institutional-level efforts should not divert funds and attention from promoting programs that bring local people into the planning and management process.

"Biological" Versus "Social" Components

People who talk about or design conservation-management plans tend to separate what administrators and evaluators refer to as "the human element" of the project, or the "sociological component" from the "biological component." Under the biological component usually falls the budgeting and planning for getting the area declared as protected by the host government, of training and deploying personnel to be ministers, administrators, wildlife managers, and guards, and of supporting biological research. The sociological component is usually accompanied by appropriate rhetoric exhorting all concerned to uphold the virtues of consulting with local people, and it normally entails a comparatively small budget for assessing the impact of the project on the local people and perhaps for developing conservation education programs at the local level. Thus program developers and evaluators see conservation planning and management as divided into two independent and even competing elements—biological versus social. Yet very little of what conservationists do is actually biological in the sense that it is research on or management of flora and fauna independent of the human sociocultural, political, or economic environment.

Making political and administrative institution building part of the biological component of management plans and relegating local peoples' concerns to little more than a sideshow at the very least suggest some confusion about how conservation planning and management measures should be organized. In this regard, reading project proposals and management plans written largely by conservation biologists and project administrators is revealing—probably not of the true motivations of these writers but of the extent to which they are molded and constrained by the culture of the conservation establishment. If they step back for a moment and take an honest look at what has consumed their time and mental energy, conservation biologists and wildlife managers must realize that protecting biodiversity has always been and will continue to be less the concern with issues of biology and wildlife ecology and much more the wrestling with human political, economic, and social processes. The ultimate goals of conservation efforts may be driven primarily by biological theory and ecological research, but the process by which conservation is achieved is overwhelmingly social and political. It has little to do with biology.

Given the pervasiveness of the top-down approach and the very structure of the process as it currently exists, it is difficult for an outside observer to conclude that conservationists are consulting local people out of a conviction that taking their experience and needs into account is either morally imperative or practically desirable. Reports, surveys, proposals, and management plans written by conservation biologists and funding agency managers lead inevitably to the conclusion that local people are considered part of the problem, not part of the solution. This is true despite widespread knowledge that projects are successful when they are culturally compatible, when they take advantage of traditional institutions, and when they address locally perceived goals. If conservation efforts in central Africa are to be effective, the process by which conservationists go about identifying, creating, developing, managing, and monitoring protected areas will have to change. To be sure, most conservation biologists are aware of this, but the impediments to integrating local people to the process appear so well entrenched that change is proving difficult.

Developing an Effective Process

Principles to Guide the Process

No one model of how to integrate local people in the planning and management of protected areas will be applicable everywhere. What is crucial is developing principles that can effectively guide and inform a process. Because the issues are so complex, one overriding principle must be that the process is adaptable and highly responsive to a broad range of social, political, and ecological systems. If the process is to be responsive to people-generated ideas and less on higher-level plans and decrees, the conservation establishment must place more emphasis on a "learning process model" in which plans and management schemes are not etched in stone from the beginning of the project but constantly revised in response to changing conditions. This means that funding agencies must be willing to take what they may perceive as greater risks by trusting host governments and contracted agencies (usually international nongovernmental conservation organizations) to effectively develop projects that are loosely structured at the funding stage.

An additional principle to be followed is that the ultimate goal of any project is that the local people should be no worse off after the project has been implemented than they were before it started. Better, and surely more effective for the long-term success of the project, is to focus from the beginning on ways to improve the lives of the people in ways they perceive as valuable for themselves.

Romer's Rule Applied to Conservation Development

In the course of evaluating sixty-eight rural development projects funded by the World Bank, Conrad Kottak (1990) found that integrating local peoples' institutions and social structures in development projects paid off economically: the "rate of return for culturally compatible projects (19%) was much higher than that of incompatible ones (less than 9%)" (Kottak 1990:723). Kottak proposed Romer's Rule as a principle to guide development projects, and it would seem to be equally applicable to conservation development. Romer's Rule refers to the theory proposed by the paleontologist Alfred Romer (1959:93–94, 327) to explain the evolution of land-dwelling vertebrates. During the Devonian, leg-bearing animals evolved from fin-bearing ones as an adaptation for getting from one ephemeral water pool to another. In other words, a feature that became essential for land life originated not to permit a fully terrestrial lifeway but to maintain an aquatic existence. Romer's Rule applied to rural development, then, is the general rule that the goal of stability is the impetus for change (Kottak 1990:723).

Within the context of development Romer's Rule suggests that projects are unlikely to be successful if they require major changes to customary social institutions and subsistence pursuits. As Kottak points out, people do want changes, but their traditional culture provides the motives to modify their behavior and the boundaries within which they are likely to accept change. This means that projects that prove to be workable promote change but not overinnovation; they have as their goal to preserve systems while making them better (Kottak 1990:724). Thus a deep understanding of existing systems is essential, and understanding is best derived from and by the local people themselves.

Recognition of the need to understand and work within local sociocultural systems leads to a commitment to provide the means by which local people can be empowered to protect the biodiversity of their lands for themselves. True empowerment of people happens when the members of a community have the opportunity to define their goals and the means to seek solutions to the underlying issues (Harcala 1992). Empowerment of local people in and around protected areas offers the challenge of responsible and ethical use of power for the common good—the good of the people themselves and, by virtue of effective conservation of the world's diversity, of the world at large. Empowerment is a methodology that seeks to commit new life to communities that more often than not are marginalized and without the means of attaining either a lifestyle beyond bare subsistence or control over their own resources. Empowerment is a strategy to activate change from the bottom up rather than the top down and an approach calling for full participation of local people in the management of their communities within a framework of sustainable biological diversity.

If this is a desirable approach—if giving people the opportunity to define their own problems and the means to seek solutions within a framework of sustainable biodiversity is a desirable goal—how do we go about it? First, we need to know something more about central African peoples, the history of their relationship with the forest, their political and social units of organization, their own perceived relations with the land and its resources, their attitudes about conservation, their demographic characteristics—these and other characteristics suggest some concrete recommendations about an effective approach to empowering local people to work toward sustainable biodiversity.

Local People and the Forest

Biodiversity as a Human Cultural Artifact

As we try to design strategies for protecting and managing biodiversity in central Africa, it is important to remember how the present diversity came about. The central African forests of today are not simply the products of thousands of years of natural processes, if by natural we mean the absence of significant anthropogenic influences. Considering the paleoenvironmental evidence indicating central Africa was significantly drier and more open as recently as nine thousand years ago (Hamilton 1976; Livingston 1975; Talbot 1984; van Noten 1977, 1982), the areas now covered by moist forest have probably been inhabited by people during most of its existence.

Purposeful clearing of central African forest for cultivation now appears to have occurred five thousand years ago (Clist 1989; David 1982; Ehret 1982; Phillipson 1985) and possibly more. Even before purposeful cultivation human foragers had significant influences on forest plant communities by affecting species' distribution and population parameters for millennia (Campbell 1965; Alcorn 1981; Balée 1989; Bailey and Headland 1991; Bailey et al. 1989). As Terese Hart and John Hart (1986) have argued for the Ituri and has been shown so convincingly by William Balée (1987, 1989, 1993) and others (Hecht and Posey 1989) for the Amazon, the present diverse composition and distribution of plants and animals in rain forests is the result of humans' introducing exotic species, creating new habitats, and chronically manipulating the forest for thousands of years. Many vegetational associations, until now considered natural, represent successional forest situated on archaeological sites, including prehistoric swiddens, settlements, and camps.

Figure 14.1   An Efe (Pygmy) camp in the Ituri Forest, Zaire. Such camps are occupied by an average of eighteen people for about three weeks. They clear a small area of forest, construct temporary huts, and discard their refuse beside and behind the huts. These foragers introduce exotic species into the forest, and the species germinate in openings the foragers create. Even minimal disturbance such as this alters the structure of the forest vegetation and creates new complexes of forest biodiversity.

Because of the long history of long-fallow shifting horticulture, along with the presence of mobile foragers, present-day central African forest plant communities are growing on soils that are the outcomes of human occupation and intervention (see figure 14.1). Although some substantial areas of forest are today not under cultivation, this has been true only for the last few decades and is the result of relatively recent colonial and governmental efforts—not always successful—to concentrate horticulturists along roadways. The widespread evidence of the effects of human land use on the character and diversity of central African forests is apparent to anyone willing to walk two days across any stretch of uncultivated forest. These forests are not pristine natural habitats; within living memory they were dotted with villages and garden plots. Today the forests of central Africa are patchworks of various successional stages shaped by generations of use by long-fallow shifting horticulturists and mobile foragers. In short, the forests many are now seeking to protect from human occupation and exploitation are human cultural artifacts. Biodiversity exists in central Africa today, not despite human habitation but because of it.

What implications does this have for the planning and management of protected areas? First, we can no longer justify the creation of a forest reserve on the basis that it is an area of "virgin forest" or is "in its natural state," or that it is, as stated in virtually every proposal and report, "pristine," "untouched," "natural," "primary," or "original" forest. Second, if we are to exclude humans from using large areas of forest, we will not be preserving the present biodiversity we hold so precious. Rather we will be altering the diversity and probably diminishing it over time. This leads us to question the practical wisdom—not to mention the ethics—of excluding indigenous peoples from lands they have been managing, albeit unwittingly, for hundreds of generations.

Who Are Local People? The Units of Organization

The strategies we use to approach local people and integrate them in planning for sustainable biodiversity will depend in part on how they are organized in cultural, political, and economic units. To be able to work through them, to apply Romer's Rule rather than replace them with new unsuitable and thus ineffective constructs, we must be conscious of the social institutions already in place among local people. At what level do the people themselves make decisions about their relations with the land and its resources? We must know this if we are to empower them to define realistic objectives and have the means to obtain them.

Most people in forest areas of central Africa reside in small villages, with residence determined by kinship affiliations. These villages are organized under chieftainships, and chiefs are usually nominated by the people, usually within specific lineages, and then appointed by state government authorities. Colonial regimes created these chieftainships for administrative purposes, often in ignorance of true cultural affinities. Traditionally, the majority of central African peoples were divided in what anthropologists call segmented lineages—quite independent clans lacking headmen, chiefs, kings, or any form of centralized authority. Cultural and political identity even today is based on extremely local language, oral history, and cultural practices—and especially kinship and identification with a specific area of forest. Thus modern chiefs and other government officials are not always effective leaders nor trusted representatives of indigenous opinion. A number of independent clans have been lumped within any one chieftainship, each clan often distrustful of the other, living apart in separate villages or clusters of villages, and identifying themselves, among other ways, by the area of forest to which they belong.

Land as a Social Entity

The way most central Africans view land is fundamentally different from the Western perception of land as a capital good that can be bought or sold or traded for other commodities. Although land is becoming a more salable commodity in areas of higher population density and rapid commercialization by immigrants (e.g., Peterson 1990), land for most central Africans remains a social entity inseparable from the identity of the community, usually the clan. Access to land cannot be limited to individuals—it is communal— and social relationships, not economic ones, structure land use and ownership. It is crucial to understand the feeling by virtually all indigenous African forest peoples that their identity is tied to a specific area of forest land, and they have a long history of identification with that area.[4] This identifying feature, along with and inseparable from kinship, is usually much stronger than any sense of larger group affiliation or identification with a chief or tribal designation.

Empowerment at the Local Level

Recognizing the long history that people have with specific areas of land and their strong affiliations at a local level, we are compelled to devise systems of organization and management for protected areas that take advantage of the sense of identity with a specific area of the forest and its resources. Groups at the clan or lineage level might be given considerable latitude in the management of areas to which they belong. The management policy for reserves should be general enough and flexible enough to allow for variation in management styles across local groups and over time. The more control handed to people at the local clan level, the greater will be the sense of stewardship and responsibility over the protection of the land and its resources by the people who use the reserve on a daily basis.

Planning the organization and management of bio reserves in central Africa will be most effective if it enlists the participation of indigenous people at levels below that of the regional government and below that of tribal chief. The unit of organization on which to focus should in most cases be the clan, or clusters of villages, which are often organized into moieties or phratries (affiliations of clans). Each clan should be given the authority to design a management plan to fit its particular social, economic, and ecological needs. This could be done within guidelines that should be general and flexible but with specific overall conservation objectives. These guidelines and objectives should be worked out with representatives from the local communities, because ideas generated by local people will be better trusted and supported.

Knowing the extent to which individuals and clans identify themselves as belonging to a specific area of forest, and understanding that people have been exploiting and manipulating these forests for millennia, lead us to question both the ethics and the efficacy of creating "wilderness areas" or "fully protected core areas" in which no people are to live and no hunting or gathering is to occur. A wilderness area in which only a handful of field biologists may tread is a Western concept that has little credence in central Africa. In most cases protection of forest areas is not incompatible with the continued presence of local people. Instead local peoples given some control over their land can enhance the efforts to protect forest flora and fauna. At low densities tribal peoples, with their mobile lifestyles, are unlikely to overexploit forest resources. The creation of protected areas should not necessitate the removal and resettlement of tribal peoples, nor should it require severe restrictions on their rights to forest resources.

Indigenous groups often are permitted to remain in protected areas so long as they remain *traditional*, a term usually defined by policy makers without consultation with or extensive historical knowledge of the peoples themselves. Such restrictions lead to "forced primitivism" (Goodland 1982:21), whereby tribal people are expected to remain traditional as the rest of the world passes them by. If limitations are placed on technology and the extent or means of forest use, special education and training for alternative lifeways should be built into the planning and budgeting of protection schemes. Much more equitable and effective over the long term would be to let local people for themselves decide how to develop their land with what technologies within specified guidelines consistent with sustainable biodiversity.

Creating opportunities for local communities to manage their own forest areas leads to a reversal of the top-down perspective that most planners and administrators have been stuck with. The authority and daily oversight of protected areas becomes decentralized—it gets away from concentrating resources and personnel around a research station, an administrative headquarters, or a guard station. As authority becomes more dispersed and mobile, more mobile management and training systems become desirable. Effective extension becomes crucial. The structure of authority becomes more lateral, less hierarchical, and each individual has broader responsibilities within smaller areas. The tendency to train and appoint well-educated nonlocal authorities and experts should give way to training local people to become the teachers, census takers, guards, wildlife monitors, health workers, trainers, and managers. In this way conservation goals should be more consistent with rural development objectives, and a greater number of local people will have knowledge of and a direct stake in achieving conservation and development goals.

Ethnicity and Conservation

Another reason to decentralize planning and management of protected areas is to take advantage of ethnic pride. Because people identify with their area of forest, if they retain control over it—if the traditional land tenure system is reinforced—they are apt to protect it. This is especially true of groups whose forest is intact but who live on the edges of regions that are being deforested rapidly. The people can see what is approaching, and they can see the changes that have been wrought by immigration and commercialization. They often realize that the people who are making profits are not the indigenous people but rather the immigrants, who extract the resources and move on, or who stay to control stores, transport of goods, and most commercial operations in the area. Thus many local people will have often witnessed the marginalization of others and will take pride in maintaining control over their own area of forest and thus their distinct cultural identity.[5]

Africans do not have a conservation ethic in the same way that some Westerners do. They often cannot foresee what development and deforestation might do to their livelihoods or those of their children. However, they do understand the value of the forest and the resources within it. They do want to preserve forest but not for its aesthetic beauty or for backpacking, camping, tourism, or wildlife biology. Rather the forest is to be exploited. Thus they want the forest and its resources to be preserved so they can continue to exploit it. However, if their forest is to be preserved but exploitation forbidden, the preserve should be established on someone else's land (Hackel 1990, n.d.). Thus Africans are generally not sympathetic to the concept of total protection. They are interested, however, in preserving the forest at a level at which they can continue to extract resources. Thus in many areas persuading local people of the importance of maintaining a sustained yield system may not be difficult.

Tourism, Ecotourism, and Anthrotourism

The recognition that local people must derive some economic benefit from the creation of a protected area on their land invariably leads to statements concerning the promise of tourism for providing both local people and host governments with revenue. Although tourism related to wildlife undoubtedly has generated substantial revenues in some African countries (Kenya and Rwanda, for example), expectations for central Africa may run too high, given the lack of easy transport, the difficulties of maintaining comfortable accommodations in tropical forests, the difficulties of finding and viewing game in forest habitats, the competition from neighboring African countries where these adverse conditions are not present, and a host of other factors. Moreover very few examples exist in Africa or elsewhere of local people benefiting from the creation and management of tourist operations on their lands. In most cases benefits accrue to government agencies and businesses not owned or operated by local people. Although these sometimes generate a few jobs for local people, few tourist revenues ever trickle down to the local people. Instead "the poorest sector of the population bear the costs of supporting wildlife when economically and aesthetically the benefits accrue nationally and internationally, to others much wealthier" (Western 1981:1).

Most proposals and management plans for protected areas in central Africa mention pygmies as potential attractions for tourists. Indeed a large selling point to garner support for reserves in central Africa from international agencies and the international public is the presence of pygmies (e.g., the Aka in the Dzanga-Sangha Reserve, the Mbuti in the Okapi Wildlife Reserve), and opportunities to "view Pygmies in their natural habitat" are promoted as ways to generate revenues to support maintenance of the reserve. Yet how such anthrotourism is to be developed—whether the people themselves have any interest in becoming humans in a zoo or becoming commodities that benefit the park administration or some tour company—is never mentioned.

References to anthrotourism by conservationists may help to gain support for protected areas, especially in central Africa where Pygmy foragers seem to hold considerable fascination for much of the world community. Indeed in central Africa indigenous people may be more of an attraction for visitors to reserves than the forest and wildlife. However, it would be a mistake for conservationists to ignore the implications of such promotions for the local people themselves. Examples of commoditization and exploitation of ethnic minorities for the purposes of tourism are many and chilling (e.g., see *Cultural Survival Quarterly* 1990a, 1990b). Conservationists must be aware of the harm they are inviting. If anthrotourism is to be a component of development of protected areas, conservationists and international agencies should design safeguards to avoid usurpation by a tourism industry of forest peoples' rights to use land and make resource management decisions. They must incorporate in their conservation plans explicit programs for integrating local people in the planning, development, management, and revenue sharing for tourism.

If developed with participation and planning by local people, tourism can be a vehicle not only for generating income for local people but also for enhancing their support for and participation in conservation efforts. If forest people are an integral part of tourism strategies, tourism can enhance cultural awareness and the knowledge of ethnic history while avoiding the people-in-a-zoo phenomenon.

When I attended the coronation of King Mswati III of Swaziland in

1986, I was struck by an incident that exemplifies how Africans invite and profit from tourism while also taking great pride in their ethnic identity and culture history. On the flight to Swaziland I sat and became friendly with three young Swazi dressed in stylish Western suits. They were listening to Sony Walkmen and discussing American basketball and European politics. They were coming to Swaziland for the coronation, on holiday from American universities. The next day I went to watch a special Swazi ceremony, held in a cattle corral in which hundreds of Swazi warriors, dressed in leopard and mongoose skins, holding spears and cattle-skin shields, were dancing to the beat of massive drums and chanting traditional songs. Standing beside me around the corral were hundreds of foreigners from all over the developed world, necks craning, cameras flashing, and videos whirring to capture the exciting spectacle. There within the corral, dressed in full traditional regalia and participating in the ceremony with their clansmen were the three Swazi I had befriended on the plane the day before. Later that evening I met with one of them for a beer. He explained to me with reverence and pride the history and cultural significance of that day's activities.

Ecotourism and anthrotourism may be effective vehicles by which local people can benefit monetarily from protected areas. But they can also be means by which local people gain knowledge and appreciation for their own ethnohistory and cultural identity, while increasing their identification with and respect for their forest lands. Segments of a local population too often become objects of tourism, members of an impoverished underclass, without a sense of their identity and ridiculed by their own. The Swaziland experience taught me that forced primitivism is not a requirement for a lucrative tourist industry. The monetary and educational value of tourism can be enhanced for both the local people and the tourists by giving indigenous people the opportunity to define their ethnicity, to develop the value of their history, and to project their customs and artifacts. The durable success of a tourist industry in any central African country can enhance conservation efforts; however, conservationists must become aware of the pitfalls of promoting eco- and anthrotourism, and they must be willing to pressure funding agencies and governments to structure reserve development so that local people participate in the design and management of the tourist industry. This will be crucial for maintaining the region's cultural and environmental integrity.

The Demographic Myth and Reality

Many areas in Africa are known to have among the highest, if not the highest, fertility rates in the world, with completed fertility in many countries averaging 6.5 to 8.0 births per woman (Robey et al. 1992). Such high fertility, even combined with relatively high mortality, results in rapid population growth and ever increasing pressure on land and natural resources. Africa's population is expected to nearly triple by the year 2025 (World Resources Institute 1990). By the year 2010, 25% more land will have to be cultivated to maintain African populations at their current nutritional levels (Food and Agriculture Organization 1989). Thus there is an urgency to calls for bringing areas of central African forest under protection before the forest is overrun by land-hungry populations. But where is pressure on the forest lands coming from? To what extent are the local populations growing and thus having to clear more land to feed their growing numbers, and to what extent is the pressure coming from outside the forest, from immigrants coming from savanna and mountain areas seeking land and exploitable resources? Although I know of no reliable demographic evidence sufficient to answer these questions directly, good evidence shows that the demographic pressure is not coming from indigenous forest populations.

Table 14.1 shows the results of demographic studies I conducted in 1987 among four horticultural populations residing in different areas in the Ituri Forest, Zaire (for a description of the people studied and methods used, see Bailey 1991; Bailey and Aunger 1995). Each population is situated in or on the edge of what has recently been declared the Okapi Wildlife Reserve, consisting of 1.37 million hectares of moist tropical forest in northeast Zaire. The completed fertility of postmenopausal women in these four populations ranges from 1.89 for the Mamvu to 3.36 for the Bila. Their child mortality rates (not shown) range from 18% to 26%, making all four populations declining or stable populations, assuming a generation time of twenty-eight years and no in- or out-migration. These low total fertility rates are accounted for in part by frequencies of primary sterility ranging from 34% (for the Budu) to 44% (for the Mamvu). In these populations more than 55% of women who have finished their reproductive careers have born one or no children.

Table 14.1

Completed Fertility and Primary Sterility of Postmenopausal Women of Four Tribal Societies in the Ituri Forest, Zaire

Tribe	Number of women	Mean no. of live births	Percentage childless
Lese	119	2.35	37%
Mamvu	105	1.89	44%
Bila	117	3.36	37%
Budu	113	3.04	34%

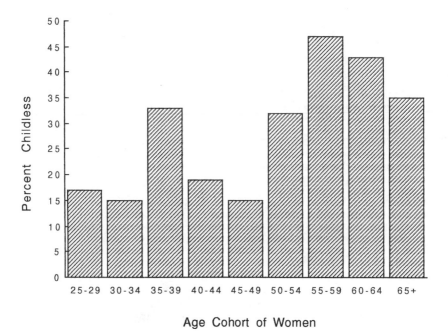

Age Cohort of Women

Figure 14.2   Primary sterility by age cohort of 993 women from four tribal societies residing in and around the Okapi Wildlife Reserve in the Ituri Forest, Zaire.

The high primary and secondary sterility observed among these forest-living populations seems not to be a recent phenomenon; low fertility has characterized these populations for many decades. Henry Morton Stanley noted the absence of children in villages in the Ituri in 1888 (Stanley 1890). The oldest people describe entire lineages dying out in their lifetimes, and elders commonly reminisce about the large size of villages and clans when they were children. More quantitative evidence is depicted in figure 14.2, which shows the parity distributions of Ituri villager women by age. The women born before 1932 (the women age fifty-five and older at the time of data collection) actually have lower completed fertility than the younger women, whose parity nevertheless is still low. This indicates that infertility is not a recent vital characteristic of these indigenous forest populations.

Although these demographic data may seem strikingly different from what we normally expect of African populations, they are actually representative of many central African forest-living populations. The Ituri is not an aberrant case for forested areas of central Africa. It is widely known and well documented that a belt across central Africa corresponds to the moist tropical forest regions and is characterized by rates of primary sterility of 30% to 45%, just as I have shown here for the four populations indigenous to the Ituri Forest (Belsey 1976; Caldwell and Caldwell 1983; Frank 1983; Page and Coale 1972; Romaniuk 1967; Voas 1981). Although many factors could cause such consistently high rates of infertility, the most likely is infection with sexually transmitted diseases (gonorrhea and chlamydia) that cause tubal occlusions and blockage (Belsey 1976; Larsen 1989). The infertility is not uniformly distributed across central Africa but occurs in pockets, often following ethnic lines (Caldwell and Caldwell 1983; Romaniuk 1968). If we examine the ethnic groups affected, they are consistently the indigenous populations, those that have resided in the forest for generations. These populations are characterized by high marital mobility and customs that permit wider sexual contacts (Retel-Laurentin 1974; Bailey and Aunger 1995). On the other hand, recent immigrants—those coming from savanna or highland areas—have significantly lower rates of primary sterility and thus are high fertility, rapidly growing populations.[6]

These data indicate that the people who have been residing in African tropical forest areas for generations are members of stationary or declining populations. These indigenous populations are not those that are causing increasing rates of forest destruction and ever increasing stress on forest resources. Immigration and commercialization by rapidly growing populations residing outside the forest are the major threats to central African forest biodiversity. This is likely to be true until at least one or a combination of the following occur: a rise in fertility rates among indigenous forest populations, perhaps by installing public health interventions that would reduce rates of primary and secondary sterility[7]; a slowing of the growth rates of populations residing on the fringes of central African forests; and development of projects that induce populations on the fringes to intensify production and remain on their homelands.

These findings suggest several strategies for developing and managing protected areas in central Africa. First, because their populations are not growing and thus not requiring new areas of land, we should work to reinforce local forest peoples' land tenure systems. This may well mean that conservation groups, funding agencies, and interested host country agencies will have to pressure central African governments to alter or bend existing laws, which often deny indigenous peoples' legal rights to land. Conservationists and international funding agencies have provided strong and ultimately successful support to indigenous groups in other forest areas of the world (e.g., Hecht and Cockburn 1989). Working with international agencies and host country counterparts, conservationists must dare to draft alternative legislation that gives local groups greater control over their lands. Other less difficult measures can also be initiated. For example, management plans for protected areas often undermine local peoples' land tenure systems by giving authority to park administrators and government officials to direct settlement and exploitation patterns. This authority could be less centralized. Local people could be given the authority to make such decisions within well-defined limits consistent with biodiversity goals.

Second, given the demographic trends within and around forest reserve areas, we should strive to ameliorate the conditions that are driving immigration into central African forests. People immigrate to the forest for many reasons. R. B. Peterson (1990, 1992) has enumerated those reasons expressed by immigrants to the Ituri Forest, Zaire. His studies strongly indicate that reversing the nonegalitarian distribution of land and promoting alternative sustainable land-use practices on the fringes of the forest would be an effective way to reduce pressure on forested areas. Such endeavors would expand the efforts of most conservation organizations beyond what they traditionally perceive as their purview. Nevertheless exerting pressure on international funding agencies and host governments to direct development resources to land tenure systems and land-use practices has become a necessary component of effective conservation planning and management.

Discussion

We have seen that despite the best intentions of conservationists to integrate local people in efforts to promote biodiversity in central Africa, the process by which protected areas are conceived, planned, managed, and monitored has worked against the involvement of local people. This is true because the process begins with a top-down approach in which outside international and national agencies are recruited in the process long before local people are even consulted, not to mention involved. If conservationists are convinced that an essential element in effective and durable conservation of African forest biodiversity is local participation in planning and management, they must turn the existing process on its head—conservationists should start with a bottom-up perspective with the aim of empowering local people to define their goals and to attain the means of seeking solutions to economic and ecological problems posed by the challenge of maintaining biodiversity.

Local people must be seen as part of the solution to the challenges of conservation. They may not be expected to initially identify an area to protect or devise some other form of conservation program (although even that may change as conservation schemes and the forces driving them become more pervasive). But they should become the very first people consulted and the very first to participate in devising ways to effect change consistent with both conservation goals and existing social institutions and subsistence pursuits. Local people must come first throughout the project cycle, from initial planning to subsequent management to ultimate control of long-term project maintenance.

New Challenges

Taking a bottom-up approach to African conservation offers several new challenges to conservationists and the international community that support biodiversity in the region. First, conservation agencies and large international funding organizations must tolerate, indeed they must encourage, projects that are designed to be fluid. They must move away from the blueprint model typically endorsed by development planners and make a commitment to a learning process model (Kottak 1990) in which projects are conceived as works in progress, responsive and continuously adapting to new ideas and fresh participation by local people. As projects apply Romer's Rule, as they strive to meet locally recognized needs and to work within existing structures, the culturally specific incentives necessary to obtain enthusiastic local participation will emerge. As they do emerge, the project plan must be sufficiently flexible to respond in directions that are culturally appropriate. As local structures are strengthened or emerge anew through local participation, the project must change to accommodate and take advantage of each new level of participation.

A second challenge presented by the bottom-up approach is to devise ways to address people's concerns at the most effective level of social and political organization. As we have seen, in most areas of forested central Africa projects designed to integrate people in planning and management are apt to be most effective if focused at the level of the clan. Most central Africans have greater identification with and loyalty to their clan (or lineage or moiety or phratry) than to a tribe, a chief, or some national officeholder; their relationship to the land and forest is defined economically, politically, and spiritually by their clan affiliation. Thus when we speak of bringing local people and local institutions into the planning and management process, we are really promoting participation by individual clans or clusters of villages. This will require a fundamental change in how projects are conceived and administered.

Decentralization of all aspects of the project will be most effective. Authority should be decentralized both structurally and geographically, rather than concentrated in the hands of a park administrator or regional officer. A larger number of project officers with broader responsibilities over a smaller area and fewer people will be more effective than one or a few powerful authorities. Administration personnel should no longer be concentrated around a research station or park headquarters but spread throughout the project area, living in the communities they serve and in touch with community opinion. This structure will place a premium on communication between administrative elements, which should meet regularly to exchange local concerns and integrate them in evolving project policy. It will also require greater mobility by all project elements. Guards, trainers, educators, monitors, regional administrators, community representatives—all project personnel will need to move between local communities to inform and be informed of project developments and community concerns. Mobile outreach and extension programs will be important for bringing every community into the process of project development and management and allow different communities to develop their own management plan. Such a decentralized, laterally organized, and mobile project structure will not only conform better with existing modes of organization but also be responsive to and take advantage of the full cultural diversity of the various communities involved with the project.

A third challenge presented to conservationists by the bottom-up approach is to consult at the earliest stages of a project with people who are already experienced with local cultures and social institutions. This will not always require a foreign expert. There may be local nongovernmental organizations or sufficient indigenous social expertise to be informative and helpful. However, most areas of central Africa have anthropologists and others trained in the social sciences who have lived in local communities and already have gained extensive knowledge of local cultures. These are experts whose methodology is expressly the bottom-up approach; they study social institutions and modes of subsistence systematically by observing while actually participating in the culture. Conservation planning, management, and assessment, if it is to be successful, will require work in the field with members of local communities. Anthropologists are trained specifically to do such work—to gather and interpret information in efficient, systematic, culturally sensitive ways with minimal bias. They have usually experienced a range of societies and read extensively about many others, giving them a comparative cross-cultural perspective that provides insights into appropriate ways to collect information and promote people's participation. These social scientists have experience and training that conservationists, despite their many talents, are lacking. Yet, because conservation is rarely recognized as being as much a social process as an ecological or biological one, the role of the anthropologist or other social expert has often been more to legitimize decisions made by others than to participate as a full and valued member of a project team.[8]

If sustainable biodiversity conservation is to be a realistic goal in central Africa or elsewhere, conservation will have to become a more interdisciplinary endeavor, integrating all the disciplines that contribute to ecology with those in the social sciences. To be successful neither the ecological nor the social can be dominant— true integration of the two should be synergistic and thus be powerful forces in the preservation of the environmental and cultural integrity of central Africa.

Acknowledgments   I am indebted to Bill Webber of the Wildlife Conservation Society for inviting me to the June 1992 African Forest Symposium in Essex, Massachusetts, where I had the opportunity to first articulate these ideas; to the dedicated authors of the many proposals, reports, and management plans I read in preparing this paper; to Robert Gustafson for his invaluable assistance searching the literature; and to the following for their helpful comments, discussions, and many insights: Serge Bahuchet, Jason Clay, Cynthia Cook, Bryan Curran, Mary Dyson, Jeff Hackel, John Hart, Terese Hart, Thomas Headland, Barry Hewlett, Conrad Kottak, Rick Peterson, Kent Redford, John Robinson, Leslie Sponsel, and David Wilkie. The field research was supported by grants from the National Science Foundation, the National Geographic Society, the Swan Fund, and the University of California—Los Angeles Academic Senate.

Notes

1. For example, the Central African Republic has recently created the Dzanga-Sangha Dense Forest Reserve and the Dzanga-Ndoki National Park supported by the World Wildlife Fund—USAID Wildlands and Human Needs Program. The Republic of the Congo is planning to manage an integrated system of six or more protected areas consisting of more than 3 million hectares of forest. Similarly, the government of Zaire recently established the Okapi Nature Reserve, consisting of 3.4 million acres of moist tropical forest in the Ituri region, and other reserves are contemplated. The government of Cameroon is considering the creation of a large nature reserve in the Lake Lobeke area in the forested southeast region of that country. International conservation agencies, especially the World Wildlife Fund, Wildlife Conservation Society, and Conservation International, have spearheaded these efforts to arrest deforestation and promote biodiversity, efforts made possible financially in large part through international development assistance agencies, especially the Global Environmental Facility of the World Bank.

2. Considerable controversy surrounds how to refer to the people living in and around proposed and existing African reserves. The term *tribal people* or *tribal groups* has for good reason fallen into disfavor. *Indigenous peoples, ethnic groups, traditional peoples,* and *residents* are all potential alternatives, but each has its problems. I use the term *local peoples* so as not to imply that all individuals living in these areas necessarily have well-defined ethnic identities. However, ethnicity in the sense of defining self as belonging with or in opposition to others is usually fundamental to social, political, and economic relations among people in central Africa, as elsewhere.

3. Although many working in conservation organizations are now lamenting that the conservation community is becoming increasingly dominated by "development specialists" not trained in ecology or resource management, those writing the reports and proposals, conducting the surveys, designing the management plans, and making the recommendations for the creation and

management of protected areas in central Africa are almost exclusively people trained in biology, zoology, ecology, or forestry.

4. A corollary of this point is that all areas of seemingly unoccupied forest in central Africa are actually perceived by the people as belonging to a specific clan or tribal group. In the Ituri Forest, for example, it would be possible to sit down with representatives from all the clans bordering the forest and plot on a map the areas to which each clan belongs. Although there may be some disputes between clans, every hectare of forest would be claimed by at least one clan. Thus for the purposes of planning the protection of any area of forested land in central Africa—no matter how untouched, pristine, primary, or mature it appears to be—we should assume a priori that any area is being used by some person, or some clan, lineage, or group that has a long history of use and identity with that area (Bailey, Bahuchet, and Hewlett 1992).

5. In the Ituri Forest immigration by BaNande from the southeast has resulted in widespread forest destruction on the lands occupied by BaBila. The more commercially minded BaNande use more intensive cultivation techniques, requiring greater labor than the more traditional swidden systems of the BaBila and other indigenous groups. This intensification, combined with ownership of shops and panning for gold, has resulted in BaNande hiring BaBila and local Mbuti foragers to supply inexpensive labor. The Lese living to the north of the BaBila have not yet experienced extensive immigration to their forest. During a conversation about the BaNande immigration the chief of the Lese Dese said to me, "If the BaNande come here, our forest will be cut down and we will end up as their slaves, just like the BaBila." The chief took great pride in the great expanse of forest still intact within his *collectivité*.

6. The causes of the differences in fertility and primary sterility among these populations are surely multiple and complex. One significant factor is access to Western medical care, especially antibiotics, as indicated by parity that increased in many areas when dispensaries were introduced by colonial administrations (Bailey 1989). However, this by no means explains all the variation.

7. U. Larsen (1989:208) has calculated that if low levels of sterility, equivalent to those of Kenya in the 1970s, prevailed in central African populations, fertility rates in Cameroon would increase to 7.3. In other words, if sterility were reduced in these areas, we would expect an increase of about 30% in the total number born to each woman and rapid population growth rates.

8. Actually, it is difficult to find conservation projects in central Africa that have had anthropologists affiliated with them. Most often impact assessments and other types of evaluations have been performed not by anthropologists trained and experienced in the methods needed to be effective and unbiased but by people trained in forestry and wildlife management or by people who have had some experience living in central Africa (e.g., an ex–Peace Corps volunteer) but no formal training. Perhaps the reason anthropologists are seldom sought to assist in conservation efforts is they are perceived as wanting too much to put peoples' interests first.

References Cited

Alcorn, J. B. 1981. Huastec noncrop resource management: Implications for prehistoric rain forest management. *Human Ecology* 9: 395–417.

Bailey, R. C. 1989. The comparative demography of Efe foragers and Lese horticulturalists in the Ituri Forest, Zaire. *Abstracts of the 88th Annual Meeting of the American Anthropological Association.* Arlington, Va.: American Anthropological Association, pp. 171–172.

———. 1991. *The Behavioral Ecology of Efe Pygmy Men in the Ituri Forest, Zaire.* Ann Arbor: University of Michigan, Museum of Anthropology.

Bailey, R. C. and R. V. Aunger. 1995. Sexuality, infertility, and sexually transmitted disease among farmers and foragers in central Africa. In *Sexual Nature/Sexual Culture*, P. R. Abramson and S. D. Pinkerton, eds., pp. 195–222. Chicago: Chicago University Press.

Bailey, R. C. and T. N. Headland. 1991. The tropical rain forest: Is it a productive environment for human foragers? *Human Ecology* 19 (2): 261–285.

Bailey, R. C., S. Bahuchet, and B. Hewlett. 1992. Development in central African rainforest: Concern for forest peoples. In *Conservation of West and Central African Rainforests*, K. Cleaver, M. Munasinghe, M. Dyson, N. Egli, A. Penker, and F. Wencelius, eds., pp. 260–269. Washington, D.C.: World Bank.

Bailey, R. C., G. Head, M. Jenike, B. Owen, R. Rechtman, and E. Zechanter. 1989. Hunting and gathering in tropical rain forest: Is it possible? *American Anthropologist* 91: 59–82.

Balée, W. 1987. Cultural forests of the Amazon. *Garden* 11 (6): 12–14.

———. 1989. The culture of Amazonian forests. In *Resource Management in Amazonia: Indigenous and Folk Strategies, Advances in Economic Botany*, D.A. Posey and W. Balée, eds., Vol. 7, pp. 1–21. New York: New York Botanical Gardens.

———. 1993. Indigenous transformation of Amazonian forests: An example from Maranhao, Brazil. *L'Homme* 33: 231–254.

Batisse, M. 1990. Development and implementation of the biosphere reserve concept and its applicability to coastal regions. *Environmental Conservation* 17 (2): 111–116.

Belsey, M. A. 1976. The epidemiology of infertility: A review with particular reference to sub-Saharan Africa. *Bulletin of the World Health Organization* 54: 319–341.

Caldwell, J. C. and P. Caldwell. 1983. The demographic evidence for the incidence and course of abnormally low fertility in tropical Africa. *World Health Statistics Quarterly* 36: 2–34.

Campbell, A. H. 1965. Elementary food production by the Australian aborigines. *Mankind* 6: 206–211.

Clist, B. 1989. Archaeology in Gabon, 1886–1988. *African Archaeological Review* 7: 59–95.

Cultural Survival Quarterly. 1990a. Breaking out of the tourist trap, Part 1. *Cultural Survival Quarterly* 14 (1): entire issue.

———. 1990b. Breaking out of the tourist trap, Part 2. *Cultural Survival Quarterly* 14 (2): entire issue.

David, N. 1982. Prehistory and historical linguistics in central Africa: Points of contact. In *The Archaeological and Linguistic Reconstruction of African History,* C. Ehret and M. Posnansky, eds., pp. 78–95. Berkeley: University of California Press.

Ehret, C. 1982. Linguistic inferences about early Bantu expansion. In *The Archaeological and Linguistic Reconstruction of African History,* C. Ehret and M. Posnansky, eds., pp. 57–65. Berkeley: University of California Press.

Food and Agriculture Organization. 1989. *The State of Food and Agriculture.* Rome: Food and Agriculture Organization.

Frank, O. 1983. Infertility in sub-Saharan Africa: Estimates and implications. *Population and Development Review* 9: 137–145.

Goodland, R. 1982. *Tribal Peoples and Economic Development: Human Ecologic Considerations.* Washington, D.C.: World Bank.

Hackel J. D. 1990. Conservation attitudes in southern Africa: A comparison between KwaZulu and Swaziland. *Human Ecology* 18 (2): 203–209.

———. n.d. Change in Rural Africa and What It Means for Nature Conservation: A Case Study from Swaziland. Manuscript, California State University, Department of Geography, San Bernadino.

Halpin, E. A. 1990. *Indigenous Peoples and the Tropical Forestry Action Plan.* New York: World Resources Institute.

Hamilton, A. C. 1976. The significance of patterns of distribution shown by forest plants and animals in tropical Africa for the reconstruction of upper Pleistocene paleoenvironments: A review. *Paleoecology of Africa* 9: 63–67.

Harcala, J. R. 1992. For recovery, empower the people. *Los Angeles Times,* June 16, p. B7.

Hart, T. B. and J. A. Hart 1986. The ecological basis of hunter-gatherer subsistence in African rain forests: The Mbuti of eastern Zaire. *Human Ecology* 14: 29–56.

Hecht, S. B. and A. Cockburn. 1989. *The Fate of the Forest: Developers, Destroyers and Defenders of the Amazon.* London: Verso.

Hecht, S. B. and D. A. Posey. 1989. Preliminary results on soil management techniques of the Kayapo Indians. In *Resource Management in Amazonia: Indigenous and Folk Strategies, Advances in Economic Botany,* D. A. Posey and W. Balée, eds., Vol. 7, pp. 174–178. New York: New York Botanical Gardens.

Kottak, C. P. 1990. Culture and "economic development." *American Anthropologist* 92: 723–731.

Larsen, U. 1989. A comparative study of the levels and the differentials of sterility in Cameroon, Kenya, and Sudan. In *Reproduction and Social Organization in Sub-Saharan Africa,* R. J. Lesthaeghe, ed., pp. 167–211. Berkeley: University of California Press.

Livingston, D. A. 1975. Late quarternary climate change in Africa. *Annual Review of Ecology and Systematics* 6: 249–280.

McNeely, J. F., K. R. Miller, W. V. Reid, R. A. Mittermeier and T. M. Werner. 1990. *Conserving the World's Biological Diversity.* Washington, D.C.: World Resources Institute.

Oldfield, M. L. and J. B. Alcorn, eds. 1991. *Biodiversity: Culture, Conservation and Ecodevelopment.* Boulder, Colo.: Westview.

Page, H. J. and A. J. Coale. 1972. Fertility and child mortality south of the Sahara. In *Population Growth and Economic Development in Africa,* S. H. Ominde and C. N. Ejiogu, eds., pp. 51–67. London: Heinneman.

Peterson, R. B. 1990. Searching for life on Zaire's Ituri Forest frontier. *Cultural Survival Quarterly* 14 (4): 56–62.

——. 1992. Conservation for Whom? A Study of Immigration onto Zaire's Ituri Forest Frontier. Paper delivered at the Symposium on Conservation of African Forests: Interdisciplinary and Applied Perspectives, June 30–July 3, Essex, Mass.

Phillipson, D. 1985. *African Archaeology.* Cambridge, England: Cambridge University Press.

Retel-Laurentin, A. 1974. Sub-fertility in black Africa: The case of the Nzakara in Central African Republic. In *Sub-Fertility and Infertility in Africa,* B. K. Adedevoh, ed., pp. 69–80. Ibadan, Nigeria: Caxton Press.

Robey, B., S. O. Rutstein, L. Morris, and R. Blackburn. 1992. The reproductive revolution: New survey findings. *Population Reports,* Series M, No. 11. Baltimore: Johns Hopkins University Population Information Program.

Romaniuk, A. 1967. *La fécundité des populations Congolaise* (The fecundity of populations in the Congo). The Hague/Paris: Mouton.

——. 1968. Infertility in tropical Africa. In *The Population of Tropical Africa,* J. C. Caldwell and C. Okonjo, eds., pp. 214–224. London: Longman.

Romer, A. S. 1959. *The Vertebrate Story.* Chicago: University of Chicago Press.

Stanley, H. M. 1890. *In Darkest Africa.* New York: Scribner's.

Talbot, M. R. 1984. Preliminary results from settlement cores from Losumtwi, Ghana. *Paleoecology of Africa and the Surrounding Islands* 16: 173–192.

van Noten, F. 1977. Excavations at Matupi cave. *Antiquity* 51: 35–45.

——. 1982. *The Archaeology of Central Africa.* Graz, Austria: Academische Druck- und Verlagsanstalt.

Voas, D. 1981. Subfertility and disruption in the Congo basin. In *African Historical Demography,* Vol. 2, pp. 777–802. Edinburgh: University of Edinburgh, Centre of African Studies.

Wells, M., K. Brandon, and L. Hannah. 1990. *People and Parks: An Analysis of Projects Linking Protected Area Management with Local Communities.* Washington, D.C.: U.S. Agency for International Development.

West, P. C. and S. R. Brechin, eds. 1991. *Resident Peoples and National Parks: Social Dilemmas and Strategies in International Conservation.* Tucson: University of Arizona Press.

Western, D. 1981. A challenge for conservation. *L. S. B. Leakey Foundation News* (Winter): 1–3.

World Resources Institute. 1990. *World Resources 1990–91.* Oxford, England: Oxford University Press.

Index